Immobilized Enzyme Principles

Immobilized Enzyme Principles

Contributors
───────────

Jinglin Fu, Jeremy Reinhold et al.

AURIS
Reference

www.aurisreference.com

Immobilized Enzyme Principles

Contributors: Jinglin Fu, Jeremy Reinhold et al.

Published by Auris Reference Limited
www.aurisreference.com

United Kingdom

Immobilized Enzyme Principles

ISBN: 978-1-78154-843-1

British Library Cataloguing in Publication Data
A CIP record for this book is available from the British Library

Printed in the United Kingdom

Exclusively distributed by CBS Publishers & Distributors Pvt. Ltd.

Sales & Distribution Rights only for India, Pakistan, Bangladesh, Sri Lanka, Nepal and Bhutan.This book is not to be sold outside these territories.

Contents

List of Abbreviations

BSA	Bovine serum albumin
CTL	Compared with the control experiment
CPG	Controlled pore glass
CLEA	Crosslinked enzyme aggregates
DMF	Dimethylformamide
EPT	Enzyme prodrug therapy
ESI	Exposes the inhibitor binding site
GPD	Glyceraldehyde-3-phosphate dehydrogenase
GK	Glycerol kinase
GPO	Glycerol-3-phosphate oxidase
HCN	Hydrogen cyanide
LDH	Lactate dehydrogenase
LDH	Layered double hydroxide
LCPA	Long-chain polyamines
PIII	Plasma immersion ion implantation
PEI	Polyethyleneimine
ROS	Reactive oxygen species
SEM	Scanning electron microscopy
SDS	Sodium dodecyl sulfate
TEM	Transmission electron microscope
TPP	Tripolyphosphate

List of Contributors

Jinglin Fu
Center for Single Molecule Biophysics, Arizona State University, Tempe, Arizona, United States of America,
Center for Innovations in Medicine, the Biodesign Institute, Arizona State University, Tempe, Arizona, United States of America,
Department of Chemistry and Biochemistry, Arizona State University, Tempe, Arizona, United States of America

Jeremy Reinhold
Center for Single Molecule Biophysics, Arizona State University, Tempe, Arizona, United States of America,
Department of Chemistry and Biochemistry, Arizona State University, Tempe, Arizona, United States of America

Neal W. Woodbury
Center for Single Molecule Biophysics, Arizona State University, Tempe, Arizona, United States of America,
Department of Chemistry and Biochemistry, Arizona State University, Tempe, Arizona, United States of America

Dong-Hao Zhang
College of Pharmacy, Hebei University, Baoding 071002, China
Key Laboratory of Pharmaceutical Quality Control of Hebei Province, College of Pharmacy, Hebei University, Baoding 071002, China

Li-Xia Yuwen
College of Pharmacy, Hebei University, Baoding 071002, China

Li-Juan Peng
College of Pharmacy, Hebei University, Baoding 071002, China

R. Malini Devi
Department of Mathematics, the Standard Fireworks Rajaratnam College for Women, Sivakasi, India

O. M. Kirthiga
Department of Mathematics, the Madura College, Madurai, India

L. Rajendran
Department of Mathematics, the Madura College, Madurai, India

Zhenhai Gan
Key Laboratory of Biomedical Information Engineering of Education Ministry, School of Life Science and Technology, Xi'an Jiaotong University, Xi'an, P. R. China

Ting Zhang
Scientific Research Center, the Second Affiliated Hospital, School of Medicine, Xi'an Jiaotong University, Xi'an, P. R. China

Yongchun Liu
Key Laboratory of Biomedical Information Engineering of Education Ministry, School of Life Science and Technology, Xi'an Jiaotong University, Xi'an, P. R. China

Daocheng Wu
Key Laboratory of Biomedical Information Engineering of Education Ministry, School of Life Science and Technology, Xi'an Jiaotong University, Xi'an, P. R. China

Lee Fung Ang
School of Pharmaceutical Sciences, Universiti Sains Malaysia, Minden, Penang, Malaysia

Lip Yee Por
Faculty of Computer Science and Information Technology, University of Malaya, Kuala Lumpur, Malaysia

Mun Fei Yam
School of Pharmaceutical Sciences, Universiti Sains Malaysia, Minden, Penang, Malaysia

Bau-Yen Hung
Department of Life Science and Institute of Biotechnology, National Dong Hwa University, Hualien-974, Taiwan

Yaswanth Kuthati
Department of Life Science and Institute of Biotechnology, National Dong Hwa University, Hualien-974, Taiwan

Ranjith Kumar Kankala
Department of Life Science and Institute of Biotechnology, National Dong Hwa University, Hualien-974, Taiwan

Shravankumar Kankala
Department of Chemistry, Kakatiya University, Telangana State-506009, India

Jin-Pei Deng
Department of Chemistry, Tamkang University, New Taipei City 251, Taiwan

Chen-Lun Liu
Department of Life Science and Institute of Biotechnology, National Dong Hwa University, Hualien-974, Taiwan

Chia-Hung Lee
Department of Life Science and Institute of Biotechnology, National Dong Hwa University, Hualien-974, Taiwan

Jian Li
School of Material Science and Chemical Engineering, Tianjin University of Science and Technology, Tianjin 300457, China

Jun Ma
School of Material Science and Chemical Engineering, Tianjin University of Science and Technology, Tianjin 300457, China

Tao Jiang
School of Material Science and Chemical Engineering, Tianjin University of Science and Technology, Tianjin 300457, China

Yanhuan Wang
School of Material Science and Chemical Engineering, Tianjin University of Science and Technology, Tianjin 300457, China

Xuemei Wen
Tianjin Synthetic Material Research Institute, Tianjin 300220, China

Guozhu Li
Key Laboratory for Green Chemical Technology of Ministry of Education,
School of Chemical Engineering and Technology, Tianjin University, Tianjin
300072, China

Jakub Zdarta
Institute of Chemical Technology and Engineering, Faculty of Chemical Technology, Poznan University of Technology, Berdychowo 4, 60965 Poznan, Poland

Łukasz Klapiszewski
Institute of Chemical Technology and Engineering, Faculty of Chemical Technology, Poznan University of Technology, Berdychowo 4, 60965 Poznan, Poland

Marcin Wysokowski
Institute of Chemical Technology and Engineering, Faculty of Chemical Technology, Poznan University of Technology, Berdychowo 4, 60965 Poznan, Poland

Małgorzata Norman
Institute of Chemical Technology and Engineering, Faculty of Chemical Technology, Poznan University of Technology, Berdychowo 4, 60965 Poznan, Poland

Agnieszka Kołodziejczak-Radzimska
Institute of Chemical Technology and Engineering, Faculty of Chemical Technology, Poznan University of Technology, Berdychowo 4, 60965 Poznan, Poland

Dariusz Moszyński
Institute of Inorganic Chemical Technology and Environmental Engineering,
West Pomeranian University of Technology, Pulaskiego 10, 70322 Szczecin,
Poland

Hermann Ehrlich
Institute of Experimental Physics, TU Bergakademie Freiberg, Leipziger Str.
23, 09599 Freiberg, Germany

Hieronim Maciejewski
Adam Mickiewicz University in Poznan, Faculty of Chemistry, Umultowska
89b, 61614 Poznan, Poland

Poznan Science and Technology Park, Adam Mickiewicz University Foundation, Rubież 46, 61612 Poznan, Poland

Allison L. Stelling
Duke University, Center for Materials Genomics, Department of Mechanical Engineering and Materials Science,144 Hudson Hall, Durham, NC 27708, USA

Teofil Jesionowski
Institute of Chemical Technology and Engineering, Faculty of Chemical Technology, Poznan University of Technology, Berdychowo 4, 60965 Poznan, Poland

Raushan Kumar Singh
Department of Chemical Engineering, Konkuk University, 1 Hwayang-Dong, Gwangjin-Gu, Seoul 143-701, Korea

Manish Kumar Tiwari
Department of Chemical Engineering, Konkuk University, 1 Hwayang-Dong, Gwangjin-Gu, Seoul 143-701, Korea

Ranjitha Singh
Department of Chemical Engineering, Konkuk University, 1 Hwayang-Dong, Gwangjin-Gu, Seoul 143-701, Korea

Jung-Kul Lee
Department of Chemical Engineering, Konkuk University, 1 Hwayang-Dong, Gwangjin-Gu, Seoul 143-701, Korea

Quan Feng
Key Laboratory of Eco-Textiles (Ministry of Education), Jiangnan University, Wuxi 214122, China
Textiles and Clothing Department, Anhui Polytechnic University, Wuhu 241000, China

Bin Tang
Textiles and Clothing Department, Anhui Polytechnic University, Wuhu 241000, China

Qufu Wei
Key Laboratory of Eco-Textiles (Ministry of Education), Jiangnan University, Wuxi 214122, China

Dayin Hou
Textiles and Clothing Department, Anhui Polytechnic University, Wuhu 241000, China

Songmei Bi
Textiles and Clothing Department, Anhui Polytechnic University, Wuhu 241000, China

Anfang Wei
Key Laboratory of Eco-Textiles (Ministry of Education), Jiangnan University, Wuxi 214122, China

Yuya Asanomi
Measurement Solution Research Center, National Institute of Advanced Industrial Science and Technology (AIST), 807-1 Shuku, Tosu, Saga 841-0052, Japan

Hiroshi Yamaguchi
Liberal Arts Education Center, Aso Campus, Tokai University, Minami-aso, Aso, Kumamoto 869 1404, Japan

Masaya Miyazaki
Measurement Solution Research Center, National Institute of Advanced Industrial Science and Technology (AIST), 807-1 Shuku, Tosu, Saga 841-0052, Japan
Department of Molecular and Material Sciences, Interdisciplinary Graduate School of Engineering Science, Kyushu University, 6-1 Kasuga-koen, Kasuga, Fukuoka 816-8580, Japan
Department of Advanced Technology Fusion, Graduate School of Science and Engineering, Saga University, 1 Honjo, Saga 840-8502, Japan

Hideaki Maeda
Measurement Solution Research Center, National Institute of Advanced Industrial Science and Technology (AIST), 807-1 Shuku, Tosu, Saga 841-0052, Japan
Department of Molecular and Material Sciences, Interdisciplinary Graduate School of Engineering Science, Kyushu University, 6-1 Kasuga-koen, Kasuga, Fukuoka 816-8580, Japan

Douglas Fernandes Silva
Department of Biological Science, University of State of São Paulo (UNESP), 19806-900 Assis, SP, Brazil

Henrique Rosa
Department of Biological Science, University of State of São Paulo (UNESP), 19806-900 Assis, SP, Brazil

Ana Flavia Azevedo Carvalho
Food Engineering Faculty, State University of Campinas (UNICAMP), 13083-970 Campinas, SP, Brazil

Pedro Oliva-Neto
Department of Biological Science, University of State of São Paulo (UNESP), 19806-900 Assis, SP, Brazil

Rakesh Sharma
Center of Nanomagnetics Biotechnology, Florida State University, Tallahassee, FL, USA
Innovations and Solutions Inc. USA, Tallahassee, FL, USA
Amity University, NOIDA, UP India

Preface

As enzymes are biological catalysts that promote the rate of reactions but are not themselves consumed in the reactions; they may be used repeatedly for as long as they remain active. An immobilized enzyme is an enzyme that is attached to an inert, insoluble material such as calcium alginate. The text *Immobilized Enzyme Principles* provides a sound basis for the design of enzymatic reactions based on kinetic principles. First chapter focuses on peptide-modified surfaces for enzyme immobilization. Second chapter provides a review of several important factors on affecting enzyme immobilization, including immobilization methods, immobilization carrier materials, and immobilization enzyme loading. In third chapter, an approximate analytical method to solve the non-linear differential equations in an immobilized enzyme film is presented. Fourth chapter introduces a novel strategy for the preparation of acetate modified cross-linked gelatin nanoparticles (CLGNs) which showed a reversible temperature-triggered swelling. In fifth chapter, we report the development, fabrication and characterization of an amperometric-based glucose biosensor with the aim of comparing the effectiveness of chitosan of different molecular weights as a matrix for enzyme immobilization using adsorption and crosslinking techniques, and to study the behavior of these enzyme-chitosan electrodes. The purpose of sixth chapter is to immobilize HRP onto IBN-4 nanoparticles via covalent linkages to facilitate the activation of the prodrug IAA (indole acetic acid) in the tumor microenvironment. In seventh chapter, a novel dual functional polymer, NH2-Alginate, is synthesized through an oxidation-amination-reduction process. The aim of eighth chapter is to use a chitin-lignin material as a novel matrix for immobilization by adsorption of the lipase from *Aspergillus niger*. Ninth chapter highlights and summarizes various studies that have aimed to improve the biochemical properties of industrially significant enzymes. In tenth chapter, PVA/PA6 composite nanofibrous membranes have been formed by electrospinning. In eleventh chapter, we summarize the recent advances of microchannel reaction technologies especially for enzyme immobilized microreactors. In twelfth chapter, the reuse of papain through the recovery of soluble enzyme by centrifugation or by its immobilization on polysaccharides has been evaluated for S. cerevisiae cells deflocculation from fuel ethanol distilleries. Last chapter introduces basic tenets of classic presumptions of enzyme inhibition, types of enzyme inhibitors, different models of enzyme inhibition, and scientific basis of emerging immobilized enzyme technology in different applications.

Chapter 1

PEPTIDE-MODIFIED SURFACES FOR ENZYME IMMOBILIZATION

Jinglin Fu[1,2,3], Jeremy Reinhold[1,3], Neal W. Woodbury[1,3]

[1]Center for Single Molecule Biophysics, Arizona State University, Tempe, Arizona, United States of America,

[2]Center for Innovations in Medicine, the Biodesign Institute, Arizona State University, Tempe, Arizona, United States of America,

[3]Department of Chemistry and Biochemistry, Arizona State University, Tempe, Arizona, United States of America

ABSTRACT

Background

Chemistry and particularly enzymology at surfaces is a topic of rapidly growing interest, both in terms of its role in biological systems and its application in biocatalysis. Existing protein immobilization approaches, including noncovalent or covalent attachments to solid supports, have difficulties in controlling protein orientation, reducing nonspecific absorption and preventing protein denaturation. New strategies for enzyme immobilization are needed that allow the precise control over orientation and position and thereby provide optimized activity.

Methodology/Principal Findings

A method is presented for utilizing peptide ligands to immobilize enzymes on surfaces with improved enzyme activity and stability. The appropriate peptide ligands have been rapidly selected from high-density arrays and when desirable, the peptide sequences were further optimized by single-point variant screening to enhance both the affinity and activity of the bound enzyme. For proof of concept, the peptides that bound to β-galactosidase and optimized its activity were covalently attached to surfaces for the purpose of capturing target enzymes. Compared to conventional methods, enzymes immobilized on peptide-modified surfaces exhibited higher specific activity and stability, as well as controlled protein orientation.

Conclusions/Significance

A simple method for immobilizing enzymes through specific interactions with peptides anchored on surfaces has been developed. This approach will be applicable to the immobilization of a wide variety of enzymes on surfaces with optimized orientation, location and performance, and provides a potential mechanism for the patterned self-assembly of multiple enzymes on surfaces.

INTRODUCTION

Surface-immobilized enzymes play an important role in many biocatalytic processes and industrial applications [1], [2]. The activity, stability and selectivity of enzymes can be improved if they are immobilized properly on surfaces [1], [3]. Many conventional protein immobilization methods [1], which rely on nonspecific absorption of proteins to solid supports or chemical coupling of reactive groups within proteins, have inherent difficulties, such as protein denaturation, poor stability due to nonspecific absorption [4], [5], variations in the spatial distance between enzymes and the enzyme-to-surface distance [6], and the inability to control protein orientation [1], [5]. New strategies for enzyme immobilization are needed which allow the precise control over orientation and position and thereby provide optimized activity[7]. Peptides represent a promising class of potential protein-anchoring/modulating molecules due to their large chemical diversity [8] and the existence of well-established methods for library synthesis [9]. There is a growing realization that, by using peptides as building blocks, it is possible to create synthetic structures with affinities and specificities comparable to natural antibodies [10], [11]. Peptide or small molecule ligands that bind to a unique region of a protein can be used for orienting the protein and modulating its activity through specific ligand-protein interactions on a solid support [11]–[13]. In this work, we present a method for creating peptide-modified surfaces that immobilize a target enzyme with optimized orientation and activity.

RESULTS

Previously, we described an approach for screening high-density peptide arrays to identify specific peptide sequences that anchor enzymes to surfaces and modulate their activity [13]. To demonstrate the utility of this approach more generally for optimized enzyme immobilization, two 20-mer peptides, YHNNPGFRVMQQNKLHHGSC (referred to as YHNN) and

QYHHFMNLKRQGRAQAYGSC (referred to as QYHH) were selected from a microarray of 10,000 peptides based on their ability to bind β-galactosidase (β-gal) and optimize its surface-immobilized activity (Table S1 in File S1). These peptides were then synthesized and covalently conjugated to aminated microwells, modifying the surface and mediating the binding of β-gal through specific peptide-enzyme interactions (Figure 1a). As controls, two inhibitory peptides, RVFKRYKRWLHVSRYYFGSC (RVFK) and PASMFSYFKKQGYYYKLGSC (PASM), and one weak-binding peptide, EFSNPTAQVFPDFWMSDGSC (EFSN), were also used to modify aminated microwells (Table S1 in File S1). β-Gal immobilized on YHNN- and QYHH-surfaces exhibited much higher activity than β-Gal immobilized on control peptide-modified surfaces (Figure 1b). The relative specific activities of β-Gal immobilized on peptide-modified microwells were shown in Table 1, which were calculated for each surface by dividing the total bound enzyme activity by the total binding intensity. Conventional surface immobilization approaches were also tested including SMCC-activated (SMCC 6) and NHS-activated (NHS 7) covalent attachment, as well as noncovalent amine-surface attachment (Amine 8). In Table 1, YHNN- and QYHH– modified surfaces resulted in a specific activity of bound enzyme that was ~2-fold greater than amine noncovalent binding and nearly 3-fold greater than NHS attachment. In addition, the YHNN- and QYHH–modified surfaces have the advantage of specifically associating with β-Gal in a protein mixture. This was shown by binding β-Gal in a solution containing 3% Bovine serum albumin (BSA). YHNN- and QYHH– modified surfaces showed 15-fold more bound enzyme activity than the amine surface and 20-fold more than the NHS surface (Figure S1 in File S1).

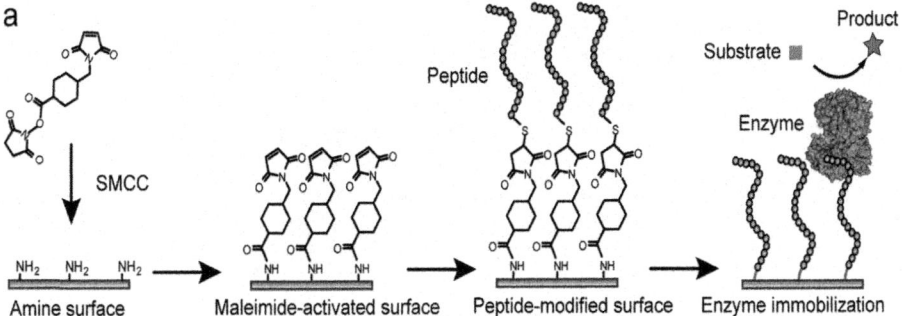

a

SMCC

Peptide

Substrate ■   Product ☆

Enzyme

NH₂   NH₂   NH₂
Amine surface   Maleimide-activated surface   Peptide-modified surface   Enzyme immobilization

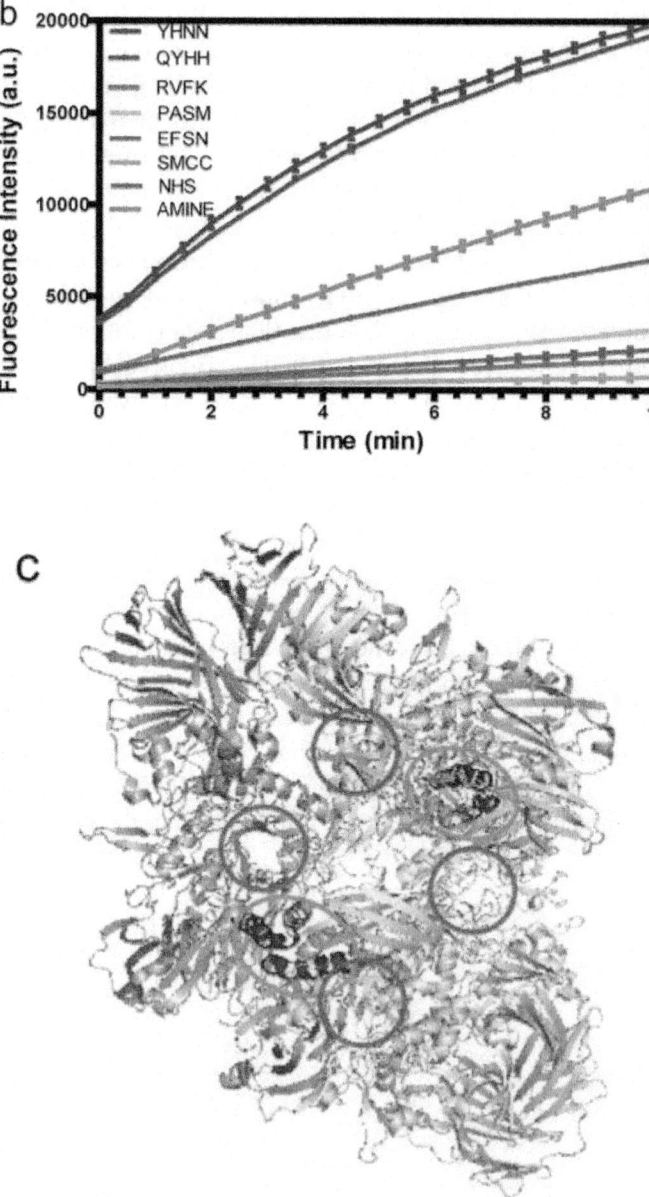

Figure 1: Enzyme immobilization on peptide-modified surfaces.

(a) The overall process for conjugating peptides to aminated microwells through specific reactions between C-terminal cysteines and maleimide-activated surfaces. (b) Activity of β-Gal immobilized on different surfaces. 25 nM β-Gal is first incubated with modified microwells for one hour and then enzyme activity is measured at 25°C as a function of time using 100 μM Resorufin β-D-galactopyranoside as the substrate. YHNN, QYHH, RVFK, PASM and EFSN represent β-Gal bound to various peptide-modified surfaces (see text). SMCC and NHS represent enzyme covalently bound via thiol and amine conjugation, respectively. AMINE represents enzyme bound noncovalently to an aminated surface. (c) Proteolytic mapping of peptide binding to tetrameric β-Gal with binding regions circled (Green). Each subunit is labeled with a unique color showing the symmetry of the β-Gal structure. The binding regions (amino acids 419–447) are highlighted in blue. The substrate-binding sites of β-Gal are circled with right color (Glu461, Met502, Tyr503 and Glu537) according to reference 16.

Table 1: Normalized activity and affinity of β-Gal immobilized on surface-modified microwells[a]

Surface for protein immobilization	Binding affinity (Norm.)	Activity (Norm.)	Specific activity (Norm.)
1 YHNNPGFRVMQQNKLHHGSC	0.9±0.03	2.1±0.1	2.2±0.1
2 QYHHFMNLKRQGRAQAYGSC	0.9±0.02	2.3±0.1	2.4±0.2
3 RVFKRYKRWLHVSRYYFGSC	0.9±0.05	0.1±0.01	0.1±0.01
4 PASMFSYFKKQGYYYKLGSC	0.5±0.1	0.3±0.06	0.5±0.1
5 EFSNPTAQVFPDFWMSDGSC	0.1±0.01	0.05±0.01	0.4±0.1
6 SMCC	0.2±0.02	0.03±0.01	0.1±0.1
7 NHS	0.8±0.1	0.6±0.1	0.8±0.1
8 Amine	1.0±0.1	1.0±0.1	1.0±0.1

[a]Types of surfaces: 1 and 2 are selected peptide-modified surfaces; 3 and 4 are control surfaces modified by inhibitory peptides; 5 is a control surface conjugated with a weak-binding peptide; 6–8 are the conventional surfaces used for covalent or noncovalent enzyme immobilization, defined as in Figure 1, legend. All of the data is normalized to that of the amine surface, 8

In addition, YHNN- and QYHH-modified surfaces were also found to improve the thermal and pH stability of immobilized β-Gal. The thermal stability of bound β-Gal was ~16-fold greater on the peptide-modified surfaces than free enzyme in solution after incubating at 55°C for one hour (Figure S2 in File S1) and more than 2-fold better than enzyme immobilized to either the NHS or amine surfaces. Immobilization of β-Gal on YHNN- and QYHH-modified surfaces also shifted the pH optimum from pH 8 in free solution to 7 on the surface. Long-term enzyme stability to storage on surfaces was greatly improved on peptide-modified surfaces, particularly when peptide modification was combined with the use of a hydrogel (5% polyvinyl alcohol) coating. β-Gal immobilized in this way and stored dry for one week at room temperature retained ~35% of its original activity. In contrast, enzyme similarly immobilized and stored on amine surfaces retained less than 5% activity and NHS surfaces retained ~14% (Figure S3 inFile S1). If one considers both the increased binding capacity of the peptide-modified surfaces and their increased stability to storage, there was 20-fold more enzyme activity per surface area after storage on the peptide-modified surfaces than either the amine surfaces or the NHS surfaces, a significant factor in the commercial immobilization and storage of enzymes.

The apparent K_d values of the YHNN- and QYHH-modified surfaces were ~5 nM and ~4 nM for β-Gal, respectively (Figure S4 in File S1). The apparent k_{cat} and K_m constants for immobilized β-Gal were measured on peptide-modified iodoacetyl resin, which has a large binding capacity and allows for the quantification of the absolute amount of bound enzyme (Figure S5 in File S1). k_{cat} values were ~46 s^{-1} for the YHNN- surface and ~53 s^{-1} for the QYHH- surface, similar to the k_{cat} of ~58 s^{-1} under the same conditions for the free enzyme. The apparent K_m values of β-Gal bound to the YHNN- and QYHH-modified surfaces were ~240 μM and 250 μM, respectively, compared ~130 μM for the free enzyme. The apparent increase in K_m for the surface-bound enzyme may be due to slow diffusion of substrate molecules to the surface and local substrate depletion [14].

Peptide-protein binding sites for YHNN and QYHH were determined by proteolytic mapping using reversible formaldehyde cross-linking [15]. YHNN and QYHH both bound to the same protein fragments (residues 419–447) at the subunit interface of β-Gal (Figure S6 in File S1). β-Gal from *E. coli* is only active in its tetrameric form [16], and it may be that YHNN and QYHH enhance the activity and stability of β-Gal by stabilizing its tetrameric structure.

Point-variant screening [17], [18] was applied to the YHNN peptide to improve both the affinity and activity of bound enzyme. 132 single-point variants, containing all substitutions of the amino acid set {Y, A, D, S, K, N, V, W} in each of the 17 randomized positions, were synthesized, printed on a microarray and analyzed for affinity and activity. Figure 2a shows the binding level vs. activity of β-Gal for each single-point variant, normalized to the YHNN- lead peptide. Several variants increased both binding level and activity, (region ii), including variant V9Y (YHNNPGFRYMQQNKLHHGSC) which increased binding by 1.5-fold and specific activity by nearly 3-fold compared to the YHNN- lead peptide (Table S2 in File S1). V9Y conjugated to an aminated microwell increased both the binding and the specific activity of immobilized β-Gal by ~2-fold compared to YHNN. This corresponds to a total bound enzymatic activity on the V9Y-modified surface that is ~12-fold greater than the NHS surface and more than 5-fold greater than the amine surface (Figure S7 in File S1). Combining two advantageous point mutations into a single peptide (e.g. V9Y and N13Y, Supplemental Table S2 in File S1) resulted in an increase in the affinity of the peptide for binding to β-Gal but did not significantly enhance the specific activity of bound enzyme compared to single-point variants.

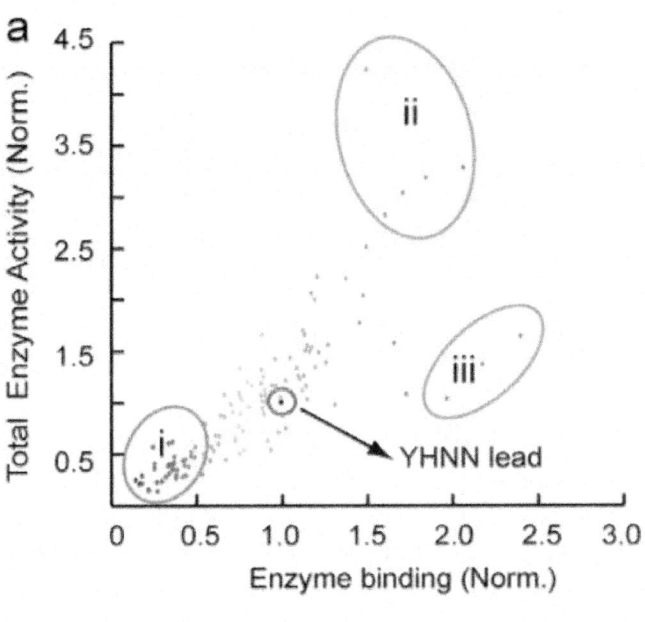

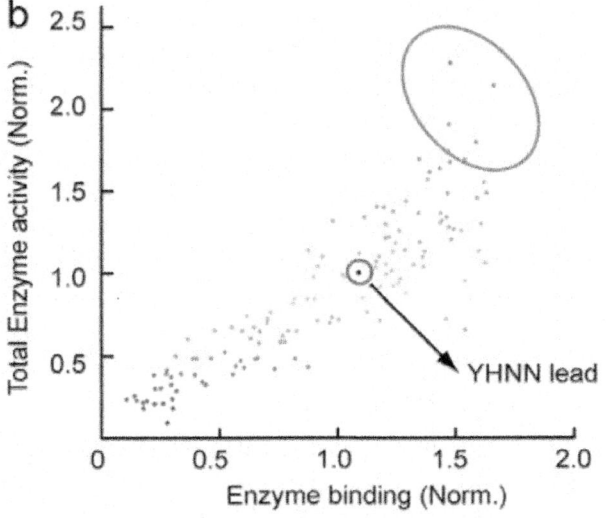

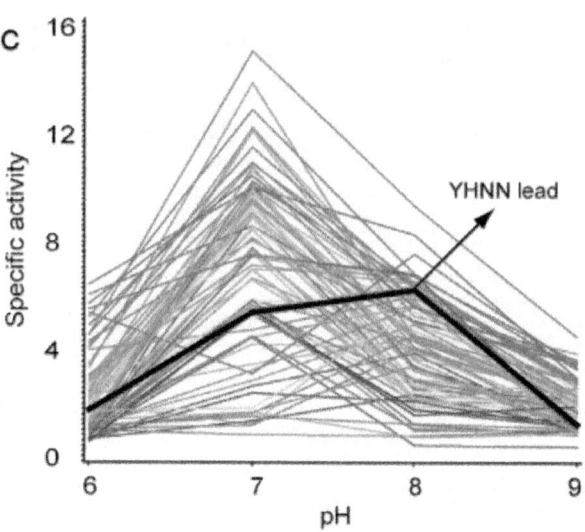

Figure 2: Point-variant screening of a lead peptide, YHNN.

β-Gal was bound to a microarray containing 132 YHNN variants and its activity was measured. (a) The activity of bound β-Gal on microarrays as a function of the amount of enzyme bound to a particular variant feature at room temperature. (i) Variants with poor affinity and activity; (ii) Variants with stronger affinity and higher activity; (iii) Variants with stronger affinity but relative lower activity. All data is normalized to the binding and activity values for the lead peptide, YHNN. (b) Thermal-stability assay. β-Gal was bound to the microarray containing YHNN variants as in (a) at room temperature, followed by incubation in phosphate buffer at 55°C for one hour. Enzyme activity was then assayed at room temperature. The selection region (circled) contains variants that bind to the enzyme with higher relative specific activity (the ratio of binding to activity) under thermal stress compared to YHNN after incubation at high temperature. (c) pH activity range assay. YHNN variant microarrays were bound to β-Gal as in (a) and incubated at room temperature in buffers with pHs ranging from 6 to 9 for one hour and then assayed for activity at the pH of incubation. The black line is the specific activity of β-Gal bound to the lead peptide, YHNN.

The library of single-point variants was also screened for enhanced thermal or pH stability of immobilized β-Gal. For thermal stability screening, enzyme was bound to microarrays containing the 132 single-point variants, at room temperature, and then the arrays were incubated at 55°C for one hour and assayed for activity at room temperature. A few point variants improved the resulting activity of bound β-Gal by nearly 50% compared to the YHNN- lead peptide (Figure 2a, circled region, and Table S3 in File S1). pH stability was screened by incubating enzyme-bound arrays in buffers ranging from pH 6 to pH 9 for one hour and then assaying activity at the pH used for incubation. In Figure 2c, some variants were found to significantly improve the specific activity of bound β-Gal at both low (pH 6) and high (pH 9) pH compared to the YHNN- lead peptide (e.g. Q12A, YHNNPGFRVMQANKLHHGSC shows a 4.1-fold activity increase at pH 6 and a 2.8-fold increase at pH 9, Table S4 in File S1).

DISCUSSION

We describe a rapid, systematic and general approach for modifying a surface in such a way that an enzyme can both bind tightly to the surface and maintain or even enhance its activity. Peptides can be rapidly selected from microarrays and covalently conjugated to surfaces for capturing target proteins. Peptide-modified surfaces improve both the specific activity and stability of bound β-Gal compared to free enzyme or to conventional enzyme surface immobilization approaches. In addition, the affinity and activity of one of the peptide-modified surfaces was further improved by single-point variant screening. Variants were found that not only improved activity under normal conditions, but enhanced thermal stability and increased enzyme activity at extreme pH. The surface modification is also specific for a particular enzyme, and thus the binding and activity enhancement can be patterned, opening the door for the development of multi-enzyme systems that are organized using a top-down patterning of surface modification combined with self-assembly of enzymatic systems on those surfaces. This approach appears to be applicable to the immobilization of a wide variety of enzymes on surfaces with optimized performance, and provides a potential mechanism for the patterned self-assembly of multiple enzymes on surfaces.

MATERIALS AND METHODS

Chemicals

Resorufin β-D-galactopyranoside (RBG) and Alexa Fluor 647 were purchased from Invitrogen (Eugene, OR). β-galactosidase (β-Gal, E.coli), polyvinyl alcohol (PVA, M.W.: 124,000~186,000), 4-nitrophenyl phosphate (PNPP), Phosphate Buffered Saline (PBS) and Tris Buffered Saline (TBS) were obtained from Sigma (St. Louis, MO). BS³(Bis[sulfosuccinimidyl] suberate), alkaline phosphatase-conjugated strepavidin and iodoacetyl resin were purchased from Pierce (Rockford, IL). Sulfo succinimidyl-4-(*N*-maleimidomethyl) cyclohexane-1-carboxylate (Sulfo-SMCC) was purchased from bioWORLD (Dublin, OH). Aminated microwell plates were ordered from Corning. A 4 mg/mL stock solution of β-Gal was prepared in 10 mM potassium phosphate buffer with 0.1 mM MgCl$_2$ at pH 7.4.

Enzyme Immobilization on modified Microwells

Peptides were conjugated to aminated microwell surfaces through the specific reaction between C-terminal cysteines and the maleimide-activated surfaces, as shown in Figure 1a. 10 mM SMCC was prepared in 1× PBS buffer, pH 7.4. Next, 30 µL of SMCC was added into each aminated microwell and incubated for one hour at room temperature. The microwell plate was then briefly washed with pure water three times. Then, 30 µL of a 300 µM peptide solution, prepared in 1× PBS pH 7.4 plus 1 mM TCEP, was added to the appropriate SMCC-activated microwells. The reaction was incubated for 4 hours at room temperature, in the dark. After the conjugation reaction was complete, the microwells were washed for 5 minutes in 1× TBST, three times, followed by three washes in water. To immobilize the enzyme on peptide-modified surfaces, 30 µL of 25 nM biotin-labeled β-Gal was incubated in the peptide-modified microwells for two hours in 10 mM phosphate buffer, pH 7.3 with 100 µM MgCl$_2$ and 0.05% Tween 20 (v/v%), at room temperature. The microwells were washed for 5 minutes in 1× TBST, three times, followed by three washes in phosphate buffer. At this point, the β-Gal-bound microwells were ready for testing. β-Gal was labeled with biotin using EZ-Link Sulfo-NHS-Biotinylation Kit purchased from Pierce (labeling ratio: ~two biotin per enzyme molecule). Figures S8–10 in File S1 show the detailed optimization procedures for peptide-modified surfaces.

Covalent attachment of β-Gal to NHS (*N*-Hydroxysuccinimide)-activated surfaces was performed using BS³ homogeneous amine-reactive cross-linker, as recommended by the manufacturer. First, 30 µL of 2 mg/mL BS³ prepared in 1× PBS, pH 7.4 was incubated with the aminated microwells for half an hour. Then, the microwells were briefly washed with nanopure water, three times, to remove unreacted BS³ molecules. Finally, 30 µL of biotin-labeled β-Gal was incubated with the microwells for one hour, which were then washed three times in 1× TBST, followed by three washes in phosphate buffer.

The activity assay of surface-bound β-Gal was performed on a SpectraMax M5 96-well plate reader (Molecular Device, Sunnyvale, CA) by adding 100 µL of 100 µM RBG into the wells. The relative amount of surface-bound β-Gal was measured using an enzyme linked immunosorbant assay (ELISA). β-Gal was first labeled with biotin. Alkaline phosphatase-conjugated strepavidin (0.4

mg/ml) was diluted at 1:1000 in 1× PBS, 0.05% (v/v) Tween 20. Next, 30 μL of streptavidin solution was added to the β-Gal-bound wells and incubated for one hour at room temperature. The streptavidin solution was then removed and the plate was washed three times with TBST buffer and three times with TBS buffer. Then, 200 μL of 1 mM PNPP was added to each well. The alkaline phosphatase activity was subsequently measured by reading the absorbance increase at 405 nm on the M5 plate reader. The β-Gal binding level was determined from the activity of alkaline phosphatase-conjugated strepavidin bound to the wells.

Determining Michaelis constants of immobilized β-Gal

The determination of the enzyme kinetic constants (K_M and k_{cat}) of immobilized β-Gal was performed on peptide-modified iodoacetyl polyacrylamide resin (UltraLink, Pierce, 50–80 μm diameter). To modify the bead surface with peptide, peptide solutions were incubated with iodoacetyl resin for one hour in 50 mM Tris buffer, 5 mM EDTA, pH 8.5. The unreacted iodacetyl groups were then capped with 50 mM L-cysteine. The amount of peptide immobilized on a bead surface was determined by comparing the peptide concentration of the unbound fraction (the remaining free peptide concentration after binding to the surface) to the starting concentration through absorbance changes at 280 nm. β-Gal was captured on the peptide-modified beads using the same protocol which immobilized the enzyme in the microwells, above. The amount of bead-immobilized β-Gal was measured by comparing the protein concentration of the unbound fraction to the starting protein concentration, determined at 280 nm. K_M and V_{max} (and thus k_{cat}, using the total enzyme concentration) of β-Gal immobilized on peptide-modified beads were determined by fitting the activity vs. substrate concentration curves in the GraphPad program using the fitting equation of "$Y=V_{max}*X/(K_m+X)$".

Peptide mapping to β-Gal

The specific regions at which the peptides YHNN and QYHH bind to β-gal were determined by reversible formaldehyde cross-linking, as described previously.[8,12] 200 μL of a 150 μM peptide solution was first conjugated to 100 μL of UltraLink iodoacetyl resin using the method described above. To promote cross-linking, the peptide-modified resin was incubated with 200 μL of 500 nM β-Gal for two hours. 200 μL of 1% formaldehyde (v/v), prepared in 1× PBS, was added to the enzyme-bound resin for 10 mins. Then, the formaldehyde solution was removed quickly by centrifugation. The resin was washed three times with 1 mM Glycine, pH 2.5 to remove enzyme that did not undergo cross-linking. Proteolytic digestion was performed by incubating the enzyme-bound resin with 34 nM Glu-c in ammonium bicarbonate buffer, pH 8.5, overnight at 37°C. Then, the resin was washed again with Glycine, pH 2.5 to remove Glu-c and any fragments that did not undergo cross-linking. The formaldehyde cross-linking was reversed by incubating the resin with 20 µL nanopure water at 70°C overnight. Following cross-link reversal, 100 µL of nanopure water was added to the resin to dissolve the free Glu-c-digested peptide fragments. The solution was spun to the bottom of the spin-column and then dried, by evaporation, in a vacuum centrifuge. The dried sample was re-dissolved with 10 µL of 1:1 acetonitrile:H_2O containing 0.1% trifluoroacetic acid and saturated alpha-cyano-4-hydroxycinammic acid matrix. The sample was spotted on a standard MALDI-MS (Matrix-assisted laser desorption/ ionization mass spectrometry) target plate, and analyzed using a Bruker Microflex MALDI-MS.

Microarray fabrication

Peptide microarrays containing 132, 20-mer single-point variants of the YHNN peptide were generated using our established, in-house printing method [19]. Each microarray was prepared by robotically spotting peptides, in triplicate, on a glass slide possessing an amino-silane surface coating. Synthesized peptides (70% purity) were purchased from Sigma. The last three carboxy-terminal positions of each peptide constituted a glycine-serine-cysteine (GSC) linker, used for conjugating the peptides to amino-silane surfaces through the C-terminal cysteine via a maleimide linker, Sulfo-SMCC (Pierce, Rockford, IL). A Telechem Nanoprint60 was used to spot approximately 500 pL of 1 mg/mL peptide prepared in 1× PBS for each feature on glass slides with 48 Telechem series SMP2 style 946 titanium pins.

Enzyme assays on PVA-coated arrays

Enzyme assays on the microarrays were performed as described in the previous work [13]. Briefly, a peptide microarray was first prewashed with surface cleaning solvent (7.33% (v/v) acetonitrile, 37% isopropyl alcohol and 0.55% trifluoroacetic acid in water) and then treated with capping buffer (3% (v/v) BSA, 0.02% (v/v) mercaptohexanol, 0.05% (v/v) Tween20 in 1×PBS) to block any active SMCC linker on the array surface. The array was incubated with a solution containing 10 nM Alexa™ 647- labeled β-Gal for two hours, allowing the enzyme to bind with peptides on the array surface. After washing off unbound enzymes, a fluorescent substrate analogue (FDG) was mixed with a 5% PVA solution and spin-coated onto the array surface for monitor the enzyme activity. The FDG molecules (substrate) in the PVA layer were converted to fluorescein (product), by the active enzymes bound to specific peptides on the array surface. The fluorescein molecules remained localized because of the PVA viscosity. Both the relative binding level of Alexa™ 647-labeled enzyme and the relative amount of fluorescein produced during the incubation period were determined by dual color scanning. Each array experiment was repeated at least three times under the same conditions for statistical analysis.

ACKNOWLEDGMENTS

We thank John Lainson for the peptide array production; Donnie Shepard, and Loren Howell for the peptide synthesis and purification, Daniel Garry for the initial test on peptide-surface conjugation, Stephen Albert Johnston for helpful discussions and Carole Flores for reading and editing the manuscript.

AUTHOR CONTRIBUTIONS

Conceived and designed the experiments: JF NW. Performed the experiments: JF JR. Analyzed the data: JF JR NW. Contributed reagents/materials/analysis tools: JF JR. Wrote the paper: JF NW. Performed peptide-modified microwell experiments: JR. Edited the paper: NW

REFERENCES

1.   Sheldon Roger A (2007) Enzyme Immobilization: The Quest for Optimum Performance. Advanced Synthesis & Catalysis 349: 1289–1307.

2.   Laurent N, Haddoub R, Flitsch SL (2008) Enzyme catalysis on solid surfaces. Trends in Biotechnology 26: 328–337.

3.   Mateo C, Palomo JM, Fernandez-Lorente G, Guisan JM, Fernandez-Lafuente R (2007) Improvement of enzyme activity, stability and selectivity via immobilization techniques. Enzyme and Microbial Technology 40: 1451–1463.

4.   Cha T, Guo A, Zhu X-Y (2005) Enzymatic activity on a chip: The critical role of protein orientation. PROTEOMICS 5: 416–419.

5.   Clarizia L-JA, Sok D, Wei M, Mead J, Barry C, et al. (2009) Antibody orientation enhanced by selective polymer–protein noncovalent interactions. Anal Bioanal Chem 393: 1531–1538.

6.   Kim J, Jia H, Wang P (2006) Challenges in biocatalysis for enzyme-based biofuel cells. Biotechnology Advances 24: 296–308.

7.   Jung Y, Kang HJ, Lee JM, Jung SO, Yun WS, et al. (2008) Controlled antibody immobilization onto immunoanalytical platforms by synthetic peptide. Analytical Biochemistry 374: 99–105.

8. Devlin J, Panganiban L, Devlin P (1990) Random peptide libraries: a source of specific protein binding molecules. Science 249: 404–406.

9. Fodor S, Read J, Pirrung M, Stryer L, Lu A, et al. (1991) Light-directed, spatially addressable parallel chemical synthesis. Science 251: 767–773.

10. Naffin JL, Han Y, Olivos HJ, Reddy MM, Sun T, et al. (2003) Immobilized Peptides as High-Affinity Capture Agents for Self-Associating Proteins. 10: 251–259.

11. Williams BAR, Diehnelt CW, Belcher P, Greving M, Woodbury NW, et al. (2009) Creating Protein Affinity Reagents by Combining Peptide Ligands on Synthetic DNA Scaffolds. Journal of the American Chemical Society 131: 17233–17241.

12. Naffin JL, Han Y, Olivos HJ, Reddy MM, Sun T, et al. (2003) Immobilized Peptides as High-Affinity Capture Agents for Self-Associating Proteins. Chemistry&Biology 10: 251–259.

13. Fu J, Cai K, Johnston SA, Woodbury NW (2010) Exploring Peptide Space for Enzyme Modulators. Journal of the American Chemical Society 132: 6419–6424.

14. Arrio-Dupont M (1988) An example of substrate channeling between co-immobilized enzymes Coupled activity of myosin ATPase and creatine kinase bound to frog heart myofilaments. FEBS Letters 240: 181–185.

15. Sutherland BW, Toews J, Kast J (2008) Utility of formaldehyde cross-linking and mass spectrometry in the study of protein-protein interactions. Journal of Mass Spectrometry 43: 699–715.

16. Jacobson RH, Zhang XJ, DuBose RF, Matthews BW (1994) Three-dimensional structure of [beta]-galactosidase from E. coli. Nature 369: 761–766.

17. Wells JA (1990) Additivity of mutational effects in proteins. Biochemistry 29: 8509–8517.

18. Greving MP, Belcher PE, Diehnelt CW, Gonzalez-Moa MJ, Emery J, et al. (2010) Thermodynamic Additivity of Sequence Variations: An Algorithm for Creating High Affinity Peptides Without Large Libraries or Structural Information. PLoS ONE. (accepted).

19. Boltz KW, Gonzalez-Moa MJ, Stafford P, Johnston SA, Svarovsk SA (2009) Peptide microarrays for carbohydrate recognition. Analyst 134: 650–652.

Chapter 2

PARAMETERS AFFECTING THE PERFORMANCE OF IMMOBILIZED ENZYME

Dong-Hao Zhang,[1,2] Li-Xia Yuwen,[1] and Li-Juan Peng[1]

[1]College of Pharmacy, Hebei University, Baoding 071002, China

[2]Key Laboratory of Pharmaceutical Quality Control of Hebei Province, College of Pharmacy, Hebei University, Baoding 071002, China

ABSTRACT

Enzyme immobilization has been investigated to improve lipase properties over the past few decades. Different methods and various carriers have been employed to immobilize enzyme. However, the application of enzymatic technology in large scale is rarely seen during the industrial process. The main obstacles are a high cost of the immobilization and the poor performance of immobilized lipase. This review focuses on the current status of enzyme immobilization, which aims to summarize the latest research on the parameters affecting the performance of immobilized enzyme. Particularly, the effect of immobilization methods, immobilization carriers, and enzyme loading has been discussed.

INTRODUCTION

Enzyme, which is produced from active cells, is a highly efficient catalyst. Compared with chemical catalyst, it has many advantages such as a high specificity, a high catalytic efficiency, and an adjustable activity, which greatly promote enzyme to be used in pharmaceutical, chemical, and food industries [1–4]. However, its poor stability, reusability, and high cost of single use limit its use in industrial production. In recent years, enzyme immobilization technology provides an effective method to circumvent these issues, which not only improves lipase catalytic properties and operational stability but also facilitates enzymes multiple reuse, separation, and continuously automatic operation in industrial production [5–7].

So far, various carriers and methodologies have been used for enzyme immobilization in order to improve the properties of free enzyme [8–11]. In fact, the performance of immobilized lipase relies on several important factors including immobilization methods, immobilization carrier materials, enzyme pretreatment before immobilization, and enzyme loading on the carriers. For instance, Knezevic et al. [12] discovered that coupling lipase via its carbohydrate moiety previously modified by periodate oxidation method had a high activity retention. Vaidya et al. [6] demonstrated that the porous AGE-EGDM polymer particles proved to be a better carrier for immobilization of Candida rugosa lipase than the porous GMA-EGDM copolymers due to the former having a larger specific surface area. Lee et al. [13] found that pretreatment of lipase with soybean oil before immobilization could prevent lipase activity loss when it was immobilized covalently on silica gel. Besides, our previous results suggested that the activity recovery and immobilized ratio had been influenced by the carrier surface properties [14]. Thus, exploiting novel immobilization methods and carrier materials have an important significance on enzyme immobilization. This paper provides a review of several important factors on affecting enzyme immobilization, including immobilization methods, immobilization carrier materials, and immobilization enzyme loading.

EFFECT OF IMMOBILIZATION METHODS

Many methods have been established in order to achieve immobilized enzyme, and each has its advantages and defects. The methods used to date include physical adsorption, entrapment, covalent binding, and the immobilization via a spacer arm.

Enzyme Immobilization via Physical Adsorption

Adsorption immobilization is a method which is used to immobilize enzyme by the attachment of enzyme on carrier surface via weak forces, such as van der Walls force, electrostatic force, hydrophobic interaction, and hydrogen bond [15]. Not surprisingly, the specific surface area of carriers is an important factor to influence the adsorption amount of enzyme. Recently, a new synthesized cyclodextrin-based carbonate nanosponge (CD-NS-1:4) was used for adsorption of Pseudomonas fluorescens lipase [16]. The results showed that the immobilized lipase presented high stabilization and good activity, even after two months of incubation at 18°C. Following the same method, Candida antarctica lipase B was adsorbed on macroporous resin NKA-9 (a polar macroporous resin produced by Chemical Plant of Nankai University, Tianjin, China) in organic medium and the immobilization conditions were optimized

[17]. Consequently, the activity recovery achieved 83% with the mass ratio of lipase to support of 1 : 80 and immobilization time of two hours at 30°C. Compared with the free lipase, the immobilized lipase showed enhanced pH value and thermal stability. Also, Ponvel et al. [18] used magnetite particles modified with alkyl benzenesulfonate as carriers to immobilizePorcine pancreas lipase via physisorption. The immobilized lipase showed enhanced specific activity (8.7 U/mg) and durability in the reuse after recovery by magnetic separations. In addition, several other publications also reported that lipase immobilization via adsorption showed higher activity than free lipase [19, 20].

Physical adsorption immobilization is one of the simplest methods and can be conducted under mild conditions. This method does not result in large loss of enzyme activity. Despite many merits of adsorption, it also presents some drawbacks. For instance, the immobilized enzyme prepared by adsorption has poor operation stability; the amount of adsorbed enzyme is more susceptible to the immobilization parameters such as temperature, ionic strength, and pH; and enzyme can be stripped off easily from the carrier because of the weak forces between them.

Enzyme Immobilization via Encapsulation

For this immobilization method, the enzyme is entrapped in the internal structure of polymer material. In the study by Yang et al. [21], lipase from Arthrobacter sp. was immobilized by encapsulation in hydrophobic sol-gel materials. Under the optimum conditions, the total activity of the prepared enzyme achieved 13.6-fold of the free form. Moreover, the encapsulated lipase exhibited higher thermal stability and operational stability than the free lipase. Compared with covalently immobilized lipase, a higher percent activity yield of the encapsulated lipase (65 U/g) was obtained when Candida rugosa lipase was encapsulated within a chemically inert sol-gel support [22]. In addition, the encapsulated lipases had higher catalytic conversion in the hydrolysis of p-nitrophenyl palmitate and higher enantioselectivity in the enantioselective hydrolysis of racemic naproxen methyl ester than that of immobilized lipase prepared by covalent binding [22], which is probably because that this encapsulation immobilization preserves the mobility of the enzyme and allows to increase its activity and enantioselectivity.

Enzyme Immobilization via Covalent Binding

Covalent binding immobilization is a method which is used to immobilize enzyme by binding the nonessential pendant group of enzyme to the functional group of carrier via chemical bonds. It must be pointed out that the immobilization reaction to form the chemical bond should be carried out under mild conditions because the vigorous reaction conditions can destroy enzyme active conformation. Recently, functionalized Fe_3O_4 nanoparticles modified with carboxymethylated chitosan were developed and used as carrier for the covalent conjugation of papain [23]. Compared with the free papain, the magnetic immobilized papain exhibited good superparamagnetism, enhanced enzyme activity, better tolerance to the variations of medium pH and temperature, and improved storage stability as well as good reusability. More recently, Bai et al. synthesized a series of mesoporous and hydrophilic bead carriers containing epoxy groups and used them to immobilize glucoamylase (Glu) by forming covalent bond between epoxy groups and enzymes [24]. The activity recovery reached 86% and the immobilized Glus exhibited excellent stability and reusability than that of the free ones. In many cases, however, some of the carriers do not have functional group or the immobilization conditions are harsh if they are coupled directly. Therefore, enzyme cannot be coupled directly to the carriers. In order to reduce the loss of enzyme activity, the researchers thus often activate the carriers using some functional reagents before immobilization, which can make the enzyme immobilization conditions mild. For instance, using carbodiimide as coupling agent to activate the hydroxyl groups of chitosan, Chiou and Wu had successfully immobilized Candida rugosa lipase on chitosan [25]. Their results showed that the operational stability and storage stability of the immobilized lipase were enhanced greatly than that of free lipase. In addition, carbonyldiimidazole was also often used to activate the hydroxyl, carboxyl, and amino group in order to immobilize enzyme under mild condition [26, 27].

Generally, enzyme immobilization by covalent binding method can combine enzyme with carrier firmly and avoid the shedding and leakage of enzyme. However, the defect of this method is that it often causes the low activity recovery, which is resulted from the destruction of enzyme active conformation during immobilization reaction, the multipoint attachment to the supports, steric hindrance of enzyme, or the strong strength of the covalent binding.

Enzyme Immobilization via a Spacer Arm

Immobilization of enzyme on carriers via a spacer arm seems to be a good way to avoid the steric hindrance and to increase enzyme activity. This type of immobilization forms a spacer arm between enzyme and carriers by means of a bifunctional reagent such as glutaraldehyde and isocyanate. With the introduction of a flexible spacer arm onto the supports, the enzyme can be allowed to stretch flexibly and catch the substrate more easily. Isgrove et al. [28] reported that β-glucosidase and trypsin immobilized on carriers via polyethyleneimine who act as spacer arm exhibited an enhanced activity compared with the corresponding carriers without spacer arm. Recently, Bayramo lu et al. studied the effect of coupling methods on the immobilized lipase [29]. They found that Candida rugosa lipase immobilized onto poly(GMA-HEMA-EGDMA) microspheres (these microspheres were prepared via suspension polymerisation by using glycidyl methacrylate (GMA), 2-hydroxyethyl methacrylate (HEMA), and ethylene glycol dimethacrylate (EGDMA) as monomers) using 1,6-diaminohexane as spacer arm showed a higher loading capacity and apparent activity compared with the lipase directly immobilized via the epoxy groups. They thought that the increase in loading capacities and lipase activity was due to the spacer-arm effect. Kim et al. also investigated the effect of three different immobilization methods on lipase immobilized on functionalized silica nanoparticles [30]. They found that high loading capacity and high activity were gained when lipase was immobilized on ethylene diamine-activated silica nanoparticles via glutaraldehyde (GA) or 1,4-phenylene diisothiocyanate (NCS) as a coupling agent. More recently, Hu et al. employed magnetic Fe_3O_4 nanoparticles modified with 3-aminopropyltriethoxysilane as carriers to immobilize lipase from S. marcescens ECU1010 (SmL) with glutaraldehyde as the coupling agent [31]. The immobilized lipase showed a higher binding efficiency and activity recovery than that of lipase adsorbed directly onto the supports. As well as, the similar results have been reported through introduction of aminopropyl spacer arm as Bacillus licheniformis L-arabinose isomerase (BLAI) was immobilized [32].

Many studies indicated that the spacer arm between enzyme and carriers could remove the enzyme away from the surface of carriers and prevent undesirable side attachment between enzyme molecules and support. This immobilization method favors the activity retention of immobilized enzyme and improves the performance of immobilized enzyme.

EFFECT OF IMMOBILIZATION CARRIERS

Another factor contributing a lot to the immobilized lipase is the immobilization carrier materials. Carrier material should be readily available, nontoxic, and should offer a good biological compatibility for enzyme [33]. As a part of the immobilized enzyme, the structure and property of the carrier have important impacts on the enzymatic properties. Several natural polymer materials and inorganic particles were commonly used as supporting materials [34–36]. Besides, many researchers had grown great concern in using various synthesized polymer materials as carriers for their good mechanical and easily adjustable properties [37–39]. Particularly, inorganic-organic composite materials have attracted deep attention [40–42].

Natural Polymer Materials

When it comes to the natural polymer materials, much attention has been paid to cellulose, chitin, chitosan, starch, and other natural polymer materials owning to their wide range of sources, easy modification, nontoxic, and pollution-free, with a variety of functional groups and good biocompatible properties. According to the previous study [43], Candida rugosa lipase which was immobilized on activated chitin by covalent attachment showed a higher thermal stability than that of the free lipase. Moreover, the immobilized lipase proved to be effective and reusable for synthesis of butyl esters. Another data reported by Hung et al. showed that immobilized lipase exhibited broader pH tolerance, higher heat stability, and still retained 74% residual activity after ten hydrolysis cycles when it was immobilized on chitosan [44]. In addition, Chiou and Wu carried out immobilization of Candida rugosa lipase on chitosan supports containing hydroxyl groups [45]. The resulting immobilized lipase enhanced the stability of lipase and exhibited excellent operational stability. And the prepared chitosan beads proved to be a good carrier for lipase immobilization.

Synthetic Polymer Materials

Synthetic polymer materials are prepared by chemical polymerization using various monomers. As a kind of important carrier, synthetic polymer materials exhibit the advantages of good mechanical rigidity, high specific surface area, easy to change their surface characteristics, and their potential for bringing specific functional group according to actual needs. Hence, they have been widely investigated and used for enzyme immobilization. Carriers which have large surface area always do a great help to obtain good immobilization efficiency. For instance, the macroporous polyacrylamide (PAM) microspheres which had a large surface area were synthesized by Lei and Jiang via inverse suspension polymerization and pectinase was covalently immobilized onto them [46]. The results showed that up to 296 mg of enzyme was immobilized per gram of the supports. Moreover, the immobilized pectinase displayed an improved thermal stability and storage stability over free lipase, and they also exhibited a better reusability than pectinase entrapped in alginate reported by Roy et al. [47]. That was possibly because PAM supports possessed many amino groups and they could provide a great number of available binding sites for immobilizing pectinase by covalent attachment. Li et al. selected polystyrene (PST) microspheres as carrier to immobilize lipase from Burkholderia cepacia and they researched the effect of the carriers' pore sizes on lipase immobilization [48]. They found that the specific activity of immobilized lipase had a close correlation with the pore size of the PST microspheres. The thermal and storage stabilities of immobilized lipase were enhanced with the increase in the support's pore sizes. It has also been reported that the property of the support was an important factor influencing the enzyme catalytic performance [49, 50]. In the study by Menaa et al., high loading capacity of protein could be obtained when carrier surface was regulated with hydrophobic groups [51]. To investigate the effect of carrier surface properties on lipase immobilization, Zhang et al. prepared a series of poly(vinyl acetate-acrylamide) microspheres with different hydrophobic/hydrophilic surface characteristics to immobilize Candida rugosalipase [52]. Their results showed that hydrophobicity/hydrophilicity of the microspheres could influence not only the immobilized ratio but also the immobilized lipase activity. This was also pointed out in several other previous reports [53, 54]. Another publication reported that Candida species 99–125 lipase was immobilized on macroporous polyglycidylmethacrylate beads via covalent binding [55]. It had been mentioned that crosslinker and porogen used in preparing the polymer beads could influence the porous structure of the polymer particles and thus affect enzyme immobilization. So the authors optimized their using amount in the preparation of poly-GMA. The prepared immobilized lipase under the optimum lipase concentration showed a good pH stability, thermal stability, and reusability. In short, synthetic polymer materials can be obtained in various forms. Moreover, certain desirable functional groups can be introduced by polymerization using different monomers or modification by different groups.

Inorganic Magnetic Particles

Many magnetic particles have been used to immobilize enzyme with good results because they can provide some advantages such as small particle size, good superparamagnetism, and large specific surface area. Thus, great attention has also been aroused using inorganic magnetic particles to immobilize enzyme. Among many types of inorganic magnetic particles, nanoiron oxide has gained much attention for its efficient carrier properties and easy recovery with the aid of a magnet. It had been mentioned that functionalized γ-Fe_2O_3 magnetic nanoparticles were used as carrier to immobilize Candida rugosa lipase via chemical bond [56]. The prepared immobilized lipase showed long-term stability, which proved the significant advantage of Fe_2O_3 nanoparticles as enzyme immobilization carriers. Recently, lipase from Serratia marcescens has successfully been immobilized onto the amino-functionalized magnetic nanoparticles [31]. The results showed that the immobilized protein load could reach as high as 35.2 mg protein g^{-1} support and the activity recovery was up to 62.0%. Moreover, the immobilized lipase demonstrated a high enantioselectivity toward (+)-MPGM and it also displayed the improved thermal stability as compared to the free lipase. More recently, it has been reported that thioredoxin (Trx1) immobilized on iron oxide superparamagnetic nanoparticles with an additional silica shell coating before being treated with 3-aminopropyltriethoxysilane (APTS) via EDC (1-ethyl-3-{3-dimethylaminopropyl}carbodiimide) method showed good activity recovery and recyclability [57]. Compared with Trx1 immobilized on iron oxide superparamagnetic nanoparticles single coated with APTS(Trx1/APTES-MagNps), the doubly coated system exhibited more stability and could be recycled many times. The authors explained this to that an inhibitory site had been created after the first catalytic assay of the Trx1/APTES-MagNps.

Magnetic Polymer Microspheres

Magnetic polymer microspheres have been widely investigated not only because they possess favorable superparamagnetic properties facilitating its recovery from the reaction mixture, but also because they have good biocompatibility and various surface functional groups which are suitable to couple enzymes. In recent years, great efforts have been put into the studies of using magnetic polymer microspheres for enzyme immobilization. Liu et al. used methacrylate and crosslinker divinylbenzene to synthesize poly(methacrylate-divinylbenzene) magnetic microsphere in the presence of magnetic fluid via modified suspension polymerization [40]. The resulting magnetic microspheres exhibited superparamagnetism and they were used as supports to immobilize Candida cylindracea lipase. The results indicated that the immobilized lipase held high enzyme loading (34.0 mg g^{-1} support), high activity recovery (72.4%), and good stability during the repeated use. Moreover, the immobilized lipase showed better heat resistance and was more stable than the free one. In another work, epoxy-functionalized magnetic chitosan beads were prepared in the presence of epichlorohydrin via phase-inversion technique and they were used as carriers to immobilize Trametes versicolor laccase by covalent immobilization [41]. The measurement indicated that the magnetic chitosan beads had a large pore volume and a porous surface structure, which would favor high immobilization capacity for the enzyme. Furthermore, the experiments revealed that the thermal stability and storage stability of laccase were enhanced when enzyme was immobilized on the magnetic chitosan beads. Besides, immobilization of lipase from Porcine pancreas on magnetic Fe_3O_4-chitosan microspheres was also successfully achieved [42]. That study implied that microspheres synthesized with and without magnetic field (MF) had different morphology and surface area. The experiments illustrated that lipase immobilized on microspheres prepared under MF showed higher activity and better operational and storage stability than those immobilized on microspheres prepared without MF. In short, magnetic polymer material is a kind of excellent support for enzyme immobilization, and it will be widely used for its good biocompatibility and magnetic feature which enables it to achieve a rapidly easy separation from the reaction medium in a magnetic field.

EFFECT OF ENZYME LOADING

The excessive enzyme loading always causes protein-protein interaction and inhibits the flexible stretching of enzyme conformation, which will result in the steric hindrance and thus the inactivation of an enzyme. That is, the enzyme molecule may be difficult to modulate its most suitable conformation for catching the substrate molecules and releasing product molecules under molecular crowding condition. Recently, several authors have investigated the effect of enzyme loading on the immobilization [46, 52, 58, 59]. For instance, in the study of pectinase immobilization on macroporous polyacrylamide microspheres by Lei and Jiang [46], they found that the activity of the immobilized pectinase decreased instead when the enzyme amount increased from 10 to 12 units/mL. A similar result was also obtained by other research groups [58]. Hu et al. also discovered that an over-high amount of lipase adsorbed onto the support would make the lipase form an intermolecular steric hindrance and thus influence the performance of lipase [31]. Zhang et al. investigated the effect of steric hindrance on the immobilized lipase [52]. The experiments further demonstrated that an excessive lipase loading had resulted in an intermolecular steric hindrance and greatly affected the lipase activity. In addition, it was also shown by Xie and Ma [59] that the activity recovery of the immobilized lipase increased at the initial stage with the increase in enzyme amount from 1 to 5 mg per 200 mg carriers and then remained nearly constant with further increase in lipase amount to 9 mg. In general, the amount of enzyme immobilized on carriers has obvious influence on the performance of the immobilized enzyme. Therefore, we should pay enough attention to it as preparing the immobilized enzyme.

CONCLUSIONS

In all, many parameters will have influence on the properties of enzyme during enzyme immobilization. Particularly, immobilization methods, carrier materials, and enzyme loading amount have proven to be important for enzyme immobilization. Therefore, we should try to select a suitable carrier as well as an appropriate method for immobilization. Moreover, the immobilized enzyme amount could not be blindly raised in immobilization with the purpose of enhancing enzyme activity.

ACKNOWLEDGMENTS

This work was supported by the Natural Science Foundation of Hebei, China (B2011201012), National Natural Science Foundation of China, and the project sponsored by the Scientific Research Foundation for the Returned Overseas Chinese Scholars, State Education Ministry (the project sponsored by SRF for ROCS, SEM).

REFERENCES

1.   S. Akgöl, Y. Kaçar, A. Denizli, and M. Y. Arca, "Hydrolysis of sucrose by invertase immobilized onto novel magnetic polyvinylalcohol microspheres," Food Chemistry, vol. 74, no. 3, pp. 281–288, 2001.

2.   D.-T. Zhao, E.-N. Xun, J.-X. Wang et al., "Enantioselective esterification of ibuprofen by a novel thermophilic Biocatalyst: APE1547," Biotechnology and Bioprocess Engineering, vol. 16, no. 4, pp. 638–644, 2011.

3.   P. Pires-Cabral, M. M. R. da Fonseca, and S. Ferreira-Dias, "Esterification activity and operational stability of Candida rugosa lipase immobilized in polyurethane foams in the production of ethyl butyrate," Biochemical Engineering Journal, vol. 48, no. 2, pp. 246–252, 2010.

4.   Y. Yücel, C. Demir, N. Dizge, and B. Keskinler, "Lipase immobilization and production of fatty acid methyl esters from canola oil using immobilized lipase," Biomass and Bioenergy, vol. 35, no. 4, pp. 1496–1501, 2011.

5.   K. Bagi, L. M. Simon, and B. Szajáni, "Immobilization and characterization of porcine pancreas lipase,"Enzyme and Microbial Technology, vol. 20, no. 7, pp. 531–535, 1997.

6.   B. K. Vaidya, G. C. Ingavle, S. Ponrathnam, B. D. Kulkarni, and S. N. Nene, "Immobilization ofCandida rugosa lipase on poly(allyl glycidyl ether-co-ethylene glycol dimethacrylate) macroporous polymer particles," Bioresource Technology, vol. 99, no. 9, pp. 3623–3629, 2008.

7.   T. Wang, H. Li, K. Nie, and T. Tan, "Immobilization of lipase on epoxy activated (1–3)-α-D-glucan isolated from Penicillium chrysongenum," Bioscience, Biotechnology and Biochemistry, vol. 70, no. 12, pp. 2883–2888, 2006.

8. C. Shu, J. Cai, L. Huang, X. Zhu, and Z. Xu, "Biocatalytic production of ethyl butyrate from butyric acid with immobilized Candida rugosa lipase on cotton cloth," Journal of Molecular Catalysis B, vol. 72, no. 3-4, pp. 139–144, 2011.

9. J. C. Moreno-Pirajàn and L. Giraldo, "Study of immobilized Candida rugosa lipase for biodiesel fuel production from palm oil by flow microcalorimetry," Arabian Journal of Chemistry, vol. 4, no. 1, pp. 55–62, 2011.

10. Z. Onderková, J. Bryjak, and M. Polakovič, "Properties of fructosyltransferase from Aureobasidium pullulans immobilized on an acrylic carrier," Chemical Papers, vol. 61, no. 5, pp. 359–363, 2007.

11. R. Lin, R. Wu, X. Huang, and T. Xie, "Immobilization of oxalate decarboxylase to eupergit and properties of the immobilized enzyme," Preparative Biochemistry and Biotechnology, vol. 41, no. 2, pp. 154–165, 2011.

12. Z. Knezevic, N. Milosavic, D. Bezbradica, Z. Jakovljevic, and R. Prodanovic, "Immobilization of lipase from Candida rugosa on Eupergit C supports by covalent attachment," Biochemical Engineering Journal, vol. 30, no. 3, pp. 269–278, 2006.

13. D. H. Lee, J. M. Kim, S. W. Kang, J. W. Lee, and S. W. Kim, "Pretreatment of lipase with soybean oil before immobilization to prevent loss of activity," Biotechnology Letters, vol. 28, no. 23, pp. 1965–1969, 2006.

14. D.-H. Zhang, L.-X. Yuwen, Y.-L. Xie, W. Li, and X.-B. Li, "Improving immobilization of lipase onto magnetic microspheres with moderate hydrophobicity/hydrophilicity," Colloids and Surfaces B, vol. 89, no. 1, pp. 73–78, 2012.

15. K. R. Jegannathan, S. Abang, D. Poncelet, E. S. Chan, and P. Ravindra, "Production of biodiesel using immobilized lipase—a critical review," Critical Reviews in Biotechnology, vol. 28, no. 4, pp. 253–264, 2008.

16. B. Boscolo, F. Trotta, and E. Ghibaudi, "High catalytic performances of Pseudomonas fluorescens lipase adsorbed on a new type of cyclodextrin-based nanosponges," Journal of Molecular Catalysis B, vol. 62, no. 2, pp. 155–161, 2010.

17. J. Sun, Y. Jiang, L. Zhou, and J. Gao, "Immobilization of Candida antarctica lipase B by adsorption in organic medium," New Biotechnology, vol. 27, no. 1, pp. 53–58, 2010.

18. K. M. Ponvel, D.-G. Lee, E.-J. Woo, I.-S. Ahn, and C.-H. Lee,

"Immobilization of lipase on surface modified magnetic nanoparticles using alkyl benzenesulfonate," Korean Journal of Chemical Engineering, vol. 26, no. 1, pp. 127–130, 2009.

19. A. Bastida, P. Armisen, R. Fernandez-Lafuente, J. Huguet, J. M. Guisan, and P. Sabuquillo, "Single step purification, immobilization, and hyperactivation of lipases via interfacial adsorption on strongly hydrophobic supports," Biotechnology Bioengineering, vol. 58, pp. 486–493, 1998.

20. R. Fernandez-Lafuente, P. Armisén, P. Sabuquillo, G. Fernández-Lorente, and J. M. Guisán, "Immobilization of lipases by selective adsorption on hydrophobic supports," Chemistry and Physics of Lipids, vol. 93, no. 1-2, pp. 185–197, 1998.

21. G. Yang, J. Wu, G. Xu, and L. Yang, "Improvement of catalytic properties of lipase from Arthrobacter sp. by encapsulation in hydrophobic sol-gel materials," Bioresource Technology, vol. 100, no. 19, pp. 4311–4316, 2009.

22. E. Yilmaz and M. Sezgin, "Enhancement of the activity and enantioselectivity of lipase by sol-gel encapsulation immobilization onto β-cyclodextrin-based polymer," Applied Biochemistry and Biotechnology, pp. 1–14, 2012.

23. Y.-Y. Liang and L.-M. Zhang, "Bioconjugation of papain on superparamagnetic nanoparticles decorated with carboxymethylated chitosan," Biomacromolecules, vol. 8, no. 5, pp. 1480–1486, 2007.

24. Y. Bai, Y. Li, and L. Lei, "Synthesis of a mesoporous functional copolymer bead carrier and its properties for glucoamylase immobilization," Applied Microbiology and Biotechnology, vol. 83, no. 3, pp. 457–464, 2009.

25. S.-H. Chiou and W.-T. Wu, "Immobilization of Candida rugosa lipase on chitosan with activation of the hydroxyl groups," Biomaterials, vol. 25, no. 2, pp. 197–204, 2004.

26. Z. S. Akdemir, S. Demir, M. V. Kahraman, and N. Kayaman Apohan, "Preparation and characterization of UV-curable polymeric support for covalent immobilization of xylanase enzyme," Journal of Molecular Catalysis B, vol. 68, no. 1, pp. 104–108, 2011.

27. S. R. Edupuganti, O. P. Edupuganti, R. O›Kennedy, E. Defrancq, and S. Boullanger, "Use of T-2 toxin-immobilized amine-activated beads as an efficient affinity purification matrix for the isolation of specific IgY," Journal of Chromatography B, vol. 923-924, pp. 98–101, 2013.

28. F. H. Isgrove, R. J. H. Williams, G. W. Niven, and A. T. Andrews, "Enzyme immobilization on nylon-optimization and the steps used

to prevent enzyme leakage from the support," Enzyme and Microbial Technology, vol. 28, no. 2-3, pp. 225–232, 2001.

29. G. Bayramo lu, B. Kaya, and M. Y. Arica, "Immobilization of Candida rugosa lipase onto spacer-arm attached poly(GMA-HEMA-EGDMA) microspheres," Food Chemistry, vol. 92, no. 2, pp. 261–268, 2005.

30. M. I. Kim, H. O. Ham, S.-D. Oh, H. G. Park, H. N. Chang, and S.-H. Choi, "Immobilization of Mucor javanicus lipase on effectively functionalized silica nanoparticles," Journal of Molecular Catalysis B, vol. 39, no. 1–4, pp. 62–68, 2006.

31. B. Hu, J. Pan, H.-L. Yu, J.-W. Liu, and J.-H. Xu, "Immobilization of Serratia marcescens lipase onto amino-functionalized magnetic nanoparticles for repeated use in enzymatic synthesis of Diltiazem intermediate," Process Biochemistry, vol. 44, no. 9, pp. 1019–1024, 2009.

32. Y.-W. Zhang, M. Jeya, and J.-K. Lee, "Enhanced activity and stability of L-arabinose isomerase by immobilization on aminopropyl glass," Applied Microbiology and Biotechnology, vol. 89, no. 5, pp. 1435–1442, 2011.

33. S. Li, J. Hu, and B. Liu, "Use of chemically modified PMMA microspheres for enzyme immobilization," BioSystems, vol. 77, no. 1-3, pp. 25–32, 2004.

34. N. Ortega, M. Perez-Mateos, M. C. Pilar, and M. D. Busto, "Neutrase immobilization on alginate-glutaraldehyde beads by covalent attachment," Journal of Agricultural and Food Chemistry, vol. 57, no. 1, pp. 109–115, 2009.

35. E. Y. Ozmen and M. Yilmaz, "Pretreatment of Candida rugosa lipase with soybean oil before immobilization on β-cyclodextrin-based polymer," Colloids and Surfaces B, vol. 69, no. 1, pp. 58–62, 2009.

36. S. Y. Chang, N.-Y. Zheng, C.-S. Chen, C.-D. Chen, Y.-Y. Chen, and C. R. C. Wang, "Analysis of peptides and proteins affinity-bound to iron oxide nanoparticles by MALDI MS," Journal of the American Society for Mass Spectrometry, vol. 18, no. 5, pp. 910–918, 2007.

37. Y. Yong, Y.-X. Bai, Y.-F. Li, L. Lin, Y.-J. Cui, and C.-G. Xia, "Characterization of Candida rugosa lipase immobilized onto magnetic microspheres with hydrophilicity," Process Biochemistry, vol. 43, no. 11, pp. 1179–1185, 2008.

38. J. Bryjak and A. W. Trochimczuk, "Immobilization of lipase and penicillin acylase on hydrophobic acrylic carriers," Enzyme and Microbial Technology, vol. 39, no. 4, pp. 573–578, 2006.

39. M. H. Sörensen, J. B. S. Ng, L. Bergström, and P. C. A. Alberius, "Improved enzymatic activity of Thermomyces lanuginosus lipase

immobilized in a hydrophobic particulate mesoporous carrier,"Journal of Colloid and Interface Science, vol. 343, no. 1, pp. 359–365, 2010.

40. X. Liu, Y. Guan, R. Shen, and H. Liu, "Immobilization of lipase onto micron-size magnetic beads,"Journal of Chromatography B, vol. 822, no. 1-2, pp. 91–97, 2005.

41. G. Bayramoglu, M. Yilmaz, and M. Y. Arica, "Preparation and characterization of epoxy-functionalized magnetic chitosan beads: laccase immobilized for degradation of reactive dyes," Bioprocess and Biosystems Engineering, vol. 33, no. 4, pp. 439–448, 2010.

42. Y. Liu, S. Jia, Q. Wu, J. Ran, W. Zhang, and S. Wu, "Studies of Fe_3O_4-chitosan nanoparticles prepared by co-precipitation under the magnetic field for lipase immobilization," Catalysis Communications, vol. 12, no. 8, pp. 717–720, 2011.

43. F. M. Gomes, E. B. Pereira, and H. F. de Castro, "Immobilization of lipase on chitin and its use in nonconventional biocatalysis," Biomacromolecules, vol. 5, no. 1, pp. 17–23, 2004.

44. T.-C. Hung, R. Giridhar, S.-H. Chiou, and W.-T. Wu, "Binary immobilization of Candida rugosa lipase on chitosan," Journal of Molecular Catalysis B, vol. 26, no. 1-2, pp. 69–78, 2003.

45. S.-H. Chiou and W.-T. Wu, "Immobilization of Candida rugosa lipase on chitosan with activation of the hydroxyl groups," Biomaterials, vol. 25, no. 2, pp. 197–204, 2004.

46. Z. Lei and Q. Jiang, "Synthesis and properties of immobilized pectinase onto the macroporous polyacrylamide microspheres," Journal of Agricultural and Food Chemistry, vol. 59, no. 6, pp. 2592–2599, 2011.

47. I. Roy, M. Sardar, and M. N. Gupta, "Evaluation of a smart bioconjugate of pectinase for chitin hydrolysis," Biochemical Engineering Journal, vol. 16, no. 3, pp. 329–335, 2003.

48. Y. Li, F. Gao, W. Wei, J.-B. Qu, G.-H. Ma, and W.-Q. Zhou, "Pore size of macroporous polystyrene microspheres affects lipase immobilization," Journal of Molecular Catalysis B, vol. 66, no. 1-2, pp. 182–189, 2010.

49. R. Torres, C. Ortiz, B. C. C. Pessela et al., "Improvement of the enantioselectivity of lipase (fraction B) from Candida antarctica via adsorpiton on polyethylenimine-agarose under different experimental conditions," Enzyme and Microbial Technology, vol. 39, no. 2, pp. 167–171, 2006.

50. N. Dizge, C. Aydiner, D. Y. Imer, M. Bayramoglu, A. Tanriseven, and

B. Keskinler, "Biodiesel production from sunflower, soybean, and waste cooking oils by transesterification using lipase immobilized onto a novel microporous polymer," Bioresource Technology, vol. 100, no. 6, pp. 1983–1991, 2009.

51. B. Mcnaa, M. Herrero, V. Rives, M. Lavrenko, and D. K. Eggers, "Favourable influence of hydrophobic surfaces on protein structure in porous organically-modified silica glasses," Biomaterials, vol. 29, no. 18, pp. 2710–2718, 2008.

52. D. H. Zhang, L. X. Yuwen, C. Li, and Y. Q. Li, "Effect of poly(vinyl acetate-acrylamide) microspheres properties and steric hindrance on the immobilization of Candida rugosa lipase," Bioresource Technology, vol. 124, pp. 233–236, 2012. ·

53. X. Hou, B. Liu, X. Deng, B. Zhang, and J. Yan, "Monodisperse polystyrene microspheres by dispersion copolymerization of styrene and other vinyl comonomers: characterization and protein adsorption properties," Journal of Biomedical Materials Research A, vol. 83, no. 2, pp. 280–289, 2007.

54. P. Reis, K. Holmberg, T. Debeche, B. Folmer, L. Fauconnot, and H. Watzke, "Lipase-catalyzed reactions at different surfaces," Langmuir, vol. 22, no. 19, pp. 8169–8177, 2006.

55. R. Zhao, J. Lu, and T. Tan, "Preparation of polyglycidylmethacrylate macropore beads and application in Candida species 99–125 lipase immobilization," Chemical Engineering and Technology, vol. 34, no. 1, pp. 93–97, 2011.

56. A. Dyal, K. Loos, M. Noto et al., "Activity of Candida rugosa lipase immobilized on γ-Fe$_2$O$_3$ magnetic nanoparticles," Journal of the American Chemical Society, vol. 125, no. 7, pp. 1684–1685, 2003.

57. C. G. C. M. Netto, E. H. Nakamatsu, L. E. S. Netto et al., "Catalytic properties of thioredoxin immobilized on superparamagnetic nanoparticles," Journal of Inorganic Biochemistry, vol. 105, no. 5, pp. 738–744, 2011.

58. Z. Lei and S. Bi, "Preparation and properties of immobilized pectinase onto the amphiphilic PS-b-PAA diblock copolymers," Journal of Biotechnology, vol. 128, no. 1, pp. 112–119, 2007.

59. W. Xie and N. Ma, "Enzymatic transesterification of soybean oil by using immobilized lipase on magnetic nano-particles," Biomass and Bioenergy, vol. 34, no. 6, pp. 890–896, 2010.

Chapter 3

ANALYTICAL EXPRESSION FOR THE CONCENTRATION OF SUBSTRATE AND PRODUCT IN IMMOBILIZED ENZYME SYSTEM IN BIOFUEL/BIOSENSOR

R. Malini Devi[1], O. M. Kirthiga[2], L. Rajendran[2]

[1]Department of Mathematics, the Standard Fireworks Rajaratnam College for Women, Sivakasi, India.

[2]Department of Mathematics, the Madura College, Madurai, India.

ABSTRACT

In this paper, an approximate analytical method to solve the non-linear differential equations in an immobilized enzyme film is presented. Analytical expressions for concentrations of substrate and product have been derived for all values of dimensionless parameter. Dimensionless numbers that can be used to study the effects of enzyme loading, enzymatic gel thickness, and oxidation/reduction kinetics at the electrode in biosensor/biofuel cell performance were identified. Using the dimensionless numbers identified in this paper, and the plots representing the effects of these dimensionless numbers on concentrations and current in biosensor/biofuel cell are discussed. Analytical results are compared with simulation results and satisfactory agreement is noted.

INTRODUCTION

Biosensors and biofuel cells are commonly used for industrial, environmental and medical applications. However there are no clear guidelines for the design of electrochemical biosensors or biofuel cells employing immobilized enzymes that will produce a targeted linear range, limit of detection and sensitivity. Such guidelines can be provided using analytical simulation tools that assess sensor feasibility prior to extensive development. Biosensors and biofuel cell face increasing demand for selective and sensitive detection of different molecules for industrial, environmental and clinical applications [1] -[4] . There are many affordable alternatives to laboratory techniques that require trained personnel, expensive equipment and possibly delayed response time. Electrochemical biosensors and biofuel cell especially desirable for use in field applications because of their compact design, ease of manufacture, real time response, sensitivity and selectivity [3] -[6] . They are used in many applications ranging from glucose detection to detection of neurotoxic agents [1] [6] [7] . Here we focus on biosensors and biofuel cell that employ immobilized enzymes and the electrochemical detection of the enzymatic reaction. Some important parameters that affect these goals are listed and include transport of the substrate and the product through the immobilized enzyme layer, oxidation/reduction kinetics at the electrode, enzyme activity and loading and operating conditions such as pH and temperature. Of these parameters optimizing the enzyme loading and activity has been a major challenge and it depends primarily on the enzyme immobilization method. Different methods such as chemical modification of the electrode surface, entrapment in a membrane and physical absorption are commonly used to create enzyme layers on electrodes [8] . A mathematical model considering reaction and diffusion processes in biofuel cell or biosensor contains a system of non-linear partial differential equations. Numerical and analytical solutions to the reaction-diffusion equations have been presented for different cases by many authors [9] -[14] . Analytical solutions are available for limiting cases, whereas numerical solutions were used to determine and optimize a wide range of experimental parameters [15]. Many of the earlier studies have focused on optimizing glucose biosensors where the enzyme was entrapped in a redox hydrogel [16] [17] . Simple Michaelis-Menten kinetics was used to model the enzyme kinetics, and first order kinetics between the mediator and the electrode were assumed [9] [17] . The effects of experimental parameters on the response at steady state and during a transient were studied [12] . Especially the behavior of the glucose sensor in the diffusion limited regime was analyzed since this leads to an extended linear range [16] [18] . Substrate and product inhibition in an enzyme with first order reaction kinetics [19] , diffusion through a semi-permeable outer membrane [20] [21] and data analysis to determine kinetic constants and enzyme activity [22] were also studied by different groups.Sachin [23] used a finite difference method for electrochemical biosensors with an immobilized enzyme layer. Sachin described the general criteria using Michaelis-Menten rate equation and effect of gel thickness on the response of this biosensor. To our knowledge no rigorous analytical solutions for non-steady-state concentration and current have been reported. In this paper, we have derived the analytical expressions of concentration and current using a new approach of Homotopy perturbation method [24] -[27] . The result of the Equations (2)-(3) in immobilized enzyme system is relevant because its solution describes important applications such as biosensors, bioreactors, and biofuel cells, among others.

MATHEMATICAL FORMULATION OF THE PROBLEM

The chemical reactions in the layer are

$$E + S \leftrightarrow ES \rightarrow E + P \tag{1}$$

where E refers to the enzyme, S is the substrate, ES is a transitory complex assumed to be at a steady concentration, and P is the product. The schematic of the system modeled in this study is shown in Figure 1. An aqueous drop containing substrate (S) is placed on the electrode with an immobilized enzyme layer. As the substrate diffuses through the enzyme layer it reacts with the enzyme to form the product (P). The product then diffuses through the layer, and if it is electroactive, is oxidized or reduced at the electrode. When modeling this system, we used Michaelis-Menten equation to describe the kinetics within the enzyme layer and coupled it with Fick's law to describe the diffusion of the substrate and product as shown in Equations (2)-(3):

$$\frac{\partial c_S}{\partial t} = D_S \frac{\partial^2 c_S}{\partial z^2} - \frac{k_{cat}[E]c_S}{c_S + K_S} \tag{2}$$

$$\frac{\partial c_P}{\partial t} = D_P \frac{\partial^2 c_P}{\partial z^2} + \frac{k_{cat}[E]c_S}{c_S + K_S} \tag{3}$$

where c_P, c_S, D_P and D_S represent the concentrations and diffusion coefficients of the product and the substrate, respectively. k_{cat} is the catalytic rate constant in the Michaelis-Menten mechanism, [E] is enzyme loading, and K_S is Michaelis constant for the substrate.

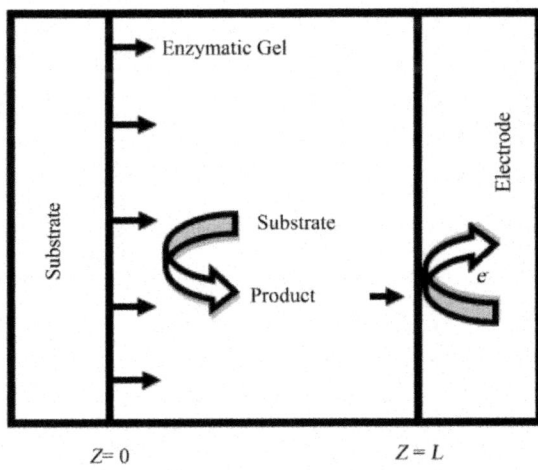

Figure 1: Schematic model of an enzyme-membrane electrodes.

In the above equations the initial and boundary conditions are given by

$$t = 0, \ 0 \leq z \leq L: \quad c_s = 0, \quad c_p = 0$$

$$t > 0, \ z = 0: \quad \frac{\partial c_s}{\partial z} = 0, \quad \frac{\partial c_p}{\partial z} = 0$$

$$t > 0, \ z = L: \quad c_s = c_{Sbulk}, \quad c_p = 0$$

$$(4)$$

where z is the distance from the electrode surface and L is the enzyme layer thickness.

c_{Sbulk} represents the concentration of substrate in bulk solution. Current i occurring at the electrode surface due to reduction or oxidation of P is given by

$$i = nFD_p \left(\frac{\partial c_p}{\partial z} \right)_{z=d}$$

$$(5)$$

Equations (2)-(3) were made dimensionless using the following dimensionless parameters:

$$c_s^* = \frac{c_s}{K_S}, \ c_p^* = \frac{c_p}{K_S}, \ t^* = \frac{D_S t}{L^2}, \ z^* = \frac{z}{L}, \ r = \frac{D_p}{D_s}, \ \phi_s^2 = \frac{k_{cat}[E]L^2}{K_S D_s}$$

$$(6)$$

The Equations (2)-(3) in dimensionless form becomes as follows:

$$\frac{\partial c_s^*}{\partial t} = \frac{\partial^2 c_s^*}{\partial z^{*2}} - \frac{\phi_s^2 c_s^*}{c_s^* + 1}$$

$$(7)$$

$$\frac{\partial c_p^*}{\partial t^*} = r \frac{\partial^2 c_p^*}{\partial z^{*2}} + \frac{\varphi_s^2 c_s^*}{c_s^* + 1}$$

$$(8)$$

From the Equation (4), the initial and boundary conditions in dimensionless form are given by

$$t^* = 0, \ 0 \leq z^* \leq 1: \quad c_s^* = 0, \quad c_p^* = 0$$

$$t^* > 0, \ z^* = 0: \quad \frac{\partial c_s^*}{\partial z^*} = 0, \quad \frac{\partial c_p^*}{\partial z^*} = 0$$

$$t^* > 0, \ z^* = 1: \quad c_s^* = \frac{c_{Sbulk}}{K_s} = c_{s0}, \quad c_p^* = 0$$

$$(9)$$

Dimensionless current density becomes

$$\psi = nFD_p \left(\frac{\partial c_p^*}{\partial z^*} \right)_{z^*=1}$$

(10)

GENERAL ANALYTICAL EXPRESSION OF CONCENTRA-TION OF SUBSTRATE AND PRODUCT UNDER NON-STEADY STATE CONDITION USING HOMOTOPY PERTURBATION METHOD (HPM)

In recent days, HPM is often employed to solve several analytical problems. In addition, several groups demonstrated the efficiency and suitability of the HPM for solving nonlinear equations in electrochemical problems [28]-[31]. He et al. [24], used HPM to solve the Lighthill equation, the Duffing equation [25] and the Blasius equation [26]. HPM has also been used to solve non-linear boundary value problems [27], integral equation [32]-[34], Klein-Gordon and Sine-Gordon equations [35], Emden-Flower type equations [36] and several other problems. Laplace transform and Homotopy perturbation method are used to solve the non-linear differential Equations (7)-(8) (Appendix A). The analytical expressions of non-steady state concentrations are as follows:

$$c_s^*\left(z^*,t^*\right) = \frac{\cosh\left(\sqrt{a}z^*\right)}{\cosh\left(\sqrt{a}\right)} - \pi \sum_{n=0}^{\infty} \frac{(-1)^n}{m}(2n+1)\cos\left(\frac{2n+1}{2}\pi z^*\right)e^{-mt^*}$$

(11)

$$c_p^*\left(z^*,t^*\right) = 1 - \frac{\cosh\sqrt{a}z^*}{\cosh\sqrt{a}} - be^{-at^*/b}\left(\frac{\cos\sqrt{(ar/b)}z^*}{\cos\sqrt{(ar/b)}} - \frac{\cosh\left(\sqrt{a(b-1)/b}z^*\right)}{\cosh\left(\sqrt{a(b-1)/b}\right)} \right)$$

$$+ \frac{4ar}{\pi}\sum_{n=0}^{\infty} \frac{(-1)^{-n}\cos\left((2n+1)\pi z^*/2\right)e^{-\left((2n+1)^2\pi^2/4r\right)t^*}}{(2n+1)\left[a-\left(b(2n+1)^2\pi^2/2r\right)\right]}$$

$$- a\pi\sum_{n=0}^{\infty} \frac{(-1)^{-n}\cos\left((2n+1)\pi z^*/2\right)(2n+1)e^{-\left[a+(2n+1)^2\pi^2/4\right]t^*}}{\left[a+(2n+1)^2\pi^2/4\right]\left[a-b\left(a+(2n+1)^2\pi^2/4\right)\right]}$$

(12)

where $$m = a + \frac{\pi^2(2n+1)^2}{4}, \quad a = \frac{\phi_s^2}{1+c_{s0}}, \quad b = (1-r)$$

(13)

Using (10) and (12), the current is given by

$$\frac{i}{nFD_p} = \left(\frac{\partial c_p^*}{\partial z^*}\right)_{z^*=1} = \left| \sqrt{abr}\, \tan\left(\sqrt{ar/b}\right) e^{-at^*/b} - \sqrt{a}\, \tanh\left(\sqrt{a}\right) + \sqrt{a(b-1)b}\, \tanh(\sqrt{a(b-1)/b}\,e^{-at^*/b} \right.$$

$$\left. + \frac{8rae^{-\pi^2(2n+1)^2t^*/4r}}{4ra - b\pi^2(2n+1)^2} - 8a\pi^2 \sum_{n=0}^{\infty} \frac{(2n+1)^2 e^{-\left(4a+\pi^2(2n+1)^2\right)t^*/4}}{\left[4a+\pi^2(2n+1)^2\right]\left[4a - b\left(4a+\pi^2(2n+1)^2\right)\right]} \right|$$

(14)

When $t^* \to \infty$ (steady state), the above equation becomes

$$\frac{i}{nFD_p} = \left| \sqrt{a}\, \tanh\left(\sqrt{a}\right) \right|$$

(15)

DISCUSSION

Equations (11) (12) and (14) are the new and simple analytical expressions of concentrations of substrate, product and current respectively. To show the efficiency of our non-steady-state result, it is compared with numerical solution in Figure 2 & Figure 3. Satisfactory agreement is noted. The SCILAB/ MATLAB program is also given in Appendix B. Figure 2 shows the time-dependent normalized concentration profiles for the substrate c_s^* in the enzyme membrane. Figures 2(a)-(c) show dimensionless concentration c_s^* versus the dimensionless

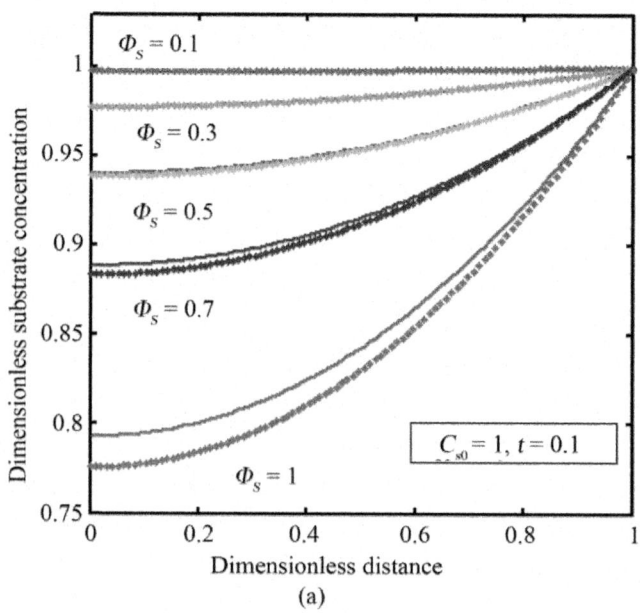

(a)

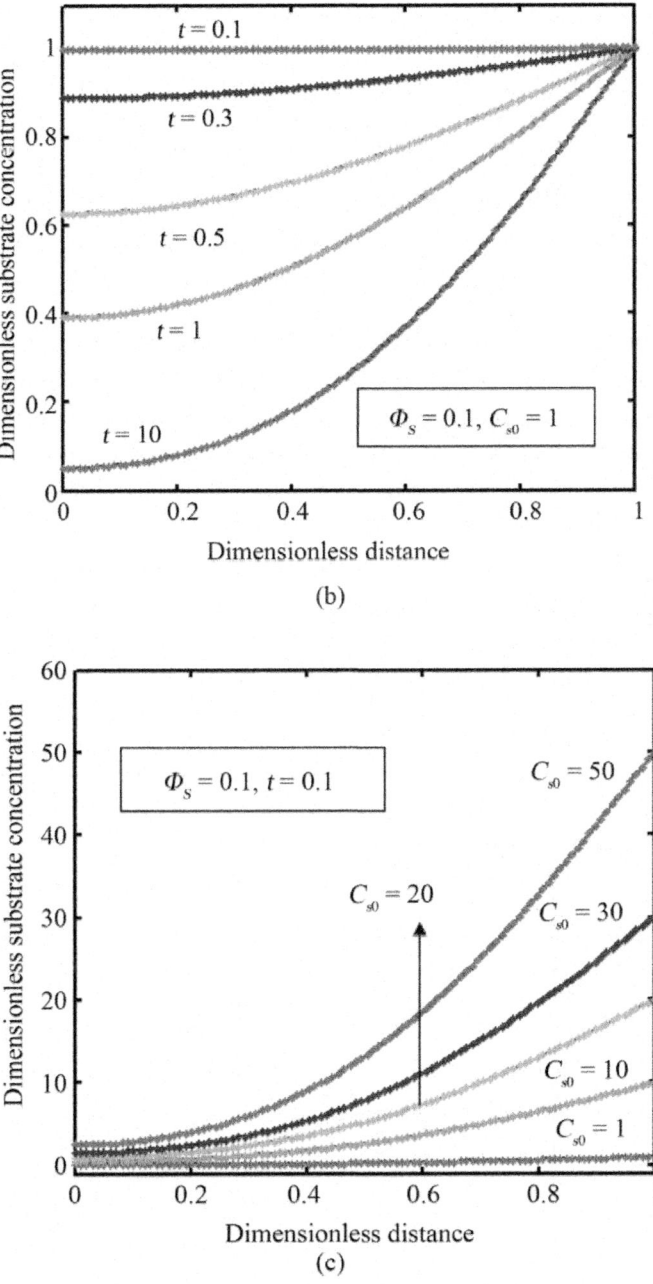

Figure 2: Dimensionless substrate concentration c_s^* versus distance from electrode surface z^* using Equation (11) for various values of parameters Φ_s, t and c_{s0}.

distance z^*. The concentration of substrate c_s^* depends upon the dimensionless parameter "a". The dimensionless parameter "a" depends upon ϕ_s and c_{s0}. When Thiele modulus ϕ_s is small, the kinetics dominate and the uptake of the substrate are kinetically controlled. From Figure 2(a), it is evident that the value of the substrate concentration c_s^* decreases when the Thiele modulus ϕ_s increases or Figure 2(b) illustrates that, when t increases, the concentration of the substrate decreases. It is obvious from Figure 2(c) that when initial substrate concentration c_{s0} increases, the concentration of substrate also decreases.

The normalized concentration of the product c_p^* for various values of Thiele modulus ϕ_s, time t and ratio of diffusion coefficient is plotted in Figures 3(a)-(c). From the Figure 3(a) & Figure 3(b), it is inferred that the normalized concentration product increases with the decrease in the value of ϕ_s and time t. The product concentration is increases when the ratio of diffusion coefficient decreases as shown in Figure 3(c).

The value of current i increases slightly when the Thiele modulus ϕ_s increases or electrode thickness increases as shown in Figure 4(a). From Figure 4(b), it is inferred that the ratio of diffusion coefficient r increases the current density is decreases. The current density increases as initial substrate concentration c_{s0}

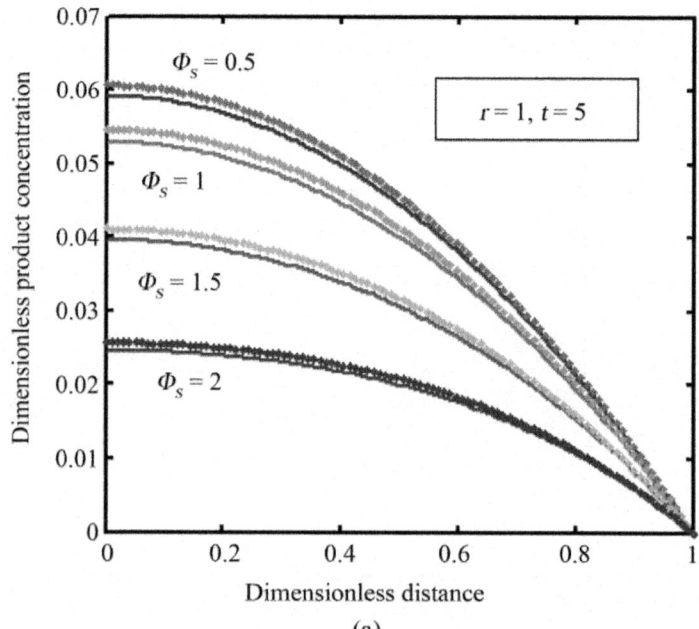

(a)

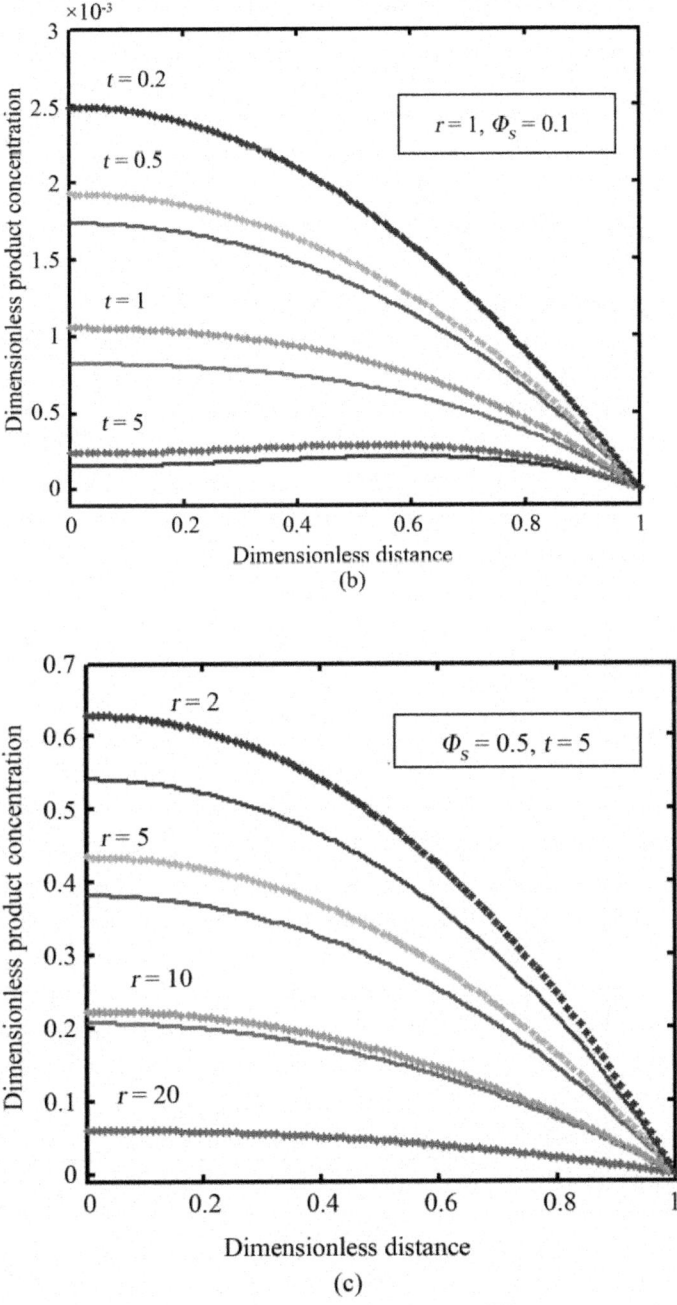

Figure 3: Dimensionless product concentration c_P^* versus distance from electrode surface z^* using Equation (12) for various values of parameters Φ_s, t and r. decreases.

ESTIMATION OF KINETIC PARAMETERS

The current is dependent upon the parameters Thiele module ϕ_s and initial substrate concentration c_{s0}. When $r = 1$, (or $D_p = D_s$) Equation (15) can be written as

$$\frac{\phi_s^2}{1+c_{s0}} = \left[\tanh^{-1}\left(-i/nFD_p \right) \right]^2$$

(16)

Substituting the value of μ and k in the above equation, we get

$$\left[\tanh^{-1}\left(-i/nFD_p \right) \right]^{-2} = \left(\frac{D_s}{k_{cat}[E]L^2} \right) C_{Sbulk} + \frac{K_S D_s}{k_{cat}[E]L^2}$$

(17)

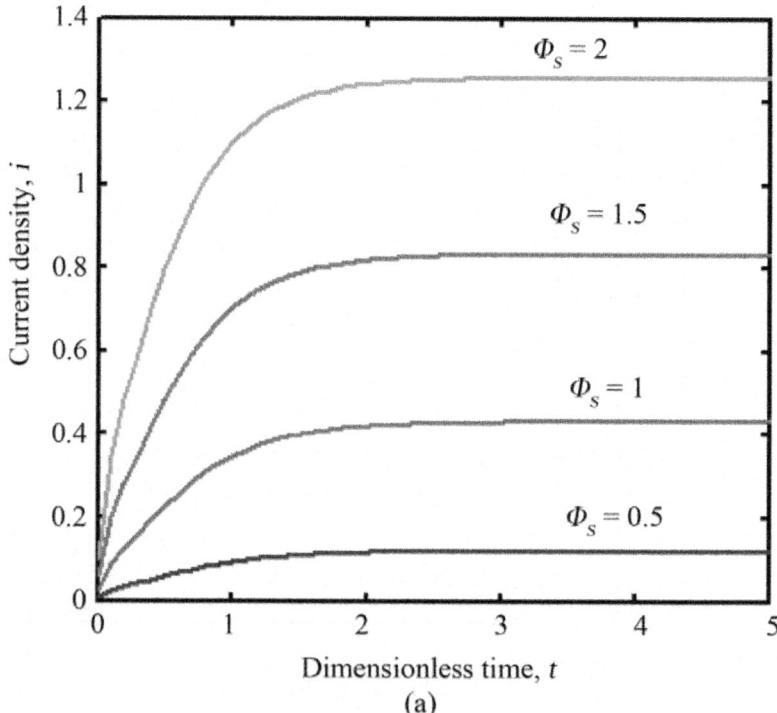

(a)

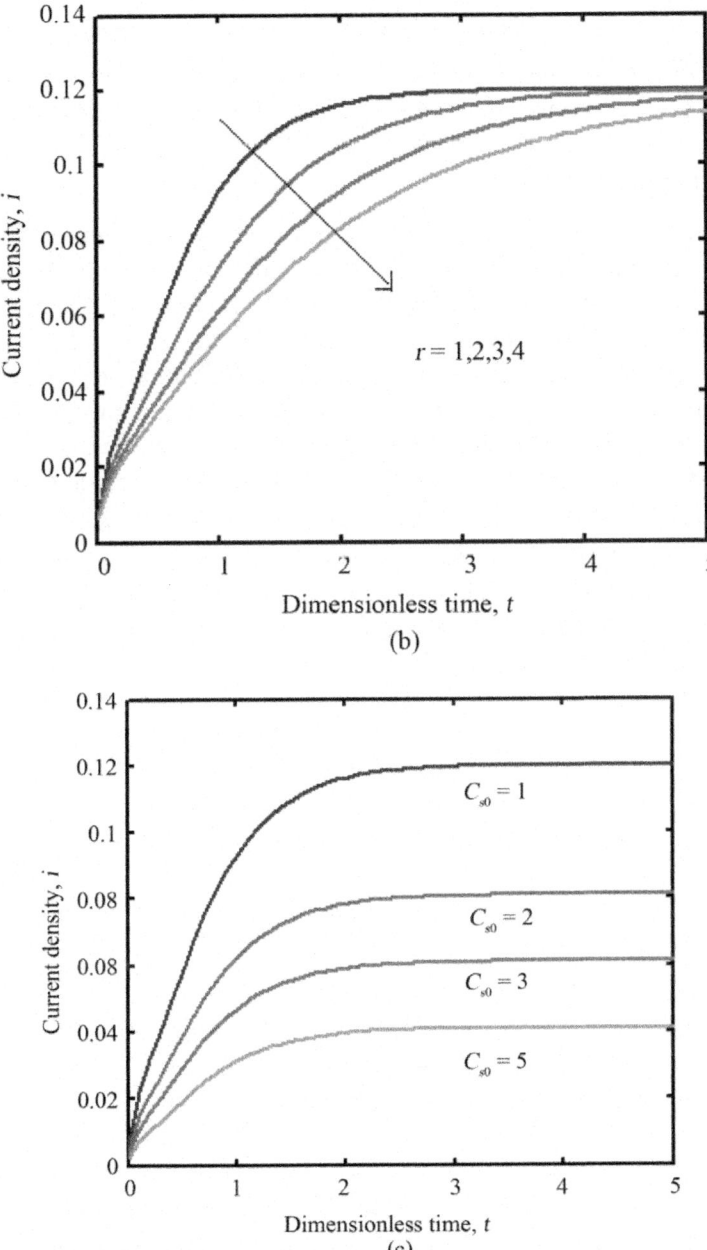

Figure 4: Dimensionless current density i/nFD_p versus time t using Equation (14) for various values of parameters Φ_s, c_{s0} and r.

The plot of $\left[\tanh^{-1}\left(-i/nFD_p\right)\right]^{-2}$ versus C_{Sbulk} gives the slope $\left(D_s/k_{cat}[E]L^2\right)$ and intercept $\left(K_S D_s/k_{cat}[E]L^2\right)$ as shown in Figure 5. From these plots, we can obtain the value of kinetic parameters K_S and $\left(D_s/k_{cat}\right)$.

CONCLUSION

The theoretical behavior of biofuel cell/biosensor was analyzed. The coupled time dependent non-steady state non-linear diffusion equations in biosensor or biofuel cell have been solved analytically and numerically. These analytical results will be used in determining the kinetic characteristics of the biofuel cell or biosensor. The analytical expressions for substrate, product concentration and transient current response are obtained using the method of Laplace transformation and HPM. A good agreement with numerical simulation data is noticed. Concentration of substrate, product and current depends upon Thiele modulus ϕ_s and initial concentration of substrate which is discussed in this communication. Evaluation of kinetic parametr from the response of the steady-state current is also completely discussed. The theoretical model presented here can be used for the optimization of the design of the biosensor.

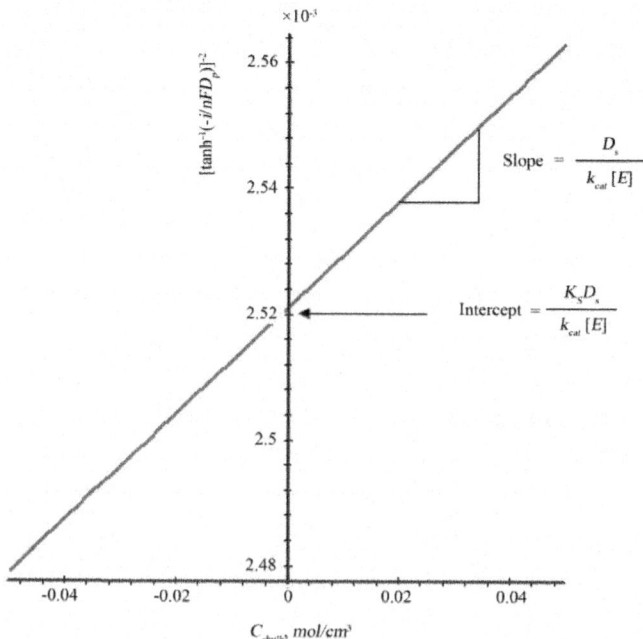

Figure 5: A plot of $\tan h^{-1}(i/nFD_p)^{-2}$ versus initial substrate concentration C_{Sbulk} using Equation (17) to estimate the kinetic parameters.

ACKNOWLEDGEMENTS

This work is supported by the Department of Science and Technology (DST) (No. SB/SI/PC-50/2012), Government of India. The authors are thankful to Shri. S. Natanagopal, Secretary, The Madura College Board and Dr. R. Murali, Principal, Mr. S. Muthukumar, Head of the Department, Department of Mathematics, The Madura College (Autonomous), Madurai, Tamilnadu, India for their constant encouragement.

REFERENCES

1.   Wang, J., Krause, R., Block, K., Musameh, M., Mulchandani, A. and Schöning, M.J. (2003) Flow Injection Amperometric Detection of OP Nerve Agents Based on an Organophosphorus-Hydrolase Biosensor Detector. Biosensors and Bioelectronics, 18, 255-260. http://dx.doi.org/10.1016/S0956-5663(02)00178-1

2.   Dennison, M.J., Hall, J.M. and Turner, A.P.F. (1996) Direct Monitoring of Formaldehyde Vapour and Detection of Ethanol Vapour Using Dehydrogenase-Based Biosensors. Analyst, 121, 1769-1773. http://dx.doi.org/10.1039/an9962101769

3.   Wu, J., Cropek, D.M., West, A.C. and Banta, S. (2010) Development of a Troponin I Biosensor Using a Peptide Obtained through Phage Display. Analytical Chemistry, 82, 8235-8243. http://dx.doi.org/10.1021/ac101657h

4.   Chen, X.J., West, A.C., Cropek, D.M. and Banta, S. (2008) Detection of the Superoxide Radical Anion Using Various Alkanethiol Monolayers and Immobilized Cytochrome c. Analytical Chemistry, 80, 9622-9629. http://dx.doi.org/10.1021/ac800796b

5.   Lei, Y., Mulchandani, P., Wang, J., Chen, W. and Mulchandani, A. (2005) A Highly Sensitive and Selective Amperometric Microbial Biosensor for Direct Determination of p-Nitrophenyl-Substituted Organophosphate Nerve Agents. Environmental Science & Technology, 39, 8853-8857 http://dx.doi.org/10.1021/es050720b

6.   Sahin, A., Dooley, K., Cropek, D.M., West, A.C. and Banta, S. (2011) A Dual Enzyme Electrochemical Assay for the Detection of Organophosphorus Compounds Using Organophosphorus Hydrolase and Horseradish Peroxidase. Sensors and Actuators B: Chemical, 158, 353-360. http://dx.doi.org/10.1016/j.snb.2011.06.034

7.   Cass, A.E.G., Davis, G., Francis, G.D., Hill, H.A.O., Aston, W.J., Higgins, I.J., Plotkin, E.V., Scott, L.D.L. and Turner, A.P.F. (1984) Ferrocene-Mediated Enzyme Electrode for Amperometric Determination

of Glucose. Analytical Chemistry, 56, 667-671.http://dx.doi.org/10.1021/ac00268a018

8.   Sheldon, R.A. (2007) Enzyme Immobilization: The Quest for Optimum Performance. Advanced Synthesis & Catalysis, 349, 1289-1307.http://dx.doi.org/10.1002/adsc.200700082

9.   Bartlett, P.N. and Pratt, K.F.E. (1995) Theoretical Treatment of Diffusion and Kinetics in Amperometric Immobilized Enzyme Electrodes Part I: Redox Mediator Entrapped within the Film. Journal of Electroanalytical Chemistry, 397, 61-78.http://dx.doi.org/10.1016/0022-0728(95)04236-7

10. Flexer, V., Pratt, K.F.E., Garay, F., Bartlett, P.N. and Calvo, E.J. (2008) Relaxation and Simplex Mathematical Algorithms Applied to the Study of Steady-State Electrochemical Responses of Immobilized Enzyme Biosensors: Comparison with Experiments. Journal of Electroanalytical Chemistry, 616, 87-98.http://dx.doi.org/10.1016/j.jelechem.2008.01.006

11. Lyons, M. (2006) Modelling the Transport and Kinetics of Electroenzymes at the Electrode/Solution Interface. Sensors, 6, 1765-1790.http://dx.doi.org/10.3390/s6121765

12. Kartono, A., Sulistan, E. and Mamat, M. (2010) The Numerical Analysis of Enzyme Membrane Thickness on the Response of Amperometric Biosensor. Applied Mathematical Sciences, 4, 1299-1308.

13. Baronas, R., Ivanauskas, F. and Kulys, J. (2009) Mathematical Modeling of Biosensors: An Introduction for Chemists and Mathematicians. Springer, Dordrecht.

14. Shunmugham, L. and Rajendran, L. (2013) Analytical Expressions for Steady-State Concentrations of Substrate and Oxidized and Reduced Mediator in an Amperometric Biosensor. International Journal of Electrochemistry, 2013, 1-12.http://dx.doi.org/10.1155/2013/812856

15. Meena, A. and Rajendran, L. (2010) Mathematical Modeling of Amperometric and Potentiometric Biosensors and System of Non-Linear Equations—Homotopy Perturbation Approach. Journal of Electroanalytical Chemistry, 644, 50- 59.http://dx.doi.org/10.1016/j.jelechem.2010.03.027

16. Cambiaso, A., Delfino, L., Grattarola, M., Verreschi, G., Ashworth, D., Maines, A. and Vadgama, P. (1996) Modelling and Simulation of a Diffusion Limited Glucose Biosensor. Sensors and Actuators B: Chemical, 33, 203-207. http://dx.doi.org/10.1016/0925-4005(96)80099-2

17. Mell, L.D. and Maloy, J.T. (1975) Model for the Amperometric Enzyme Electrode Obtained through Digital Simulation and Applied to the

Immobilized Glucose Oxidase System. Analytical Chemistry, 47, 299-307. http://dx.doi.org/10.1021/ac60352a006

18. Mell, L.D. and Maloy, J.T. (1976) Amperometric Response Enhancement of the Immobilized Glucose Oxidase Enzyme Electrode. Analytical Chemistry, 48, 1597-1601.http://dx.doi.org/10.1021/ac50005a045

19. Šimelevičius, D. and Baronas, R. (2010) Computational Modelling of Amperometric Biosensors in the Case of Substrate and Product Inhibition. Journal of Mathematical Chemistry, 47, 430-445. http://dx.doi.org/10.1007/s10910-009-9581-x

20. Puida, M., Ivanauskas, F. and Laurinavičius, V. (2010) Mathematical Modeling of the Action of Biosensor Possessing Variable Parameters. Journal of Mathematical Chemistry, 47, 191-200. http://dx.doi.org/10.1007/s10910-009-9541-5

21. Schulmeister, T. and Pfeiffer, D. (1993) Mathematical Modelling of Amperometric Enzyme Electrodes with Perforated Membranes. Biosensors and Bioelectronics, 8, 75-79.http://dx.doi.org/10.1016/0956-5663(93)80055-T

22. Rinken, T. (2003) Determination of Kinetic Constants and Enzyme Activity from a Biosensor Transient Signal. Analytical Letters, 36, 1535-1545.http://dx.doi.org/10.1081/AL-120021535

23. Sachin, A. (2012) Development of Electrochemical Methods for Detection of Pesticides and Biofuel Production. Columbia University, New York.

24. He, J.H. (1999) Homotopy Perturbation Technique. Computer Methods in Applied Mechanics and Engineering, 178, 257-262. http://dx.doi.org/10.1016/S0045-7825(99)00018-3

25. He, J.H. (2003) Homotopy Perturbation Method: A New Nonlinear Analytical Technique. Applied Mathematics and Computation, 135, 73-79. http://dx.doi.org/10.1016/S0096-3003(01)00312-5

26. He, J.H. (2003) A Simple Perturbation Approach to Blasius Equation. Applied Mathematics and Computation, 140, 217-222. http://dx.doi.org/10.1016/S0096-3003(02)00189-3

27. He, J.H. (2006) Homotopy Perturbation Method for Solving Boundary Value Problems. Physics Letters A, 350, 87-88. http://dx.doi.org/10.1016/j.physleta.2005.10.005

28. Ghori, Q.K., Ahmed, M. and Siddiqui, A.M. (2007) Application of Homotopy Perturbation Method to Squeezing Flow of a Newtonian Fluid. International Journal of Nonlinear Sciences and Numerical Simulation, 8, 179-184.http://dx.doi.org/10.1515/ijnsns.2007.8.2.179

29. Ozis, T. and Yildirim, A. (2007) Relation of a Nonlinear Oscillator with Discontinuities. International Journal of Nonlinear Sciences and Numerical Simulation, 8, 243-248.http://dx.doi.org/10.1515/IJNSNS.2007.8.2.243

30. Li, S.J. and Liu, Y.X. (2006) An Improved Approach to Nonlinear Dynamical System Identification Using PID Neural Networks. International Journal of Nonlinear Sciences and Numerical Simulation, 7, 177-182. http://dx.doi.org/10.1515/IJNSNS.2006.7.2.177

31. Mousa, M.M. and Ragab, S.F. (2008) Application of the Homotopy Perturbation Method to Linear and Nonlinear Schrodinger Equations. Zeitschrift für Naturforschung A, 63, 140-144. http://dx.doi.org/10.1515/zna-2008-3-404

32. Golbabai, A. and Keramati, B. (2008) Modified Homotopy Perturbation Method for Solving Fredholm Integral Equations. Chaos, Solitons & Fractals, 37, 1528-1537.http://dx.doi.org/10.1016/j.chaos.2006.10.037

33. Ghasemi, M., Kajani, T.M. and Babolian, E. (2007) Numerical Solutions of the Nonlinear Volterra-Fredholm Integral Equations by Using Homotopy Perturbation Method. Applied Mathematics and Computation, 188, 446-449.http://dx.doi.org/10.1016/j.amc.2006.10.015

34. Biazar, J. and Ghazvini, H. (2009) He's Homotopy Perturbation Method for Solving System of Volterra Integral Equations of the Second Kind. Chaos, Solitons & Fractals, 39, 770-777. http://dx.doi.org/10.1016/j.chaos.2007.01.108

35. Odibat, Z., Momani, S., Odibat, Z. and Momani, S. (2007) A Reliable Treatment of Homotopy Perturbation Method for Klein-Gordon Equations. Physics Letters A, 365, 351-357. http://dx.doi.org/10.1016/j.physleta.2007.01.064

36. Chowdhury, M.S.H. and Hashim, I. (2007) Solutions of Time-Dependent Emden-Fowler Type Equations by Homotopy-Perturbation Method. Physics Letters A, 368, 305-313.http://dx.doi.org/10.1016/j.physleta.2007.04.020

Chapter 4

TEMPERATURE-TRIGGERED ENZYME IMMOBILIZATION AND RELEASE BASED ON CROSS-LINKED GELATIN NANOPARTICLES

Zhenhai Gan[1], Ting Zhang[2], Yongchun Liu[1], Daocheng Wu[1]

[1]Key Laboratory of Biomedical Information Engineering of Education Ministry, School of Life Science and Technology, Xi'an Jiaotong University, Xi'an, P. R. China,

[2]Scientific Research Center, the Second Affiliated Hospital, School of Medicine, Xi'an Jiaotong University, Xi'an, P. R. China

ABSTRACT

A glucoamylase-immobilized system based on cross-linked gelatin nanoparticles (CLGNs) was prepared by coacervation method. This system exhibited characteristics of temperature-triggered phase transition, which could be used for enzyme immobilization and release. Their morphology and size distribution were examined by transmission electron microscopy and dynamic light scattering particle size analyzer. Their temperature-triggered glucoamylase immobilization and release features were also further investigated under different temperatures. Results showed that the CLGNs were regularly spherical with diameters of 155±5 nm. The loading efficiencies of glucoamylase immobilized by entrapment and adsorption methods were 59.9% and 24.7%, respectively. The immobilized enzyme was released when the system temperature was above 40°C and performed high activity similar to free enzyme due to the optimum temperature range for glucoamylase. On the other hand, there was no enzyme release that could be found when the system temperature was below 40°C. The efficiency of temperature-triggered release was as high as 99.3% for adsorption method, while the release of enzyme from the entrapment method was not detected. These results indicate that CLGNs are promising matrix for temperature-triggered glucoamylase immobilization and release by adsorption immobilization method.

INTRODUCTION

Enzyme immobilization is a technique in which an enzyme is made to attach to an inert material. It can provide increased resistance for the immobilized enzyme to the changes in conditions such as pH value or temperature [1]–[5]. Compared with free enzyme systems, immobilized enzymes offer the advantages of having batch or continuous processes operations, rapid termination of reactions, controlled product formation, ease of removal from the reaction mixture, and adaptability to various engineering designs [6]. Thus, enzyme immobilization has been widely used in analysis [7], medicine [8], food [9], chemical industry[10], [11] and biotechnology [12], [13] areas. For example, in recently reported studies, Terres et al. immobilized chloroperoxidase on silica matrix both physically and covalently for the biocatalytic desulfurization of fossil fuels [14]. Laveille et al. reported mesoporous silica immobilization for hemoglobin. In this system, the adsorbed hemoglobin was carried out for the detection of polycyclic aromatic hydrocarbons pollutant in the presence of H_2O_2. The catalytic performance of hemoglobin was enhanced for the real sample treatment under the pH range 6.5–8.5, while optimum pH for free hemoglobin is 5.0 [15].

Enzyme immobilization may be broadly divided into two main categories. One is based on the formation of covalent bonds between enzyme and matrix. The other method is physical method which is based on molecular interactions between the enzyme and support, such as adsorption and entrapment. Covalently immobilized enzyme showed an inactivating catalytic ability because of the reaction between groups necessary for enzyme activity or groups responsible for tertiary structure [16], [17]. Comparatively, enzyme immobilization by adsorption appears to offer better commercial potential because it is simpler, less expensive, and retains higher catalytic activity [18], [19]. With this method, no coupling reagent or reactive group of any amino acid residue of the enzyme is required to form specific covalent bonds with the supports. However, several studies have found that physically adsorbed enzymes on most supports were generally not strong enough, causing slow enzyme leakage during washing, operation and reaction processes [20]–[22]. Hence, strong adsorption between the enzyme and support should be achieved to prevent enzyme desorption from immobilization supports [23]. Furthermore, enzymes immobilized on solid-supports refer to two-phase systems with inherent mass transfer limitations. Immobilized enzyme systems suffer from producing unfavorable effects on their overall catalytic performances, such as the catalysis happened beyond the optimum temperature of enzyme or enzyme could not be released at desirable time [1], [24],[25]. Thus, developing a new strategy for enzyme immobilization is urgently needed to address the above-mentioned problems.

Smart hydrogels, which have the capability respond to small external stimulus changes, have attracted substantial attention in recent years, particularly in the pharmaceutical and material fields [26]. Many physical and chemical stimuli have been applied to induce various responses of the smart hydrogel systems. The physical stimuli include temperature, electric fields, solvent composition, light, pressure, sound and magnetic fields, while the chemical or biochemical stimuli include pH, ions and specific molecular recognition events [27]–[29]. If these systems were used for enzyme immobilization, enzyme would be released at needful time and optimum conditions under which enzyme could perform its highest catalysis ability. Among the smart hydrogels available, temperature sensitive hydrogels are probably the most commonly studied, especially in drug delivery and enzyme immobilization research. Qiu et al. summarized a series of temperature-sensitive systems, including poly(N-isopropylacrylamide), poly(N, N-diethylacrylamide), poly(ethylene oxide) and poly(propylene oxide) [26]. However, the biocompatibility of currently available polymers for these temperature sensitive hydrogels is not satisfactory, which limits their applications in fields such as the food and beverage industry. The natural biopolymer gelatin, obtained through partial hydrolysis of collagen, exhibits good biocompatibility, non-toxic biodegradation *in vivo* and readily excreted products. It also features highly effective drug encapsulation and can be fabricated into various forms of carriers for controlled drug and DNA delivery [30], [31]. Based on our previous research, a cross-linked gelatin nanoparticles (CLGNs) system prepared by coacervation method showed the ability to shrink at room temperature and swell when the temperature increased. Thus, a temperature-sensitive system made of gelatin could provide a promising strategy for enzyme immobilization and release, which has the characteristic of controlled-release at desirable time and optimum temperature for the enzyme. To the best of our knowledge, few reports on the gelatin hydrogel system for the enzyme immobilization existed and no controlled-release enzyme immobilization system based on gelatin hydrogel could be found [16], [32]–[34]. Glucoamylase is an industrially important enzyme used for the large-scale saccharification of malto-oligosaccharides into glucose and various syrups required in the food, beverages and fuel ethanol industry [23]. The suitable temperature for this enzyme is 40–70°C. Thus, its wide application in the biotechnology field is limited. Immobilize glucoamylase into a temperature-sensitive smart hydrogel system can either protect it from environmental damage when the temperature is beyond the suitable range, or release it when the temperature is suitable for the enzyme perform high catalytic ability, thereby yielding optimum application conditions.

In this work, a strategy for designing a temperature-sensitive cross-linked gelatin nanoparticles-based immobilized enzyme system is proposed.

Specifically, this system has the characteristics of temperature-triggered enzyme release ranging from 40 to 80°C, which is established on the swelling of the CLGNs as shown in Figure 1. Glucoamylase was selected as the model enzyme because of its appropriate activity temperature range around 40–70°C and the considerable academic interest it has gained [16], [17], [35], [36]. During glucoamylase immobilization, non-covalent physical adsorption and entrapment methods were compared based on their immobilization and release efficiencies. The results indicate that this kind of temperature-sensitive enzyme immobilization system may be a new technology for enzyme immobilization, and could provide a new and effective alternative for enzyme immobilization in biotechnology.

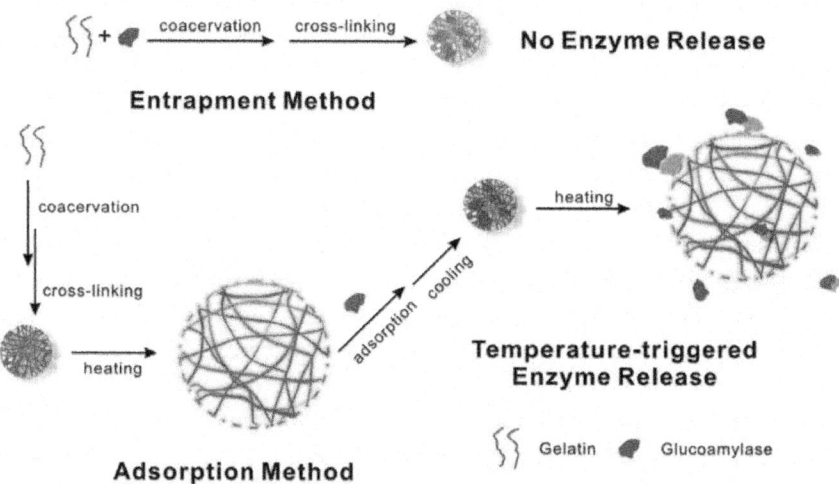

Figure 1: Schematic illustrations for the preparation of CLGNs, glucoamylase immobilization, and temperature-triggered enzyme release.

The gelatin and glucoamylase were mixed together before the cross-linking step in the entrapment method. In the adsorption method, the enzyme was adsorbed after the formation of the CLGNs.

MATERIALS AND EXPERIMENTS

Materials

Gelatin (Type B, 238–282 Bloom, isoelectric point 4.0–5.7) was purchased from Amresco Inc. (Solon, OH, U.S.A.). Tween-20 was obtained from Sigma-Aldrich Corporation (St. Louis, MO. U.S.A.). Glutaraldehyde was obtained from Sinopharm Chemical Reagent (Shanghai, China). Sodium sulfate, isopropanol, sodium pyrosulfite, sodium hydroxide, 3,5-dinitrosalicylic acid (DNS), and starch were analytical grade and used as purchased. Glucoamylase (EC 3.2.1.3) was purchased from Shanghai Kayon Biological Technology Co., Ltd. (Shanghai, China) and dissolved in acetate buffer (0.1 mol/L, pH 4.6). The concentration of the glucoamylase solution was determined to be 1.0 mg/mL using UV-Vis absorption spectra. Sodium salts with different anions used for the surface modification of the CLGNs were all analytical grade and used as purchased. The concentrations of these series solutions were dependent on the charge number of the anions to obtain the same ionic strengths for all these solutions. A 14KD dialyzer was purchased from Sino-American Biotechnology (Shanghai, China). All chemicals were used without further purification. Double-distilled water (18 MΩ cm) was used as solvent.

Preparation of Cross-linked Gelatin Nanoparticles

The preparation method involved the process of coacervation with sodium sulfate and cross-linking by glutaraldehyde. In our previously published paper, the fluoric anion was used for the modification of the CLGNs in an ultrasound-triggered drug release system because fluoride anions have the ability to attract electrons and can establish hydrogen bonds between oxygen and nitrogen atoms in the gelatin nanoparticles [37]. In present study, a series of anions with different charge numbers, such as F^-, Cl^-, NO_3^-, CH_3COO^-, SO_4^{2-} and PO_4^{3-}, were used to investigate the optimum temperature response property. Briefly, CLGNs were obtained as follows: 2.0 mL of 5% gelatin stock solution and 1.0 mL of anion solution with the same ionic strengths for surface modification was mixed together in a 50 mL beaker, followed by the addition of double-distilled water to the final volume of 10 mL. The mixture solution was stirred under 37°C for several minutes with the addition of 100 μL of Tween-20. A 20% (w/w) of sodium sulfate solution was added dropwise to the mixture until a small amount of precipitate formed. Isopropanol was then added dropwise to dissolve the precipitate by sodium sulfate and obtain a clear solution again. The temperature of the mixture was lowered to 25°C, and 0.1 mL glutaradehyde was introduced into the mixture as a cross-linking reagent. After 5 min stirring, 2.0 mL of 12% (w/w) sodium pyrosulfite was added to stop the cross-linking, and the system was maintained under continuous stirring for another 90 min. The final clear product was dialyzed with double-distilled water for 24 hours to remove the unreacted reagents and stored at 4°C prior to use.

Characterization of Cross-linked Gelatin Nanoparticles

The temperature responses of the different anion modified CLGNs solutions were examined by immersing the solutions in a thermostated water bath with the temperature set to rise gradually. The temperatures and response times were recorded when the clear solutions turned translucent.

The morphology of the CLGNs was characterized by transmission electron microscopy (JEM-2000EX, JEOL, Tokyo, Japan) using phosphotungstic acid as negative staining reagent. TEM imag of glucoamylase immobilized CLGNs prepared by adsorption method was also taken (H-7650, Hitachi, Tokyo, Japan). Their sizes and size distributions were determined using dynamic light scattering (DLS, Malvern Zetasizer Nano ZS90, Malvern instruments Ltd., U.K.). The zeta potential of the CLGNs under different pH and temperature were also measured with the same instrument.

The temperature-triggered swelling behavior of the CLGNs was investigated at different temperatures, increasing from 25 to 60°C, then decreasing back to 25°C. The swelling ratio was calculated using following equation:

$$SD(\%) = \frac{D_T - D_{T0}}{D_T} \times 100\%$$

(1)

Here SD, D_T, and D_{T0} are the swelling ratio, the diameter of CLGNs at certain temperature, and the diameter at 25°C, respectively.

The functional groups on the surface of the CLGNs were evaluated by a Fourier transform infrared spectrometer (FTIR, IR presitge-21, Shimazu, Kyoto, Japan). In a typical procedure, 0.25 mg of lyophilized CLGNs powder was mixed with IR-grade KBr (0.1 g) and pressed into a tablet form, and then the spectrum was recorded.

Immobilization of Glucoamylase

In the present study, two strategies for glucoamylase immobilization were chosen. The first one was the entrapment of glucoamylase for immobilization. In this method, a certain volume of glucoamylase stock solution was added to the gelatin mixture before the addition of sodium sulfate at the coacervation step. The following steps were the same as the preparation of CLGNs mentioned above. After dialysis for 24 hours, the mixture was stored at 4°C prior to use. The other method was adsorption of the glucoamylase. Briefly, after the preparation of pure CLGNs as described above, the dialyzed nanoparticles solution was lyophilized to powder form. A certain amount of gelatin powder was added to 3 mL of glucomylase solution and stirred at different temperature (25, 37, and 60°C) for four hours, and then at 25°C for one hour. The glucoamylase adsorbed CLGNs solution was also stored at 4°C prior the measurement of temperature-triggered release of enzyme.

Efficiency of Glucoamylase Immobilization

The activities of glucoamylase were assayed by the measurement of glucose produced by the hydrolysis of starch solution catalyzed by glucoamylase. The amount of glucose was determined using the DNS method according to reported reference [23]. Briefly, 0.5 mL of diluted free enzyme or the solution containing immobilized glucoamylase were added to 0.5 mL of soluble starch solution which contained 1.0% of (w/v) soluble starch in water as the substrate, followed by the addition of 0.5 mL of acetate buffer solution (0.1 mol/L, pH 4.6) under different temperatures. After stirring for 15 min, the reaction was stopped by the addition of 0.5 mL of NaOH solution (10% w/v). The glucose content in the reaction medium was then determined using the DNS method by the measurement of the UV-Vis absorption spectra at the wavelength of 520 nm. Each sample was measured three times, after that the average value was used for further calculations.

To determine the amount of immobilized glucoamylase, the mixture solution for the glucoamylase immobilization, both entrapped and adsorbed, was centrifuged for 10 min at 18000 rpm for the sedimentation of CLGNs. The activities of glucoamylase in the suspensions were measured by DNS method. The efficiency of immobilization was calculated by the decreased glucoamylase activity in the suspension after the immobilization compared to the free enzyme solution used for immobilization according to following equation:

$$\eta_1 = \frac{\text{immobilized enzyme}}{\text{added enzyme}} = \frac{U_f - U_s}{U_f} \times 100\%$$
$$= \frac{C_f - C_s}{C_f} \times 100\%$$

$$(2)$$

Here η_1 is the efficiency of the immobilization. U_f and U_s are the glucoamylase activities of the free enzyme solution used for immobilization and the suspension after centrifugation, respectively. C_f and C_s are the corresponding glucose concentrations determined by the DNS method for the starch solution added the free enzyme solution and the suspension after centrifugation, respectively.

Temperature-Triggered Glucoamylase Release

To determine the temperature-triggered glucoamylase release efficiency, the release of enzyme was carried out at different temperature. After centrifugation, the glucoamylase immobilized CLGNs sedimentation was washed three times with 0.5 mL acetate buffer solution (0.1 mol/L, pH 4.6) and centrifugation again. After the last washing, certain volumes of solution containing the enzyme immobilized CLGNs were added to series cuvettes, in which 0.5 mL of 1.0%(w/v) starch solution and acetate buffer (0.1 mol/L, pH 4.6) were previously mixed. These cuvettes were incubated under different temperatures for exactly 5 min for the release of glucoamylase. 0.5 mL of NaOH solution (10% w/v) was then added to stop the glucoamylase catalyzed hydrolysis of starch. The DNS method was used to determine the activity of the released glucoamylase as mentioned above. The release efficiency was calculated as the proportion of the released enzyme to the immobilized enzyme, which could be calculated according to the equation:

$$\eta_R = \frac{\text{released enzyme}}{\text{immobilized enzyme}} = \frac{U_r V_r}{(U_f - U_s)V_i} \times 100\%$$

(3)

Here η_R is the release efficiency. U_r is the activity of the released glucoamylase. V_i and V_r are the volumes of the solutions in the enzyme immobilization and release steps, respectively. We set the V_i and V_r the same value in the experiment, then equation (3) could be simplified as following:

$$\eta_R = \frac{U_r}{U_f - U_s} \times 100\% = \frac{C_r}{C_f - C_s} \times 100\%$$

(4)

Here C_r referred to the glucose concentration in the release experiment and determined by DNS method.

To investigate the effect of temperature on the activity of glucoamylase at the immobilization and release steps, a certain volume of enzyme immobilized CLGNs solution was immersed in a thermostated water bath at 60°C for four hours to complete the release of glucoamylase. The solution was then centrifuged, and the activity of enzyme in the suspension was measured at different temperature using the DNS method as mentioned above. The free enzyme activities at different temperature were also measured for comparison.

Statistical Analysis

One-way ANOVA was performed to determine the differences among the size distribution of CLGNs in the temperature-sensitive swelling experiment. The differences of glucoamylase immobilization and release efficiency at different temperatures were also determined. Newman–Keuls test was performed as post-hoc analysis for one-way ANOVA. Differences between the compared data with P-values <0.05 were considered statistically significant.

RESULTS AND DISCUSSION

Temperature Response of Cross-linked Gelatin Nanoparticles

Different anion modifications could affect the temperature response of CLGNs. With the increase of temperature, the clear solutions of CLGNs turned translucent. The images of CLGNs at 25°C, 35°C and 45°C were taken and shown in Panel A of Figure 2. As the temperatures increased, the clear solutions turned translucent. While the temperatures decreased, the solutions turned clear again. These phenomena clearly showed the swelling and contracting of CLGNs. With several cycles of temperature rising and cooling, these phenomena could still be observed. This result indicated that the reversible swelling and contracting of CLGNs. The time taken for the phase transition of CLGNs was recorded as response time. Then the temperature decreased gradually to room temperature to observe if the translucent solution turned clear again. The results are listed in Table 1.

Table 1: Temperature-Sensitive Response for Different Anions Modified CLGNs

Anion	Phase Transition Temperature (°C)	Response Time	Reversibility
X⁻ (2.0 mol/L)			
F^-	40	2'26''	√
Cl^-	46	5'59''	
Br^-	47	1'28''	
I^-	45	6'	
NO_3^-	48	4'	√
CH_3COO^-	40	1'53''	√
X²⁻ (1.0 mol/L)			
CO_3^{2-}	41	3'08''	
SO_4^{2-}	45	1'46''	
X³⁻ (0.7 mol/L)			
HPO_4^{3-}	40	2'27''	

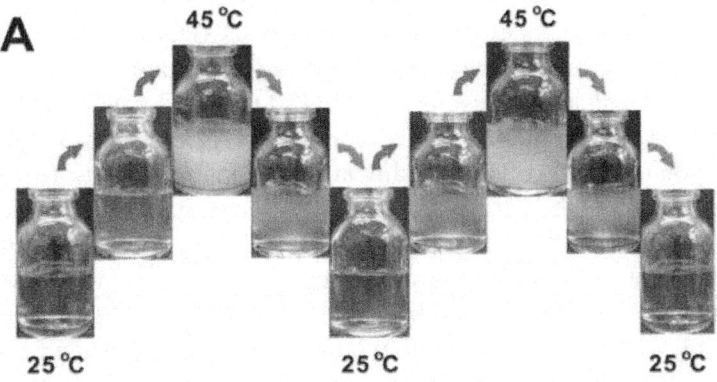

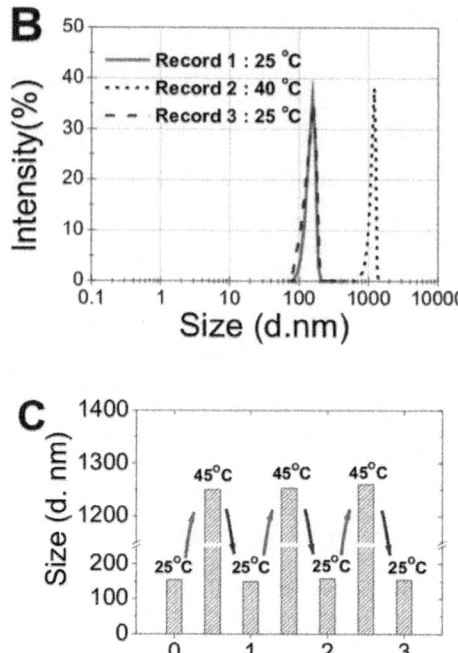

Figure 2: Temperature responses effect of CLGNs.

(A) Pictures of CLGNs at different temperatures. (B) Size distribution of CLGNs at different temperatures measured using DLS. (C) Size changes of CLGNs at 25°C and 45°C within three cycles.

As the results shown in Table 1, the fluoride and acetate anion modified CLGNs clearly bear an excellent reversible temperature response. Some anions could precipitate gelatin in the CLGNs preparation step, such as CO_3^{2-} and PO_4^{3-}. While others could be used to get CLGNs but made the CLGNs irreversible to the temperature-triggered phase transition, such as Cl^-, Br^-, I^- and SO_4^{2-}. To explain the reasons that CLGNs modified with F^- and CH_3COO^- could exhibit excellent reversible temperature response, we hypothesized that the polarizability of the modification anions contribute to the surface properties of CLGNs. The polarizability of ions was reported increase in the following order: $F^-<CH_3COO^-<Cl^-<Br^-<I^-$ [38]. This means the F^- and CH_3COO^- could hold electrons more tightly than other anions, while all of the counter cations for the modification were sodium. Due to the fact that anions with weak ion polarizability would have strong repulsion between each other, we hypothesized that the modification of CLGNs with F^- and CH_3COO^- would result in a stronger repulsion at the surface of the CLGNs. When the

temperature was increased, the stronger repulsion could more easily counteract the hydrogen bonds between oxygen and nitrogen atoms in CLGNs [39]. In other words, the CLGNs modified with F⁻ and CH_3COO^- should have a lower phase transition temperature. Moreover, the acetate anion modified CLGNs showed a quicker response time, Thus, 1.0 mL of 2.0 mol/L sodium acetate was used as the modification solution in the preparation of reversible temperature-sensitive CLGNs.

Morphology and Size Distribution of Cross-linked Gelatin Nano-particles

Panel B of Figure 2 showed that the size distribution measured by dynamic light scattering at different temperatures. The results showed that at room temperature, CLGNs were monodispersed with diameter of 155±5 nm and polydispersity index (PdI) around 0.14. With the small PdI value, it is clearly known that the CLGNs were homogeneous [40], [41]. As the temperature elevated from 25 to 40°C, the diameter of CLGNs increased to 1257+24 nm while the PdI remained the same value. TEM image also showed that CLGNs were monodispersed and almost spherical in shape (SEM photo of CLGNs also confirmed this as Figure S1 shown). Most of the CLGNs had a diameter around 150 nm (Figure 3, Panel A). TEM image of glucoamylase loaded CLGNs prepared by adsorption method was also shown (Figure 3, Panel B). After the enzyme immobilization, the size of CLGNs didn't change significantly. The PdI value measured from the TEM image was 0.18, which correspondence with the data of unloaded CLGNs. These results confirmed that during the enzyme immobilization process, the CLGNs exhibited an excellent reversible swelling property.\.

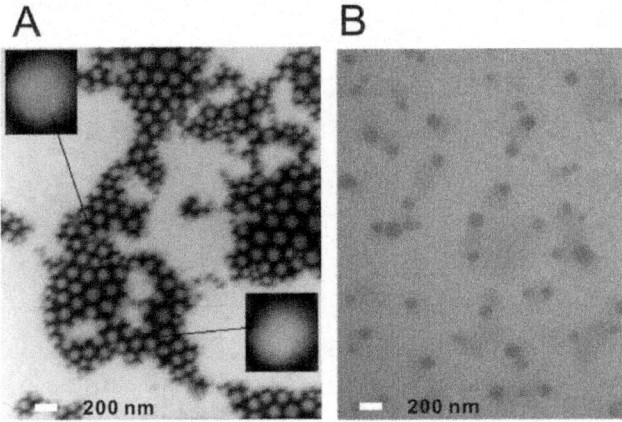

Figure 3: TEM image of the CLGNs.

(A) negative stained by phosphotungstic acid and (B) gluocoseamylase immobilized CLGNs prepared by adsorption method.

From the results of temperature-sensitive size distribution of CLGNs, it was also confirmed that the CLGNs exhibited an excellent reversible temperature-sensitive response. The increased diameter could be attributed to the swelling of the CLGNs rather than the aggregation caused by raised temperature because the constant PdI and the contracting of CLGNs as the temperature decreased to room temperature. The Zeta potential measurements of CLGNs at different temperature were also confirmed reversible swelling of the nanopaticles. With the increasing of temperature, the Zeta potential became more negative among acidic, neutral and basic conditions (Panel A in Fugre 4). These results indicated that with increased temperature, the repulsion of at the surface of the gelatin became stronger, which were according with the results listed in Table 1.

In addition, after storing under common conditions, such as in a refrigerator or at room temperature for six months, the CLGNs did not exhibit any significant change in morphology and particle size (data not shown). These indicate the stability and good fluidity of acetate modified CLGNs.

The swelling degree results of CLGNs are listed in Table 2. As the temperature elevated, the CLGNs exhibit swelling process, observed as the increased diameter at the temperature above 40°C. When the temperature decreased below 40°C, the diameter of the CLGNs got back to 155±5 nm. The results showed that the CLGNs have excellent reversible temperature-triggered swelling properties.

Table 2: Reversible Temperature-Triggered Swelling Degree of the CLGNs, gulocoseamylase immobilized CLGNs prepared by entrapment and adsorption methods respectively.

T (°C)	CLGNs		Enzyme Entrapped		Enzyme Adsorbed	
	Average Diameter (nm)	SD%	Average Diameter (nm)	SD%	Average Diameter (nm)	SD%
25	155.0	0	154.0	0	156.0	0
40	1257[a]	87.7	1261[a]	87.8	1258[a]	87.6
50	1300[a]	88.1	1298[a]	88.1	1325[a]	88.2
60	1381[a]	88.8	1315[a]	88.3	1330[a]	88.3
50	1326[a]	88.3	1296[a]	88.1	1312[a]	88.1
40	1230[a]	87.4	1250[a]	87.7	1256[a]	87.6
25	155.0	0	156.2	1.4	158.3	1.5

In Panel B of Figure 4, the IR spectra of lyophilized CLGNs are shown. Glutaraldehyde was used as a cross-linker to for a Schiff base with the amino group of gelatin. This method was widely used in biochemistry and gelatin nanoparticles preparations [42]. Comparison of the IR spectra of the CLGNs and pure gelatin shows that the strengthened IR absorption around 2900 nm^{-1} could be attributed to the stretching vibration of $-(CH_2)_3-$, which was derived from the glutaraldehyde used as cross-linking reagent. On the other hand, the strong peak of glutaraldehyde around 1750 nm^{-1} was not observed, while the intensity around 1110 and 1660 nm^{-1} increased. This confirmed the formation of Schiff base between the carbonyl group of glutaraldehyde and amino group of gelatin [43]–[45].

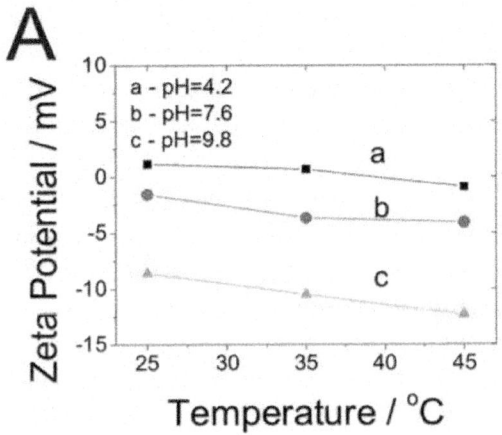

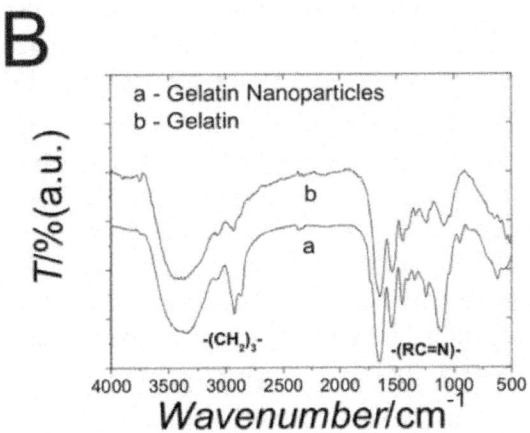

Figure 4: Characterization of CLGNs.

(A) Zeta potential measurements values of the CLGNs under different temperature. (B) IR spectra of lyophilized CLGNs. Spectra of pure gelatin were also presented for comparison.

Efficiency of Glucoamylase Immobilization

Glucoamylase (EC 3.2.1.3) is a biocatalyst capable of hydrolyzing α-1, 4 glycosidic linkages in raw or soluble starches and related oligosaccharides to produce β-glucose. Recent research on glucoamylase immobilization mainly focused on the entrapment of the cross-linked enzyme and covalent binding on different matrices in order to achieve industrial applications [23]. The disadvantage of these immobilization methods is the inactivation of the enzyme [16], [17]. In current study, glucoamylase was immobilized by the entrapment and adsorption methods to prevent inactivation. Furthermore, the optimum temperature range for glucoamylase is 40–70°C. This property makes it possible for the enzyme to maintain its activity during the immobilization process and perform its highest catalytic ability after its release, which was carried out above the CLGNs' phase transition temperature (40°C).

In the first attempt to immobilize glucoamylase by the entrapment method in distilled water, the efficiency of the immobilization was 22.5% and the amount of the enzyme loaded was 2.48 mg per 100 mg CLGNs. The isoelectric point (pI) of the gelatin used in this study is 5.0, whereas the pI of the glucoamylase is around 3.0~4.0. When the gelatin and glucoamylase were oppositely charged, the efficiency of the immobilization was predicted to increase. The experimental results showed that if gelatin was dissolved in an acetate buffer solution (0.1 mol/L, pH 4.6) instead of distilled water at the beginning step of CLGNs preparation, the efficiency of immobilization increased to 59.9% and the amount of enzyme loaded was 6.59 mg per 100 mg CLGNs.

Immobilization of glucoamylase by the adsorption method was carried out at different temperatures (25, 37 and 60°C). The efficiency of the immobilization at 25 and 37°C were 19.8% and 26.7% respectively, using the CLGNs without dialysis. When the adsorption temperature was raised to 60°C, the immobilization efficiency was as high as 82.7%, indicating that the temperature-triggered CLGNs phase transition could increase the amount of adsorbed enzyme. But in the following glucoamylase release experiment, there was no release observed for adsorbed enzyme from the CLGNs.

Another strategy using CLGNs with dialysis to adsorb glucoamylase was carried out at different temperature. The results listed in Table 3 showed that immobilization efficiency with dialyzed CLGNs were not as high as non-dialyzed ones. But in the following glucoamylase release experiments, the dialyzed CLGNs showed excellent temperature-triggered enzyme release property.

Table 3: Glucoamylase Immobilization Efficiency by Adsorption Method Using Dialyzed and Non-dialyzed CLGNs at Different Temperature

Temperature (°C)	Non-dialyzed CLGNs η_i(%)	Dialyzed CLGNs η_i(%)
25	19.8	8.3
37	26.7	6.7
60	82.7 [a, b]	24.7 [a, b]

The reasonable explanation for the immobilization efficiencies of dialyzed and non-dialyzed CLGNs also rely on that the glutaraldehyde used as cross-linking reagent was remained in the solution of non-dialyzed CLGNs. In the immobilization step, glutaraldehyde could react with the glucoamylase forming covalent linkage between gelatin and glucoamylase, which resulted in the higher immobilization efficiency. But the covalently linked enzyme was hardly to be released or the covalent linkages deactivated the glucoamylase. On the other hand, when perform the immobilization using dialyzed CLGNs, the glucoamylase was physically adsorbed at the gelatin matrix. Although the immobilization efficiency was only around 25%, the temperature-triggered enzyme release property of the dialyzed CLGNs was satisfactory.

Various amounts of dialyzed and lyophilized CLGNs powder (20, 30 and 60 mg) were used to investigate the optimum conditions for immobilization. The immobilization efficiency was determined as 17.8%, 24.7% and 24%, respectively. The results showed that the increased amount of the CLGNs powder improved the immobilization efficiency. However, when the added gelatin powder exceeded 30 mg, the efficiency did not increase any further. Thus the optimum amount of CLGNs powder for enzyme adsorption was determined to be 30 mg.

The concentration of glucoamylase also had an effect on the immobilization efficiency. 3 mL of solutions with different glucoamylase concentrations were used to carry out the adsorption studies. The concentrations of these solutions varied from 0.5, 1.0 and 2.0 mg/mL with the resulting immobilization efficiencies as 23.8%, 24.7% and 12.2%, respectively. The decreased immobilization efficiency with 2.0 mg/mL glucoamylase solution indicated that the adsorption of enzyme saturated at the higher concentration. This means at low level of enzyme concentration, the more enzyme added, the more it could be adsorbed. When the enzyme concentration was high enough, the CLGNs adsorbed enzyme in its capacity completely. So the concentration of glucoamylase was set at 1.0 mg/mL in the subsequent experiments for the temperature-triggered enzyme release.

Temperature-Triggered Glucoamylase Release

Enzyme immobilization by entrapment showed no release of enzyme even when the temperature was raised to 60°C. So this immobilization method could not be used for temperature-triggered enzyme release. The reason for the high immobilization efficiency and no release in the entrapment method for enzyme immobilization could be attributed to the glutaraldehyde added in the CLGNs preparation. Glutaraldehyde not only cross-links the amino acid residues on the gelatin chain, but also with those on the glucoamylase. The cross-linking between gelatin and glucoamylase resulted in the high immobilization efficiency. On the other hand, the firm cross-linking could also inhibit the release of glucoamylase.

Temperature-triggered enzyme release for the immobilized glucoamylase by adsorption method with dialyzed CLGNs was also examined. The results are shown in Panel A of Figure 5. For comparison, the activities of the free enzyme at the same temperature are also shown. At temperatures below 40°C, the activities of glucoamylase in the CLGNs solution were very low, indicating that the immobilization was effective and there was no enzyme released. When the temperature was set above 40°C, higher than the phase transition temperature of the CLGNs, the immobilized enzyme was immediately released and showed recovered activities. The release efficiency examined at 60°C was 99.3%, indicating that the temperature-triggered enzyme release was complete.

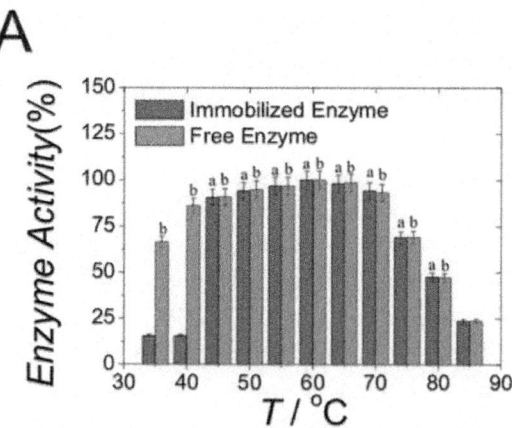

B

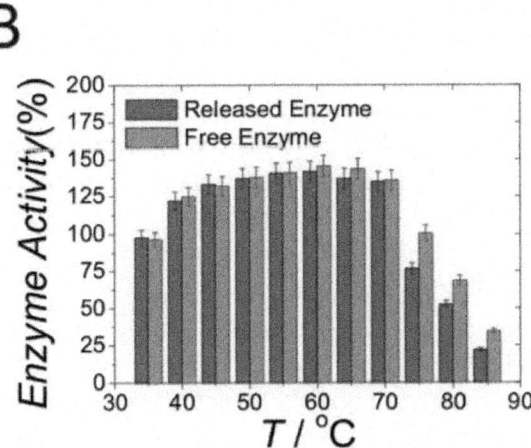

Figure 5: Activities of the glucoamylase at different temperature.

(A) 30 mg of dialyzed and lyophilized powder was added into 3 mL of 1.0 mg/mL glucoamylase solution to perform the adsorption of enzyme at 60°C for four hours and then at 25°C for one hour. After centrifugation and washing process to remove free enzyme, the sedimentation of CLGNs were incubated at different temperature to determine the activities of released enzyme. The activities of free glucoamylase were also calculated at the same condition for comparison. (B) Comparison of enzyme activities for the released glucoamylase at 60°C and free enzyme at temperature range of 30–85°C a: Data are compared with the data of released enzyme at 40°C and significantly different ($P<0.05$) from that data; b: Data are compared with the data of free enzyme at 85°C and significantly different ($P<0.05$) from that data.

In order to confirm the fact that the immobilized enzyme could not perform its catalytic ability, the experiment of mixing CLGNs and glucose were carried out using blank CLGNs and enzyme immobilized ones at the temperature blow 40°C. The results showed that neither the blank CLGNs nor the glucoamylase immobilized ones show evident enzyme activities. The conclusion is the immobilized glucoamylse could not catalyze the hydrolysis of glucose. In other words, the immobilization by CLGNs prevented the catalytic abilities.

Based on the experiments of temperature-triggered enzyme release using dialyzed CLGNs by adsorption method, we hypothesized that compact CLGNs could be formed when the temperature is below 40°C because of the hydrogen bonds between oxygen and nitrogen atoms in the amino acid residues. When the temperature was raised above 40°C, the hydrogen bonds broke and the CLGNs swelled, which was observed as the phase transition. The size of the CLGNs dramatically increased, and the mesh size at the surface of the nanoparticles also enlarged. At this phase, the glucoamylase could easily enter and come out from the CLGNs. When the temperature was decreased below 40°C, the hydrogen bonds were formed again and the CLGNs quickly shrank back to their original size. This process completes the adsorption of glucoamylase. After immobilization was completed, the centrifugation and washing step removed the free enzymes remaining in the solution. When the temperature was elevated again, the swelling of the CLGNs made it possible for the complete release of immobilized glucoamylase.

By comparison of the activities for the released glucoamylase at 60°C and free enzyme at temperature range of 30–85°C, there were no significant differences (Panel B in Figure 5). These results indicated that the immobilization and release process had no major effect on the activities of glucoamylase. The different enzyme activities for the released and free glucoamylase observed below 40°C in Panel A of Figure 5 evidently derived from the immobilization of glucoamylase in the CLGNs which could not catalyzed the hydrolyzation of starch while the free glucoamylase could.

Attempts to achieve reversible immobilization of glucoamylase after the temperature-triggered release were also carried out. After the release of glucoamylase above the phase transition temperature, the temperature was reduced to 25°C for four hours for reversible immobilization. The solution was centrifuged for 10 min under 18000 rpm for the sedimentation of CLGNs, as described previously. The sediment was then re-dissolved in an acetated buffer (0.1 mol/L, pH 4.6) and the temperature was raised to 60°C for enzyme release determination. Unfortunately, reversible immobilization and release were not observed. This may be because the immobilization efficiency by adsorption was not high enough (24.7%) for the re-adsorption. Future researches will focus on improvement of immobilization efficiency to explore the reversible immobilization and release for enzymes.

CONCLUSION

The presented study introduced a novel strategy for the preparation of acetate modified CLGNs which showed a reversible temperature-triggered swelling. Morphology results showed that the CLGNs were regularly spherical, with average diameters of about 155±5 nm. As well, the CLGNs exhibited favorable dispersivity. When the ambient temperature was above 40°C, the diameters of CLGNs could increase to over 1 μm in a few minutes, and their reversible temperature-sensitive properties are fine. By the adsorption method, glucoamylase could be easily immobilized in the CLGNs with loading efficiency of 24.7% and 3.14 mg glucoamylase per 100 mg CLGNs, respectively. When the system temperature was higher than 40°C, which was the phase transition temperature of acetate modified CLGNs, the absorbed glucoamylase was released completely. The activity of released glucoamylase was not affected by the immobilizationn and it could function as a free enzyme during the catalytic reaction. This system may be an effective alternative for enzyme immobilization used in biotechnology.

REFERENCES

1. Krajewska B (2004) Application of chitin- and chitosan-based materials for enzyme immobilizations: a review. Enzyme and Microbial Technology 35: 126–139. doi: 10.1016/j.enzmictec.2003.12.013

2. Wang B, Zhu W, Zhang Y, Yang ZG, Ding JD (2006) Synthesis of a chemically-crosslinked thermo-sensitive hydrogel film and in situ encapsulation of model protein drugs. Reactive & Functional Polymers 66: 509–518. doi: 10.1016/j.reactfunctpolym.2005.10.003

3. Shakya AK, Sami H, Srivastava A, Kumar A (2010) Stability of responsive polymer-protein bioconjugates. Progress in Polymer Science 35: 459–486. doi: 10.1016/j.progpolymsci.2010.01.003

4. Milasinovic N, Milosavljevic N, Filipovic J, Knezevic-Jugovic Z, Krusic MK (2010) Synthesis, characterization and application of poly(N-isopropylacrylamide-co-itaconic acid) hydrogels as supports for lipase immobilization. Reactive & Functional Polymers 70: 807–814. doi: 10.1016/j.reactfunctpolym.2010.07.017

5. Mondal K, Mehta P, Mehta BR, Varandani D, Gupta MN (2006) A bioconjugate of Pseudomonas cepacia lipase with alginate with enhanced catalytic efficiency. Biochimica Et Biophysica Acta-Proteins and Proteomics 1764: 1080–1086. doi: 10.1016/j.bbapap.2006.04.008

6. Gai LL, Wu DC (2009) A Novel Reversible pH-Triggered Release Immobilized Enzyme System. Applied Biochemistry and Biotechnology 158: 747–760. doi: 10.1007/s12010-008-8373-2

7. Asanomi Y, Yamaguchi H, Miyazaki M, Maeda H (2011) Enzyme-Immobilized Microfluidic Process Reactors. Molecules 16: 6041–6059. doi: 10.3390/molecules16076041

8. Ge J, Lu DN, Liu ZX, Liu Z (2009) Recent advances in nanostructured biocatalysts. Biochemical Engineering Journal 44: 53–59. doi: 10.1016/j.bej.2009.01.002

9. Torres-Salas P, del Monte-Martinez A, Cutino-Avila B, Rodriguez-Colinas B, Alcalde M, et al. (2011) Immobilized Biocatalysts: Novel Approaches and Tools for Binding Enzymes to Supports. Advanced Materials 23: 5275–5282. doi: 10.1002/adma.201101821

10. .Sheldon RA (2007) Enzyme immobilization: The quest for optimum performance. Advanced Synthesis & Catalysis 349: 1289–1307. doi: 10.1002/adsc.200700082

11. Zhang BH, Weng YQ, Xu H, Mao ZP (2012) Enzyme immobilization for biodiesel production. Applied Microbiology and Biotechnology 93: 61–70. doi: 10.1007/s00253-011-3672-x

12. Garcia-Galan C, Berenguer-Murcia A, Fernandez-Lafuente R, Rodrigues RC (2011) Potential of Different Enzyme Immobilization Strategies to Improve Enzyme Performance. Advanced Synthesis & Catalysis 353: 2885–2904. doi: 10.1002/adsc.201100534

13. Jun SH, Lee J, Kim BC, Lee JE, Joo J, et al. (2012) Highly Efficient Enzyme Immobilization and Stabilization within Meso-Structured Onion-Like Silica for Biodiesel Production. Chemistry of Materials 24: 924–929. doi: 10.1021/cm202125q

14. Terres CME, Dominguez JM, Aburto J (2007) Immobilization of chloroperoxidase on silica-based materials for 4,6-dimethyl dibenzothiophene oxidation. Journal of Molecular Catalysis B-Enzymatic 48: 90–98. doi: 10.1016/j.molcatb.2007.06.012

15. Laveille P, Falcimaigne A, Chamouleau F, Renard G, Drone J, et al. (2010) Hemoglobin immobilized on mesoporous silica as effective material for the removal of polycyclic aromatic hydrocarbons pollutants from water. New Journal of Chemistry 34: 2153–2165. doi: 10.1039/c0nj00161a

16. Tanriseven A, Olcer Z (2008) A novel method for the immobilization of glucoamylase onto polyglutaraldehyde-activated gelatin. Biochemical Engineering Journal 39: 430–434. doi: 10.1016/j.bej.2007.10.011

17. Bai YX, Li YF, Lei L (2009) Synthesis of a mesoporous functional copolymer bead carrier and its properties for glucoamylase immobilization. Applied Microbiology and Biotechnology 83: 457–464. doi: 10.1007/s00253-009-1864-4

18. Mahkam M, Vakhshouri L (2010) Colon-Specific Drug Delivery Behavior of pH-Responsive PMAA/Perlite Composite. International Journal of Molecular Sciences 11: 1546–1556. doi: 10.3390/ijms11041546

19. Gaur R, Lata, Khare SK (2005) Immobilization of xylan-degrading enzymes from Scytalidium thermophilum on Eudragit L-100. World Journal of Microbiology & Biotechnology 21: 1123–1128. doi: 10.1007/s11274-005-0080-3

20. Saleem M, Rashid MH, Jabbar A, Perveen R, Khalid AM, et al. (2005) Kinetic and thermodynamic properties of an immobilized endoglucanase from Arachniotus citrinus. Process Biochemistry 40: 849–855. doi:

10.1016/j.procbio.2004.02.026

21. Wang MF, Qi W, Yu QX, Su RX, He ZM (2010) Cross-linking enzyme aggregates in the macropores of silica gel A practical and efficient method for enzyme stabilization. Biochemical Engineering Journal 52: 168–174. doi: 10.1016/j.bej.2010.08.003

22. Tanriseven A, Dogan S (2001) Immobilization of invertase within calcium alginate gel capsules. Process Biochemistry 36: 1081–1083. doi: 10.1016/s0032-9592(01)00146-7

23. .Wang F, Guo C, Liu HZ, Liu CZ (2007) Reversible immobilization of glucoamylase by metal affinity adsorption on magnetic chelator particles. Journal of Molecular Catalysis B-Enzymatic 48: 1–7. doi: 10.1016/j. molcatb.2007.05.003

24. Krajewska B (2009) Ureases. II. Properties and their customizing by enzyme immobilizations: A review. Journal of Molecular Catalysis B-Enzymatic 59: 22–40. doi: 10.1016/j.molcatb.2009.01.004

25. Bayramoglu G, Karakisla M, Altintas B, Metin AU, Sacak M, et al. (2009) Polyaniline grafted polyacylonitrile conductive composite fibers for reversible immobilization of enzymes: Stability and catalytic properties of invertase. Process Biochemistry 44: 880–885. doi: 10.1016/j. procbio.2009.04.011

26. Qiu Y, Park K (2001) Environment-sensitive hydrogels for drug delivery. Advanced Drug Delivery Reviews 53: 321–339. doi: 10.1016/s0169-409x(01)00203-4

27. .Kaetsu I, Uchida K, Sutani K, Sakata S (2000) Intelligent biomembrane obtained by irradiation techniques. Radiation Physics and Chemistry 57: 465–469. doi: 10.1016/s0969-806x(99)00416-8

28. Satarkar NS, Hilt JZ (2008) Magnetic hydrogel nanocomposites for remote controlled pulsatile drug release. Journal of Controlled Release 130: 246–251. doi: 10.1016/j.jconrel.2008.06.008

29. Sakata S, Uchida K, Kaetsu I, Kita Y (2007) Programming control of intelligent drug releases in response to single and binary environmental stimulation signals using sensor and electroresponsive hydrogel. Radiation Physics and Chemistry 76: 733–737. doi: 10.1016/j. radphyschem.2004.12.010

30. Saraogi GK, Gupta P, Gupta UD, Jain NK, Agrawal GP (2010) Gelatin nanocarriers as potential vectors for effective management of tuberculosis. International Journal of Pharmaceutics 385: 143–149. doi: 10.1016/j. ijpharm.2009.10.004

31. Kommareddy S, Amiji M (2007) Poly(ethylene glycol)-modified thiolated gelatin nanoparticles for glutathione-responsive intracellular DNA delivery. Nanomedicine-Nanotechnology Biology and Medicine 3: 32–42. doi: 10.1016/j.nano.2006.11.005

32. Mao JS, McShane MJ (2006) Transduction of volume change in pH-sensitive hydrogels with resonance energy transfer. Advanced Materials 18: 2289–2293. doi: 10.1002/adma.200600040

33. Naganagouda K, Mulimani VH (2006) Gelatin blends with alginate: Gel fibers for alpha-galactosidase immobilization and its application in reduction of non-digestible oligosaccharides in soymilk. Process Biochemistry 41: 1903–1907. doi: 10.1016/j.procbio.2006.03.040

34. Sheelu G, Kavitha G, Fadnavis NW (2008) Efficient immobilization of lecitase in gelatin hydrogel and degumming of rice bran oil using a spinning basket reactor. Journal of the American Oil Chemists Society 85: 739–748. doi: 10.1007/s11746-008-1261-7

35. Kovalenko GA, Perminova LV (2008) Immobilization of glucoamylase by adsorption on carbon supports and its application for heterogeneous hydrolysis of dextrin. Carbohydrate Research 343: 1202–1211. doi: 10.1016/j.carres.2008.02.006

36. Tardioli PW, Vieira MF, Vieira AMS, Zanin GM, Betancor L, et al. (2011) Immobilization-stabilization of glucoamylase: Chemical modification of the enzyme surface followed by covalent attachment on highly activated glyoxyl-agarose supports. Process Biochemistry 46: 409–412. doi: 10.1016/j.procbio.2010.08.011

37. Wu DC, Wan MX (2008) A Novel Fluoride Anion Modified Gelatin Nanogel System for Ultrasound-Triggered Drug Release. Journal of Pharmacy and Pharmaceutical Sciences 11: 32–45.

38. Solomatin SV, Bronich TK, Bargar TW, Eisenberg A, Kabanov VA, et al. (2003) Environmentally responsive nanoparticles from block ionomer complexes: Effects of pH and ionic strength. Langmuir 19: 8069–8076. doi: 10.1021/la0300151

39. Bajpai AK, Goswami S, Bajpai J (2010) Designing Gelatin Nanocarriers as a Swellable System for Controlled Release of Insulin: An In-Vitro Kinetic Study. Journal of Macromolecular Science Part a-Pure and Applied Chemistry 47: 119–130. doi: 10.1080/10601320903458556

40. Yates MZ, ONeill ML, Johnston KP, Webber S, Canelas DA, et al. (1997) Emulsion stabilization and flocculation in CO2.2. Dynamic light scattering. Macromolecules 30: 5060–5067. doi: 10.1021/ma961694s

41. Pereira-Lachataignerais J, Pons R, Panizza P, Courbin L, Rouch J, et al.

(2006) Study and formation of vesicle systems with low polydispersity index by ultrasound method. Chemistry and Physics of Lipids 140: 88–97. doi: 10.1016/j.chemphyslip.2006.01.008

42. Beauchamp RO, St Clair MB, Fennell TR, Clarke DO, Morgan KT, et al. (1992) A critical review of the toxicology of glutaraldehyde. Crit Rev Toxicol 22: 143–174. doi: 10.3109/10408449209145322

43. Gunasekaran S, Wang T, Turhan M (2004) Selected properties of pH-sensitive, biodegradable chitosan-poly(vinyl alcohol) hydrogel. Polymer International 53: 911–918. doi: 10.1002/pi.1461

44. Dubey VK, Singh AN, Singh S, Suthar N (2011) Glutaraldehyde-Activated Chitosan Matrix for Immobilization of a Novel Cysteine Protease, Procerain B. Journal of Agricultural and Food Chemistry. 59: 6256–6262. doi: 10.1021/jf200472x

45. Singh M, Tarannum N (2010) Synthesis and Characterization of Zwitterionic Organogels Based on Schiff Base Chemistry. Journal of Applied Polymer Science 118: 2821–2832. doi: 10.1002/app.32393

Chapter 5

STUDY ON DIFFERENT MOLECULAR WEIGHTS OF CHITOSAN AS AN IMMOBILIZATION MATRIX FOR A GLUCOSE BIOSENSOR

Lee Fung Ang[1], Lip Yee Por[2], Mun Fei Yam[1]

[1]School of Pharmaceutical Sciences, Universiti Sains Malaysia, Minden, Penang, Malaysia,

[2]Faculty of Computer Science and Information Technology, University of Malaya, Kuala Lumpur, Malaysia

ABSTRACT

Two chitosan samples (medium molecular weight (MMCHI) and low molecular weight (LMCHI)) were investigated as an enzyme immobilization matrix for the fabrication of a glucose biosensor. Chitosan membranes prepared from acetic acid were flexible, transparent, smooth and quick-drying. The FTIR spectra showed the existence of intermolecular interactions between chitosan and glucose oxidase (GOD). Higher catalytic activities were observed on for GOD-MMCHI than GOD-LMCHI and for those crosslinked with glutaraldehyde than using the adsorption technique. Enzyme loading greater than 0.6 mg decreased the activity. Under optimum conditions (pH 6.0, 35°C and applied potential of 0.6 V) response times of 85 s and 65 s were observed for medium molecular weight chitosan glucose biosensor (GOD-MMCHI/PT) and low molecular weight chitosan glucose biosensor (GOD-LMCHI/PT), respectively. The apparent Michaelis-Menten constant (K_M^{app}) was found to be 12.737 mM for GOD-MMCHI/PT and 17.692 mM for GOD-LMCHI/PT. This indicated that GOD-MMCHI/PT had greater affinity for the enzyme. Moreover, GOD-MMCHI/PT showed higher sensitivity (52.3666 nA/mM glucose) when compared with GOD-LMCHI/PT (9.8579 nA/mM glucose) at S/N>3. Better repeatability and reproducibility were achieved with GOD-MMCHI/PT than GOD-LMCHI/PT regarding glucose measurement. GOD-MMCHI/PT was found to give the highest enzymatic activity among the electrodes under investigation. The extent of interference encountered by GOD-MMCHI/

PT and GOD-LMCHI/PT was not significantly different. Although the Nafion coated biosensor significantly reduced the signal due to the interferents under study, it also significantly reduced the response to glucose. The performance of the biosensors in the determination of glucose in rat serum was evaluated. Comparatively better accuracy and recovery results were obtained for GOD-MMCHI/PT. Hence, GOD-MMCHI/PT showed a better performance when compared with GOD-LMCHI/PT. In conclusion, chitosan membranes shave the potential to be a suitable matrix for the development of glucose biosensors.

INTRODUCTION

A biosensor is commonly described as an analytical device incorporating a biological or biologically derived recognition element, either intimately associated or integrated within a physicochemical transducer to produce a signal proportional to the target analyte concentration[1]. What is important is a direct relationship between the biosensor signal and the quantity of the analyte. Since the invention of the first oxygen electrode by Clark and Lyons (1962), enzymes have been the most regularly employed biorecognition elements encountered in catalytic biosensors for the analysis of small molecules such as glucose, which is widely monitored in medicine, biotechnology and the food industry [2], [3].

Glucose biosensors are categorized based on the type of transducer used, such as electrochemical, piezoelectric, thermoelectric, acoustic and optical sensors (Wilkins &Atanasov, 1996) [4]. In electrochemical sensors, the electrical signal is a direct result of a chemical process occurring at the transducer/analyte interface [4]. Electrochemical transducers include potentiometric, voltammetric (amperometric), conductometric and field-effect transistor (FET)-based electrodes [5]. To date, electrochemical sensors dominate in glucose sensing owing to the simplicity of electrochemical measuring principles and the low cost [5]. Glucose oxidase was selected as a model enzyme in this study. Glucose oxidase is highly specific for β-D-glucose. It is widely used for the determination of glucose in body fluids and in removing residual glucose and oxygen from beverages and foodstuffs. Moreover, glucose oxidase is suitable for biosensor applications.

Chitosan was selected as a matrix for immobilization of the enzyme because of its biocompatibility, non-toxicity, high mechanical strength and excellent membrane forming ability. Chitosan can be divided into three categories, namely low molecular weight, medium molecular weight and high molecular weight. Chitosan of higher molecular weight possesses longer molecular chains with the availability of more hydroxyl groups. There is also a higher possibility that there are more amino groups, although the number of amino groups is determined by the degree of deacetylation. These amino groups are responsible for crosslinking. However, a review of the current literature revealed leaked documented reports on a comparative study of differences in the molecular weight of chitosan as enzyme matrix for biosensor. Therefore, in the present study, we investigated whether higher molecular weight chitosan could improve enzyme retention activity and loading, and thus function as a suitable matrix for enzyme immobilization. By using this chitosan membrane as a biosensor, it was hypothesized that it would provide a better performance in terms of sensitivity and stability. In this study, we report the development, fabrication and characterization of an amperometric-based glucose biosensor with the aim of comparing the effectiveness of chitosan of different molecular weights as a matrix for enzyme immobilization using adsorption and crosslinking techniques, and to study the behavior of these enzyme-chitosan electrodes.

RESULTS AND DISCUSSION

Investigation of the Intermolecular Interactions of Immobilized GOD-chitosan Membranes using FTIR

FTIR is a powerful tool for identifying the types of chemical bonds in a molecule by producing an infrared absorption spectrum that is like a "molecular fingerprint".

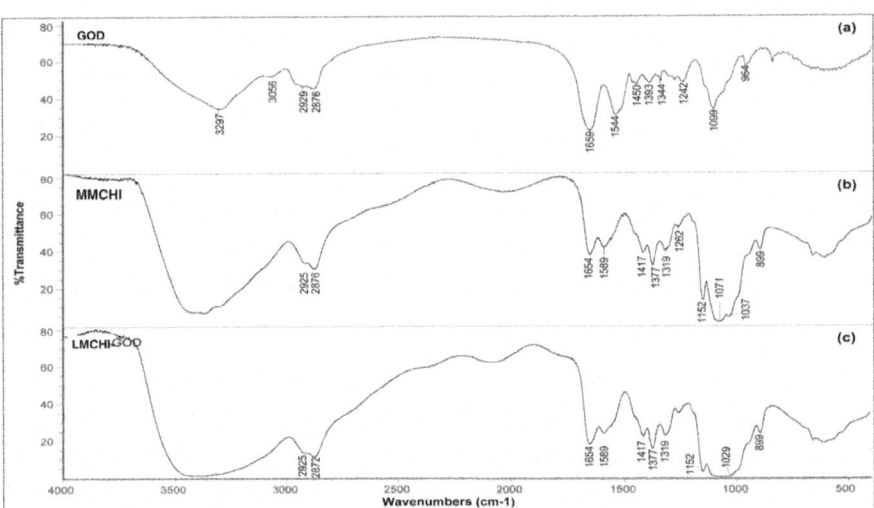

Figure 1: FTIR spectra of (a) crystalline GOD; (b) MMCHI membrane; (c) GOD-MMCHI membrane showing the interactions between GOD and chitosan membrane after immobilization.

In this study, FTIR spectroscopy was used to investigate the interactions between chitosan and glucose oxidase.Figure 1 displays the Fourier transform infrared spectroscopy (FTIR) spectra of glucose oxidase (GOD) crystals, medium molecular weight chitosan (MMCHI) membrane and the GOD-MMCHI membrane.

Detailed information on the secondary structure of a polypeptide chain is provided by the shape of the amide I and amide II infrared absorbance bands of the protein [6]. In the crystalline GOD spectrum, the C=O stretching of the amide bond was observed at 1659 cm^{-1}. The peak at 1544 cm^{-1} was assigned to strong N-H bending vibration of the secondary amide[7]. The absorption bands between 1340 cm^{-1} and 1450 cm^{-1} were attributed to symmetric and asymmetric vibrations of COO^{-} groups [8]. The bands for symmetric and asymmetric vibrations of CH$_2$ groups were found at 2876 cm^{-1} and 2929 cm^{-1}, respectively [1], [9]. The strong band observed at 3297 cm^{-1} and a weaker band at 3056 cm^{-1}were attributed to amide A and amide B bands of GOD, respectively. The amide A band arises due to N-H stretching vibrations, whereas the amide B band arises due to the first overtone of the amide II vibration intensified by Fermi resonance with amide A vibrations [1].

Although the overall FTIR spectra of the MMCHI membrane (cast from acetic acid), crystalline GOD and the GOD-MMCHI membrane were similar to each other, they showed subtle differences in the absorption intensities and range of frequencies. The presence of residual moisture content and glycerol in the GOD-MMCHI membrane resulted in a broad peak from 3300 to 3500 cm^{-1}. The amide I and II bands of the GOD-MMCHI membrane increased in intensity compared to that of the MMCHI membrane. The peak at 1544 cm^{-1} in the spectrum of crystalline GOD shifted to 1589 cm^{-1} in the GOD-MMCHI membrane. This could be due to intermolecular interactions between the enzyme and membrane via the formation of a Schiff base linkage between the aldehyde groups of glutaraldehyde and the amine groups of GOD and chitosan. Musale and Kumar (2000) reported that increased intensity of an amide II band may be due to the overlapping of peaks corresponding to -N-H stretching in –NHCOCH$_3$ of chitosan and C=N stretching of the newly formed Schiff base between the –NH$_2$ groups of chitosan and the =CHO groups of glutaraldehyde [10]. In the spectrum of the GOD-MMCHI membrane, the heme vibrational modes (1340–1400 cm^{-1}) [11] of GOD are overlapped with the bands of CH$_3$ bending of the MMCHI membrane. The 1000–1100 cm^{-1} region in the GOD-MMCHI membrane spectrum appeared as a characteristic absorption spectrum typical of the phosphate buffer in the enzyme solution. Furthermore, the broad band of O-H stretching resulted from intermolecular hydrogen bonds between chitosan and GOD molecules. Figure 2shows the FTIR spectra of the GOD-MMCHI membrane and the GOD-LMCHI (low molecular weight chitosan) membrane displaying a similar pattern.

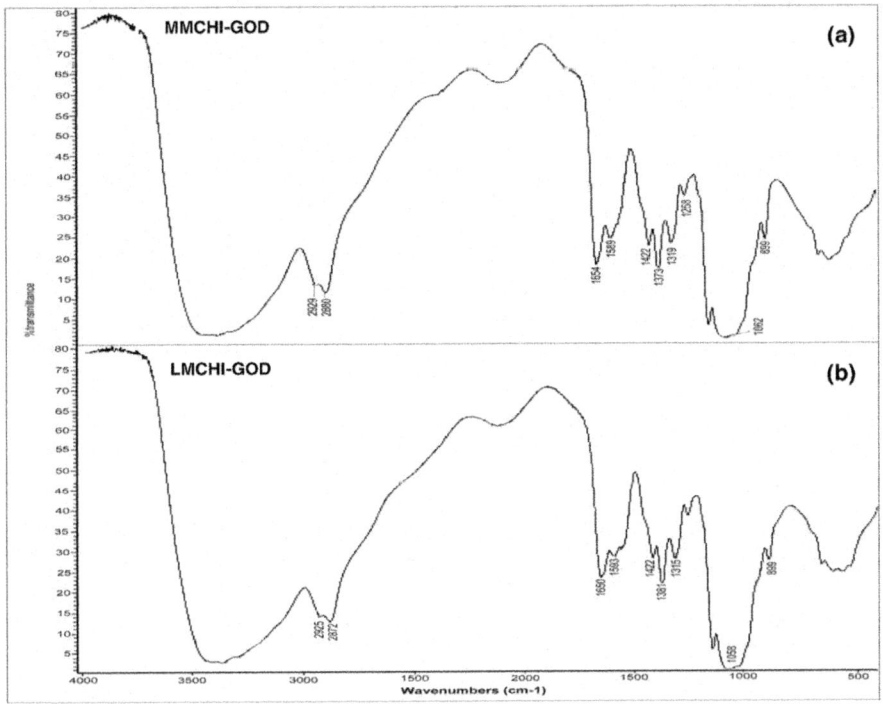

Figure 2: FTIR spectra of (a) GOD-MMCHI membrane and (b) GOD-LMCHI membrane.

Catalytic Activity Measurements of the Soluble and Immobilized Enzyme

The GOD-peroxidase (POD) enzymatic system is an assay for determining glucose concentrations based on a spectrophotometric method. In the catalyzed reaction of GOD, H_2O_2 is produced as a side product, which is then reduced in the POD enzymatic reaction. 2,2′-azino-di-(3-ethylbenzthiazoline)-6-sulfonic acid (ABTS) was used as a hydrogen donor in the H_2O_2 reduction process. It was oxidized to a stable, non-toxic, water-soluble brilliant blue-green derivative (ABTS+). The intensity of the solution color is proportional to the ABTS+concentration and indirectly to the H_2O_2 concentration. The amount of ABTS+ generated was also a measure of the glucose oxidized with two moles of ABTS oxidized per mole of glucose. The molar extinction coefficient of flavin adenine dinucleotide (FAD) permits direct a calculation of the concentration of H_2O_2 produced from the absorbance as a function of time plot. The angular coefficient of the straight line at the initial rates of H_2O_2 production gives the enzyme activity expressed as µmol/min. In this way, the specific enzyme activity (SEA) in units/mg can be calculated using the following equation:

$$SEA\ (units/mg) = \frac{\Delta A_{450}/min}{11.3 \times mg\ enzyme\ per\ ml\ in\ reaction\ mixture}$$

where 11.3 refers to the millimolar extinction coefficient of oxidized ABTS at 450 nm. One unit (U) of activity of glucose oxidase is defined as the amount of enzyme that catalyzes the oxidation of one µmol of ABTS per minute under the above mentioned conditions. The retention activity of the immobilized enzyme was then obtained by expressing the SEA of immobilized enzyme as a percentage of the SEA of the soluble enzyme (Table 1).

Table 1: Catalytic activity of different immobilized enzyme-membranes

Amount of GOD (mg)	Specific enzyme activity (Units/mg)					Retention activity (%)			
	Soluble GOD	cGOD-MMCHI	cGOD-LMCHI	aGOD-MMCHI	aGOD-LMCHI	cGOD-MMCHI	cGOD-LMCHI	aGOD-MMCHI	aGOD-LMCHI
0.05	1.402±0.002	1.042±0.007	0.964±0.007	0.566±0.002	0.525±0.002	74.32	68.76	40.37	37.45
0.10	1.174±0.003	0.901±0.006	0.857±0.005	–	–	76.75	73.00	–	–
0.20	1.025±0.002	0.810±0.007	0.789±0.006	–	–	79.02	76.98	–	–
0.40	0.737±0.001	0.623±0.004	0.599±0.003	–	–	84.53	81.28	–	–
0.60	0.622±0.003	0.621±0.003	0.561±0.003	–	–	99.84	90.19	–	–
0.80	0.482±0.002	0.429±0.001	0.416±0.002	–	–	89.00	86.31	–	–

The most important aspect of enzyme immobilization is the retention of its activity upon immobilization. As shown in the tabulated results, the immobilized enzyme retained a relatively high activity with varying amounts of enzyme for the crosslinking method employed, presumably due to strong crosslinking with the formation of a Schiff base linkage. MMCHI membranes showed slightly higher activity when compared to LMCHI membranes. This might be due to the higher molecular weight (MW) of MMCHI. In a study by Xu and Du (2003) on the molecular structure of chitosan, they reported that chitosan with higher MW possess longer polysaccharide chains which can entrap a greater amount of protein [12]. Alsorra et al. (2002) reported that chitosan with high MW could improve enzyme loading and reduce enzyme release [13]. In the present study, MMCHI with higher MW showed a higher retention activity as a result of greater enzyme loading. On the other hand, the weak binding of the enzyme to the matrix in the adsorption method resulted in low enzyme activity.

The effect of enzyme loading on the activity of the immobilized enzyme (via glutaraldehyde crosslinking) was studied by varying the amount of GOD (2.5–40.0 mg/mL or 0.05–0.80 mg) for immobilization (Figure 3). The enzymatic activity increased with increasing the amount of GOD up to 0.6 mg (30.0 mg/mL), but decreased at higher GOD content. Similar results were reported by Onda *et al.* (1999) and Bindhu & Abraham (2003) [14], [15]. The drop in enzyme activity at high enzyme concentrations might be due to the increased hindrance in substrate diffusion on the surface of the membrane.

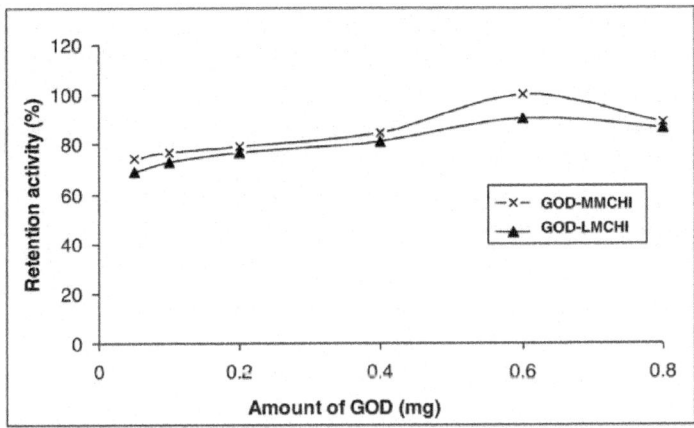

Figure 3: Effect of enzyme loading on the retention activity of GOD on chitosan.

Determination of the Michaelis-Menten Constant for the Soluble Enzyme

The Michaelis-Menten constant, K_M^{app} for the soluble GOD determined from Eadie-Hofstee plot (Figure 4) was found to be 0.5135 mM, indicating a strong affinity between enzyme and substrate.

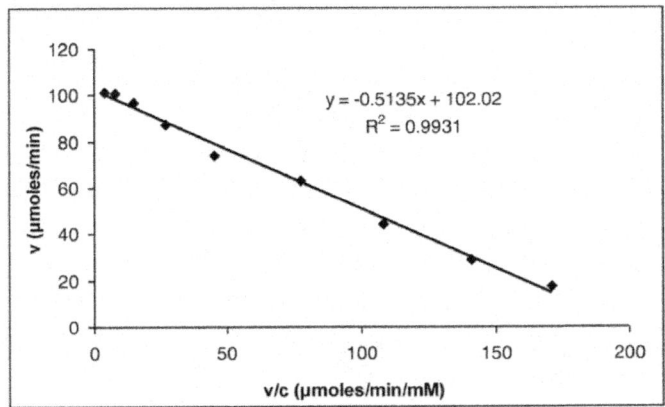

Figure 4: Eadie-Hofstee plot for the determination of Michaelis-Menten constant of the soluble GOD.

Characteristics of the Glucose Biosensor

Response time

The response time is defined as the time required reaching a steady-state current value on sensing the substrate. Figures 5, shows that the response times of medium molecular weight chitosan glucose biosensor (GOD-MMCHI/PT) and low molecular weight chitosan glucose biosensor (GOD-LMCHI/PT) to 2 mM glucose were less than 85 s and 65 s, respectively. The fast response time indicated good permeability of the chitosan membrane in the biosensors to the enzymatically-generated H_2O_2.

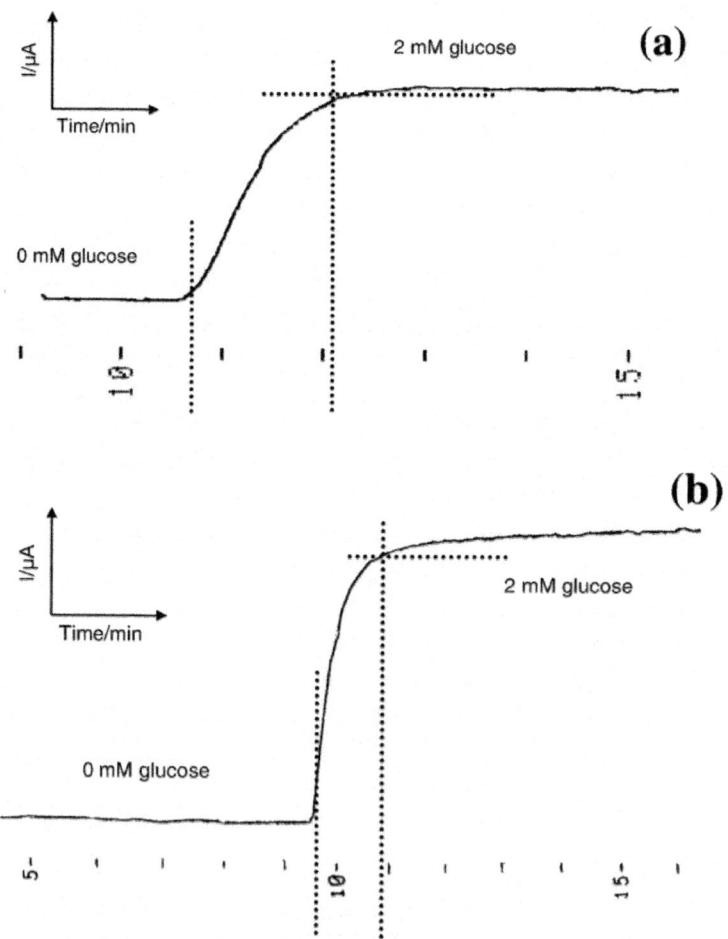

Figure 5: (a) Response time curve for GOD-FCHIT/PT to glucose; (b) Response time curve for GOD-SCHIT/PT to glucose.

Calibration of Glucose Biosensor

GOD-MMCHI/PT and GOD-LMCHI/PT were calibrated under optimal experimental conditions with glucose concentrations from 9.99×10^{-6} to 1.29×10^{-1} M and 1.00×10^{-5} to 9.90×10^{-2} M, respectively (Figures 6). The anodic current increased with increasing glucose concentration. The GOD-MMCHI/PT biosensor exhibited a linear calibration curve up to 10.8 mM. The slope of the initial range was 0.0620 µA/mM with a correlation coefficient of 0.9942 (Figure 6 (a), inset). A detection limit (S/N>3) of 0.1 mM glucose with a sensitivity of 52.3666 nA/mM was obtained. In the case of GOD-LMCHI/PT, the linear range was from 10.0 µM to 11.4 mM with a slope of 0.0366 µA/mM and a correlation coefficient of 0.9969 (Figure 6 (b), inset). The detection limit of 0.1 mM glucose at S/N>3 with a sensitivity of 9.8579 nA/mM was obtained.

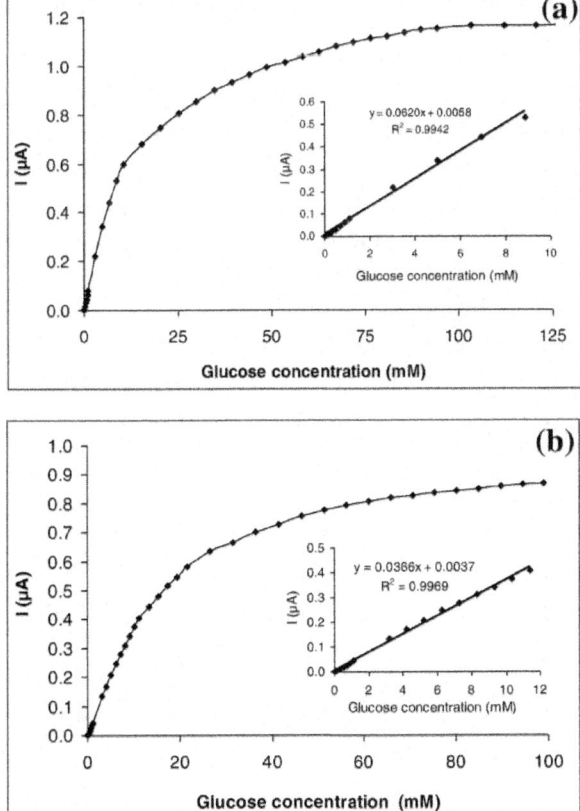

Figure 6: (a) Calibration curve of the GOD-FCHIT/PT under optimal experimental conditions.

Inset: Linear range from 10.0 μM to 10.8 mM glucose with linear regression equation y=0.0620x +0.0058; R²=0.9942, n=6.(b) Calibration curve of the GOD-SCHIT/PT under optimal experimental conditions. Inset: Linear range from 10.0 μM to 11.4 mM glucose with linear regression equation y=0.0366x +0.0037; R²=0.9969, n=5.

Singhal *et al.* (2002) reported that the normal blood glucose concentration for humans is in the range of 80 to 140 mg/dL (4.44–7.77 mM). In the present study, both biosensors exhibited a wide linear range covering the range of normal blood glucose concentrations [1]. GOD-MMCHI/PT showed a greater sensitivity in the amperometric measurement when compared to GOD-LMCHI/PT. The limit of detection of GOD-MMCHI/PT was lower compared to GOD-LMCHI/PT, indicating that GOD-MMCHI/PT was more sensitive in the detection of glucose.

Determination of the Apparent Michaelis-Menten Constant

The value of the apparent Michaelis-Menten constant (K_M^{app}) for an enzyme is an indication of its affinity for the substrate; the lower the value, the higher the affinity for immobilized GOD[16].The K_M^{app} for the immobilized enzyme can be determined by amperometry as suggested by Shu and Wilson (1976) [17]. A linear line was obtained in the Eadie-Hofstee plot for GOD-MMCHIT/PT from 4.99 to 44.17 mM glucose (Figure 7). The linear regression of the plot was y=−12.7370x +1.2282 with R²=0.9936. Thus, the K_M^{app} for glucose and I_{max} value for the biosensor were found to be 12.7370 mM and 1.2282 μA respectively. Similarly, the Eadie-Hofstee plot for GOD-LMCHIT/PT was also a linear line from 11.39 to 75.47 mM (Figure 7). The linear regression of the plot was y=−17.6920x +1.0406 with R²=0.9969. The K_M^{app} value for GOD was 17.6920 mM and the I_{max} was 1.0406 μA.

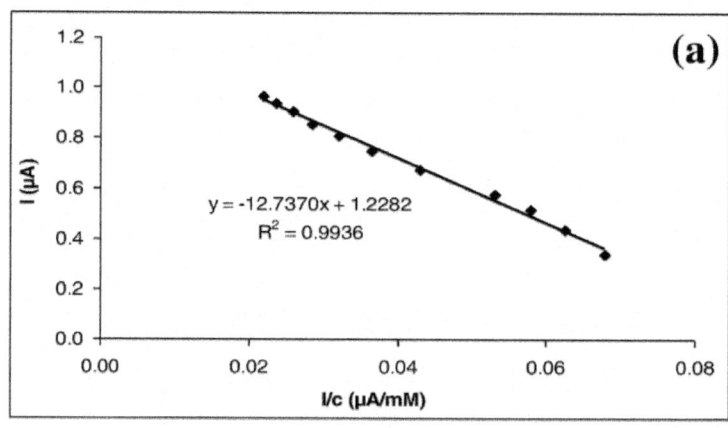

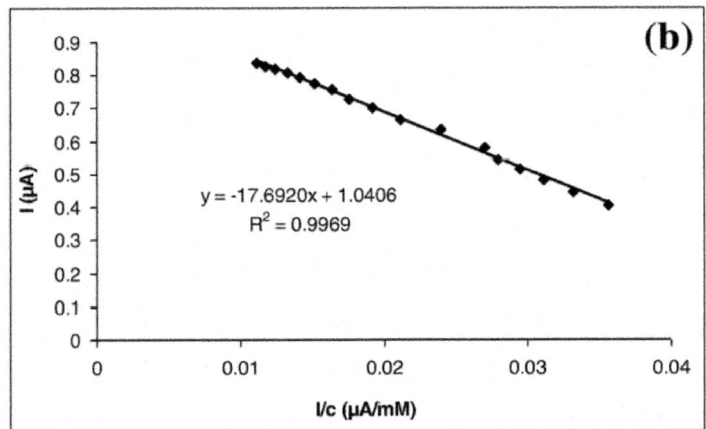

Figure 7: Eadie-Hofstee plot of (a) GOD-MMCHI/PT (b) GOD-LMCHI/PT.

The glucose concentration range chosen was optimal for the determination of K_M^{app} and I_{max}.

The smaller the value of K_M^{app}, the stronger the affinity between the enzyme and the substrate[18]. The smaller K_M^{app} value of GOD-MMCHIT/PT indicated that the immobilized GOD on MMCHIT membrane possessed higher enzymatic affinity. However, when the kinetic parameters of the immobilized GOD were compared to those of soluble GOD (K_M^{app}=0.5135 mM), an increase in K_M^{app} for immobilized GOD was observed. Spagna et al. (1997) and Bhatia et al. (2000) reported similar behavior for the immobilized enzyme [19], [20]. The increase in K_M^{app} value in the immobilized enzyme could be due to problems associated with the diffusion of the substrate or products through the support [21]. The K_M^{app} values of both the biosensors were lower when compared to the value of 23.30 mM reported by Chen et al. (2006)and 32.71 mM reported by Xu and Chen (2000) [18], [22].

Repeatability and Reproducibility

A reliable glucose biosensor should show good precision (repeatability and reproducibility). Repeatability refers to the agreement between successive measurements of the same sample whereas reproducibility is used to describe the closeness of agreement between results (signals) obtained with the same method under different conditions (using different glucose biosensors) [23].

Both the glucose biosensors showed both good repeatability and reproducibility in the glucose measurements as indicated by their RSD value.

The repeatability and reproducibility of GOD-MMCHI/PT were 2.30% and 3.70% and that of GOD-LMCHI/PT were 4.12% and 9.46%, respectively (Table 2). The precision of the enzyme immobilized on MMCHI was higher compared to that of LMCHI.

Table 2: The repeatability and reproducibility of the biosensors

Electrodes	Repeatability		Reproducibility	
	I (µA) (mean±SD, n=20)	RSD (%)	I (µA) (mean±SD, n=6)	RSD (%)
GOD-MMCHI/PT	0.22±0.01	2.30	0.23±0.01	3.70
GOD-LMCHI/PT	0.24±0.01	4.12	0.22±0.02	9.46

Stability Study

Enzyme stability in the matrix is a vital factor to consider in the development of a biosensor. As such, the stability of the GOD bioelectrodes was evaluated over a period of two months (Figure 8).

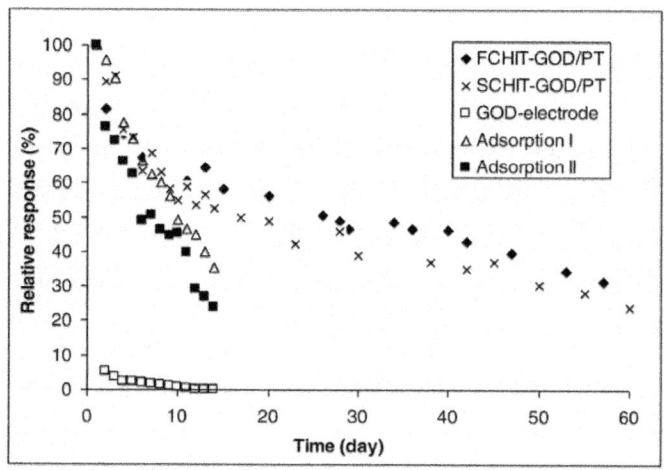

Figure 8: Stability of glucose biosensors over a period of 60 days.

During the initial period, both GOD-MMCHI/PT and GOD-LMCHI/PT were comparably stable. After one week, both biosensors retained about 60–70% of their original activity. On Day 14, the current was still more than half the initial value. Thereafter, the response dropped gradually to about 20–30% of the initial value at the end of two months. The activity of GOD-MMCHI/PT was higher when compared to GOD-LMCHI/PT throughout the period of investigation. The relatively longer stability of GOD-MMCHI/PT could be due to the higher MW of MMCHI, which could prevent enzyme from leaking [13].

The response of the electrodes designated "Adsorption I" and "Adsorption II" dropped by 50% of their initial value on Day 10 and Day 7 respectively. On Day 14, their response was less than 35% of the original value. Weak binding and desorption might occur over a period of storage and analysis [24]. The loss in activity was greater than that of the biosensors obtained via crosslinking with glutaraldehyde. The enzyme immobilized onto the MMCHI membrane was found to be more stable compared to that immobilized on the LMCHI membrane. The response of the free GOD electrode dropped drastically by 95% on the second day and was less than 1% of its initial value after 10 days. The free enzyme-electrode was not suitable for reuse after one cycle.

The retention activity was influenced by the MW of chitosan. The gradual decrease of in the current might have been due to the temperature change experienced by the enzyme electrodes between the storage and experimental temperatures. Partial enzyme denaturation might have occurred over a period of time. The possibility of electrode fouling could also affect the sensitivity of the biosensors. The results of the GOD electrode suggest that the immobilization process permits the enzyme to be reused, resulting in operational stability over a period of time.

Effect of Electroactive Compounds on Biosensor Response

Electroactive compounds present in the matrix can pose a problem in biosensors employing amperometric assays. The interference may be due to direct electrode oxidation with an increase in the anodic current or enzyme inhibition resulting in a reduced response current [25]. The selectivity of the biosensors therefore has to be evaluated in their presence.

The fourteen substances used to evaluate the selectivity of the two biosensors (GOD-MMCHI/PT and GOD-LMCHI/PT) included DL-glutamic, D(+) galactose, lactose, sucrose, oxalate, urea, citric acid, L-leucine, L-histidine, lactic acid, L-cysteine, ascorbic acid, acetaminophen and uric acid. The current obtained for each interfering substance at 0.1 mM was compared in the presence and absence of 5.0 mM glucose for the biosensors. No observable current shift in the presence of DL-glutamic, D(+) galactose, lactose, sucrose, oxalate, urea, citric acid, L-leucine, L-histidine and lactic acid. Hence, the concentration of these substances was increased to 0.5 mM in the presence of 5.0 mM glucose. Despite the increase in concentration, these substances did not affect the response of both biosensors in analyzing glucose. On the other hand, there were interfering signals from 0.1 mM acetaminophen (PCM), uric acid (UA), ascorbic acid (AA) and L-cysteine (Cys) when analyzing glucose at an applied potential of +0.6 V. This created a response deviation for the fabricated biosensors with easy co-oxidation of these substances at similar potentials [26]. The extent of interference encountered by both biosensors was not significantly different.

To overcome the effect of interferents, the biosensors were coated with a layer of Nafion to yield Nafion-GOD-MMCHI/PT and Nafion-GOD-LMCHI/PT. The effectiveness of this polyelectrolyte matrix to reduce the permeability of negatively charged substrates had been previously reported [27]. In addition, Nafion can prevent electrode fouling by proteins in the blood serum [28]. The presence of Nafion in Nafion-GOD-MMCHI/PT and Nafion-GOD-LMCHI/PT was found to reduce the effect of interference to different extents when compared to GOD-MMCHI/PT and GOD-LMCHI/PT (Table 3), because the negatively charged sulfonate groups in Nafion can prevent anions from partitioning onto the electrode surface [28], [29]. The deviation of current in the presence of PCM and UA sensed by Nafion-GOD-MMCHI/PT was significantly lower when compared with that of GOD-MMCHI/PT (Figure 9). There was no significant difference between the results obtained from GOD-LMCHI/PT and Nafion-GOD-LMCHI/PT in sensing these electroactive compounds. On the contrary, the response of Nafion-GOD-MMCHI/PT and Nafion-GOD-LMCHI/PT to glucose was significantly reduced ($P<0.001$) by about 17% and 13% respectively when compared to GOD-MMCHI/PT and GOD-LMCHI/PT (Figure 10). The addition of Nafion to the membrane acted as a diffusion barrier on top of the enzyme-chitosan membrane.

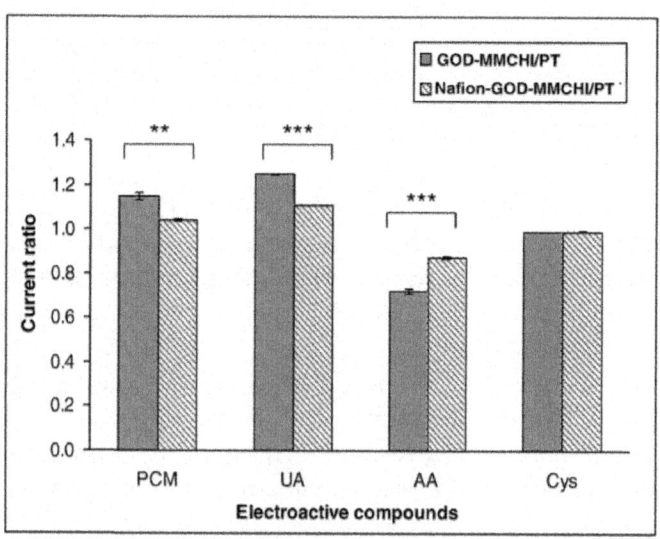

Figure 9: Ratio of currents for mixtures containing 0.1 mMelectroactive compound and 5.0 mM glucose to 5.0 mM glucose alone (mean±SEM, n=3).

** and *** indicate significance level among the comparison groups at $P<0.01$ and $P<0.001$, respectively.

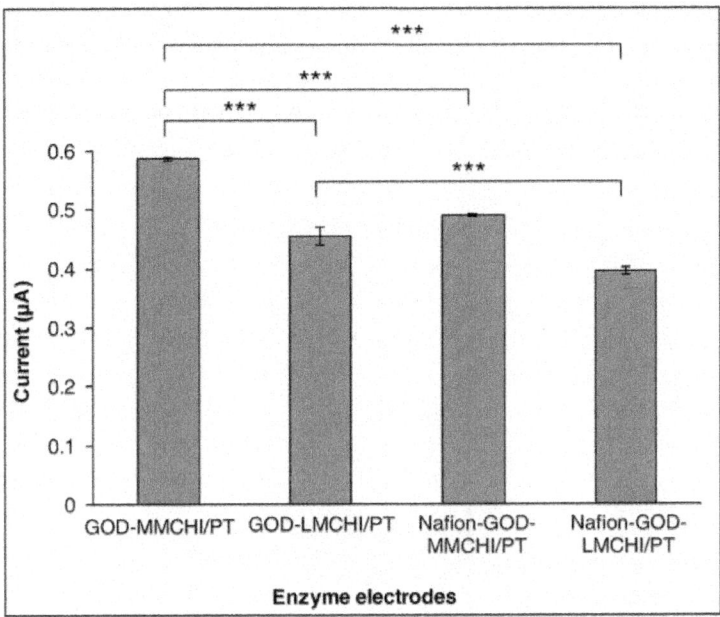

Figure 10: Ratio of currents for mixtures containing 0.1 mMelectroactive compound and 5.0 mM glucose to 5.0 mM glucose alone (mean±SEM, n=3).

** and *** indicate significance level among the comparison groups at $P<0.01$ and $P<0.001$, respectively.

Table 3: Influence of some electroactive compounds on glucose biosensor response. Mean±S.E.M, n=3.

Enzyme electrodes Compounds	Current ratios' GOD-MMCHI/PT	GOD-LMCHI/PT	Nafion-GOD-MMCHI/PT	Nafion- GOD-LMCHI/PT
Acetaminophen	1.15±0.017	1.18±0.006	1.04±0.005	1.12±0.035
Uric acid	1.25±0.003	1.20±0.109	1.11±0.001	1.11±0.004
Ascorbic acid	0.72±0.013	0.73±0.004	0.87±0.005	0.81±0.006
L-cysteine	0.99±0.001	0.99±0.004	0.99±0.003	0.99±0.001

In short, there was little interference encountered by the biosensors despite the absence of Nafion presumably due to the anti-interferents property of chitosan [29], [30]. In addition, GOD (pI =4.2) is negatively charged at pH 6.0 [31], [32] which can prevent anionic interferents from interaction with the surface of the platinum (Pt) electrode.

Accuracy and Recovery

The glucose levels in rat serum samples were assayed using both glucose biosensors (GOD-MMCHI/PT and GOD-LMCHI/PT) and compared with those determined by an ABTS-based spectrophotometric method (Table 4).

Table 4: Comparison of glucose level in rat serum determined using glucose biosensors and ABTS-spectrophotometric method.

Sample no.	[a]Concentration of glucose (mM)		Concentration of glucose added (mM)	[b]Concentration of glucose found (mM)		[c]Concentration of glucose recovered (mM)		Recovery (%)	
	Spectrophotometric method	GOD-MMCHI/PT	GOD-LMCHI/PT	GOD-MMCHI/PT	GOD-LMCHI/PT	GOD-MMCHI/PT	GOD-LMCHI/PT	GOD-MMCHI/PT	GOD-LMCHI/PT
1	5.9472	5.9740	6.3997	4.9751	10.2994	11.3736	4.3254	4.9739	86.94	99.98
2	5.6908	5.1729	5.7066	4.9751	9.7295	10.6571	4.5566	4.9505	91.50	99.51
3	6.5721	6.0281	5.3659	4.9751	10.0061	10.1133	4.8780	4.7474	98.05	95.42
4	6.4751	6.9658	6.5908	7.4442	13.9239	13.9101	6.9572	7.3193	93.46	98.32
5	8.1342	8.5940	8.2489	7.4442	15.7004	15.4889	7.1064	7.2400	95.46	97.26

The accuracy of these biosensors was also evaluated by a standard addition method in determining the recoveries of glucose in rat serum samples. A comparison of the glucose levels in rat serum is given in Table 4. The results indicated that GOD-MMCHI/PT and GOD-LMCHI/PT exhibited reasonably satisfactory results with an average recovery of 93.10% and 98.10% respectively. The results therefore suggest that chitosan membranes could be used as a matrix in the development of glucose biosensors.

CONCLUSION

There was anintermolecular interaction between chitosan and GOD. GOD-MMCHIpossessedahighercatalyticactivitythanGOD-LMCHI.Immobilization via crosslinking exhibited a higher enzymatic activity than through adsorption. The immobilized enzyme-chitosan membranes were studied by incorporation into glucose biosensors. MMCHI possessed higher enzymatic activity and improved kinetic characteristics compared to that of LMCHI. GOD-MMCHI/PT exhibited higher sensitivity, repeatability, reproducibility, retention activity and stability. The response of the two biosensors to interferences, was almost similar. GOD-MMCHI/PT demonstrated better performance than GOD-LMCHI/PT for the determination of glucose in rat serum. In conclusion, chitosan membranes could be a suitable matrix in the development of glucose biosensors.

MATERIALS AND METHODS

Preparation of Enzyme-chitosan Membranes

One gram of chitosan (Low molecular weight (LMCHI) and medium molecular weight (MMCHI)) was dispersed in 100 ml of acetic acid at concentration of 0.8% w/v (The physical properties of chitosan solution and membrane were studied by dissolving the chitosan in different organic acids (acetic acid, lactic acid and maleic acid). Both the chitosan samples were most soluble in aqueous acetic acid, followed by lactic acid and maleic acid). Chitosan membranes prepared from acetic acid were flexible, transparent, smooth and quick-drying. They exhibited good mechanical strength and elongation at break and the values were significantly higher than those prepared in lactic acid and maleic acid).The chitosan membranes were prepared into Petri dish with casting measurement 0.21 mL/cm^2 membrane thickness for GOD-MMCHI/PT and 0.35 mL/cm^2 membrane thickness for GOD-LMCHI/PT. After the membrane was formed, it was neutralized with 1% w/v sodium hydroxide (NaOH) for 30 minutes followed by rinsing with distilled water to remove excess NaOH. The neutralized membrane was cut into small squares (1.5×1.5 cm^2) for immobilization. The extra membranes were kept in distilled water at 4°C until further use.

The glutaraldehyde activation method reported byMagalhães and Machado (1998) was adopted with a slight modification [33]. A smaller dimension of the chitosan squares (1.5×1.5 cm²) was used instead of 2.0×2.0 cm². This was because the diameter of platinum electrode used was less than 1 cm. One side of the membrane square was coated with 20 µl of 1% v/v glutaraldehyde and allowed to dry at room temperature (25°C). Subsequently, 20 µl of glucose oxidase (GOD) (concentration ranges of 2.5 mg/ml to 40.0 mg/ml in 0.1 M of phosphate buffer, pH 7.0) containing 5% v/v glycerol was spread evenly onto the same surface of the membrane with the aid of a L-shaped glass rod. The immobilized membrane was left to dry at room temperature. The small amount of glycerol in the enzyme solution acts as an emollient to facilitate even spreading of GOD on the membrane surface.

For the adsorption method, (1.5×1.5 cm²) squares of chitosan membranes (MMCHI and LMCHI) were separately incubated in 1 ml of GOD solution (2.5 mg/ml) at 4°C for 24 hours. The membrane was then washed with distilled water and kept in phosphate buffer (0.1 M, pH 7.0) at 4°C until further use.

Investigation of Intermolecular Interactions of Immobilized GOD-chitosan Membranes

Intermolecular interactions of chitosan membranes (MMCHI and LMCHI) with GOD were investigated by FTIR analysis. The FTIR spectra of GOD as potassium bromide (KBr) disks were obtained. Approximately 10 mg of GOD crystals and 100 mg of KBr was ground with an agate mortar and pestle for 5 min. About 40 mg of the mixture was then loaded onto an evacuable potassium bromide die and compressed for 1 min by applying an IR hydraulic press (Hydraulic Unit Model #3192, Carver Laboratory Equipment, Wabash, Indiana, USA) of 10 tons to obtain a yellowish, transparent and thin disk. It was placed in an oven at 40°C for 2 hours before analysis. GOD-MMCHI and GOD-LMCHI membranes (20 µl of 30 mg/ml GOD was used to crosslink onto chitosan membrane) were washed with distilled water to remove the unbound enzyme on the surface followed by air-drying at room temperature. They were kept in the oven at 40°C for 2 hours before scanning.

Colorimetric Determination of Glucose

The activities of soluble and immobilized GOD were assayed using ABTS [2,2'-azino-bis(3-ethylbenzthiazoline)-6-sulfonic acid] method adapted from Bergmeyer and Bernt (1974) [34]. Reagent A was consisted of 0.1 M phosphate buffer (pH 7.0), 0.92 mM ABTS, 1.5 U/mL POD and 9 U/mL GOD, whereas reagent B was similar to reagent A but without GOD. For glucose calibration, the glucose concentration used was from 0.1221 mM to 1.0 M. For each run, 0.1 mL of mutarotated glucose solution was added to 5.0 mL of reagent A (oxygenated) and mixed using a vortex mixer. Initial reaction rate was determined by monitoring the rate of change of absorbance at 450 nm recorded by UV/VIS spectrometer (Lambda 45, Perkin Elmer Instruments, USA). The Michaelis-Menten constant for the soluble enzyme was determined following this procedure from Eadie-Hofstee plot (initial velocity of enzymatic activity, v versus $\frac{v}{c}$, where c is concentration of glucose). For blood samples, 0.1 ml of serum was used instead of glucose standard. Thereafter, 0.1 mL of standard glucose solution (1.0 M) was added to 5.0 ml reagent B (oxygenated) and mixed well. For the determination of GOD activity, 20 μLof enzyme solution was added to reagent B. When GOD-immobilized membranes were used as samples, they were placed into reagent B for activity determination. In this case, the GOD membranes were sonicated for 1 hour before adding glucose solution. The glucose consumption was determined by the increase in absorbance at 450 nm.

The molar extinction coefficient (E) of FAD at 450 nm is 11.3 ± 10^{-3} $M^{-1}cm^{-1}$ [35]. The absorbance (A) was converted to concentration, according to the equation $A=ELc$, where L is the cell thickness or path length of light through the sample (usually 1-cm) and c the concentration of absorbing material in the sample.

Construction of the GOD-chitosan Electrode

The platinum electrode was first polished with 0.05 μm alumina on a polishing pad, washed with distilled water and finally sonicated for few minutes to remove the alumina particles. The GOD-chitosan membrane placed over a moist dialysis membrane as a protective layer was fastened onto surface of the platinum electrode with an O-ring (Figure 11). Amperometric detection of glucose was performed using a potentiostat (cyclic voltammograph CV-1B, Bioanalytical Systems Inc. (BAS), West Lafayette, Indiana, USA) poised at +0.6 V connected to an integrator-plotter (Chromato-Integrator D-2500, Hitachi, Tokyo, Japan) and a digital multimeter (8022A, Fluke, USA). The experimental set-up for electrochemical measurements is illustrated in Figure 11. The conventional three electrodes consisted of a silver/silver chloride (Ag/AgCl) reference electrode (MF-2079, BAS, West Lafayette, Indiana, USA), a platinum wire (0.25 mm diameter, 99.99%, Aldrich, Milwaukee, Wisconsin, USA) as counter electrode and a platinum working electrode (MF-2013, BAS, West Lafayette, Indiana, USA) with the enzyme-chitosan layer. Unless stated otherwise, all experiments were carried out in 10 ml of phosphate buffer (0.1 M, pH 7.0) maintained at 25.0±0.1°C using a digital temperature controller (Model 9001, Poly Science, USA).

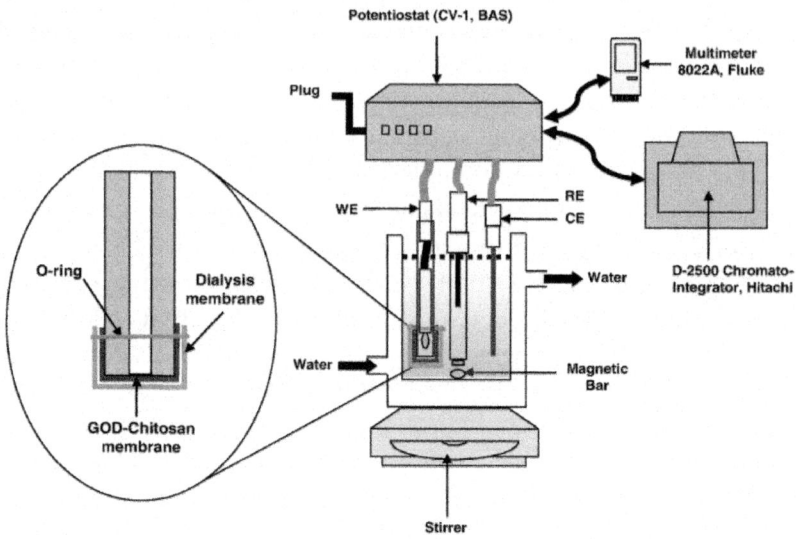

Figure 11: Schematic representation of the experimental set-up (WE: working electrode; RE: reference electrode; CE: counter electrode).

Characteristics of the Glucose Biosensor

Response time

The response time of the glucose biosensor was determined from the time of the addition of the analyte to the time the anodic current produced achieved steady-state.

Calibration of glucose biosensor.

Calibration of the glucose biosensors was performed under optimized experimental conditions. Aliquots of 10 µl of β-D-glucose stock solution were successively added into 10 ml of stirred phosphate buffer in an electrochemical cell. Three different glucose stock solutions at concentrations of 0.01 M, 0.1 M and 1.0 M were prepared to obtain the hydrodynamic response for glucose concentrations from 9.99×10^{-6} to 1.29×10^{-1} M. Calibration curves from the anodic current-glucose concentration plot for each biosensor were used to obtain the linear detection range.

Determination of apparent Michaelis-Menten constant

The apparent Michaelis-Menten constant (K_M^{app}) for the immobilized enzyme was determined from Eadie-Hofstee plot with the equation shown below: $I = I_{max} - K_M^{app}\left(\dfrac{I}{c}\right)$ where I is the steady state current after the addition of substrate, c is the bulk concentration of the substrate (glucose) and I_{max} is the maximum current measured under saturated substrate condition. K_M^{app} and the upper detection limit, I_{max} of the biosensor were then determined from the slope and intercept on the y-axis of the plot.

Repeatability and reproducibility

The repeatability generated by the glucose biosensors (GOD-MMCHI/PT and GOD-LMCHI/PT) was studied by measuring the anodic current generated by 3.98 mM glucose in 10 ml phosphate buffer (pH 6.0) for a total of 20 times in a single day. On the other hand, the reproducibility of the biosensors was studied by measuring the current generated by 3.98 mM glucose in 10 ml phosphate buffer (pH 6.0) by using six different glucose biosensors. The signals obtained were calculated and summarized as relative standard deviation (RSD). The RSD was calculated using the following equation:

$$RSD(\%) = \frac{\text{Standard deviation}}{\text{Average}} \times 100\%$$

The stability of three types of biosensors was explored under optimal experimental conditions. The three types of biosensors were prepared via i) glutaraldehyde crosslinking (GOD-MMCHI/PT and GOD-LMCHI/PT), ii) physical adsorption method [Adsorption I (GOD-MMCHI) and Adsorption II (GOD-LMCHI)] and iii) coating the surface of the platinum electrode with 20 µL of GOD solution (30 mg/ml) at room temperature (25°C). The membranes were then placed onto surface of the platinum electrode covered with a layer of dialysis membrane and fastened with an O-ring.

The responses of GOD-MMCHI/PT and GOD-LMCHI/PT to 3.98 mM glucose were measured daily in triplicate for the first two weeks. After 2 weeks, the biosensors were tested every 3–5 days over a period of 60-days. The mean of the relative current to initial current sensed by these biosensors was plotted as function of time. The responses of the other two types of biosensors were measured under similar experimental conditions to compare the stability of these biosensors. All enzyme electrodes were kept in phosphate buffer and stored at 4°C when not in use.

Effect of electroactive compounds on biosensor response

The effect of possible interferences from electroactive substances were investigated by adding the interfering compounds of 0.1 mM to 0.5 mM into a standard solution of 5.0 mM glucose. The current obtained was then compared to that generated by glucose alone. A layer of Nafion (0.2% v/v diluted with alcohol) coated over the GOD-chitosan layer was prepared to study if the effect of interferences can be minimized or overcome.

Accuracy and recovery

Blood samples were collected intracardially according to the method reported by Orphan *et al.*(2003) from six non-fasting Sprague-Dawley malerats (~250 g) aged between 3–5 months (Approval by Animal Ethic Committee, Universiti Sains Malaysia) [36]. The blood samples were allowed to coagulate for 1 hour after withdrawal and were then centrifuged at 3500 rpm for 10 min at 4°C. The supernatant (serum) was collected and glucose content was determined. In the electrochemical measurement, 200 µL of serum samples were added to 10 ml of 0.1 M phosphate buffer (pH 6.0) and the response was obtained at an applied potential of +0.6 V and temperature of 35°C. A separate analysis of glucose concentrations was carried out using the ABTS method based on spectrophotometric detection described under section "Colorimetric determination of glucose". The recovery of glucose in serum was performed by adding known amount of glucose to the serum samples. The amount of added glucose was then determined using the two glucose biosensors, GOD-MMCHI/PT and GOD-LMCHI/PT. The concentration of glucose recovered was calculated from the difference in glucose concentration between the spiked and unspiked serum samples.

STATISTICAL ANALYSIS

All the results were expressed as mean±standard error mean (SEM). The results were analyzed using one-way analysis of variance (ANOVA). When a statistically significant difference was obtained ($P<0.05$), LSD test was then performed. The relationship between parameters was analyzed using the Pearson correlation test.

AUTHOR CONTRIBUTIONS

Conceived and designed the experiments: LFA MFY. Performed the experiments: LFA MFY. Analyzed the data: LFA LYP MFY. Contributed reagents/materials/analysis tools: LFA LYP MFY. Wrote the paper: LFA LYP MFY.

REFERENCES

1. Singhal R, Takashima W, Kaneto K, Samanta SB, Annapoorni S, et al. (2002) Langmuir-Blodgett films of poly(3-dodecyl thiophene) for application to glucose biosensor. Sensors and Actuators B 86(1): 42–48. doi: 10.1016/s0925-4005(02)00145-4

2. Clark LCJ, Lyons C (1962) Electrodes systems for continuous monitoring in cardiovascular surgery. Annals of the New York Academy of Sciences 102: 29–45. doi: 10.1111/j.1749-6632.1962.tb13623.x

3. Freitag R (1999) Utilization of enzyme-substrate interactions in analytical chemistry. Journal of Chromatography B 722(1–2): 279–301. doi: 10.1016/s0378-4347(98)00507-6

4. Wilkins E, Atanasov P (1996) Glucose monitoring: state of the art and future possibilities. Medical Engineering and Physics 18(4): 273–288. doi: 10.1016/1350-4533(95)00046-1

5. Scheller F, Schubert F (1992) *Biosensors*. The Netherlands: Elsevier Science Publishers B.V.

6. Kauppinen JK, Moffatt DJ, Mantsch HH, Cameron DG (1981) Fourier self-deconvolution: a method for resolving intrinsically overlapped bands. Applied Spectroscopy 35(3): 271–276. doi: 10.1366/0003702814732634

7. Liang W, Zhuobin Y (2003) Direct electrochemistry of glucose oxidase at a gold electrode modified with single-wall carbon nanotubes. Sensors 3(12): 544–554. doi: 10.3390/s31200544

8. Zhao CZ, Egashira N, Kurauchi Y, Ohga K (1998) Electrochemiluminescence sensor having a pt electrode coated with a $Ru(bpy)_3^{2+}$ modified chitosan/silica/gel membrane. Analytical Sciences 14(2): 439–441. doi: 10.2116/analsci.14.439

9. Dhanikula AB, Panchagnula R (2004) Development and characterization of biodegradable chitosan films for local delivery of paclitaxel. The AAPS Journal 6(3): 88–99. doi: 10.1208/aapsj060327

10. Musale DA, Kumar A (2000) Effects of surface crosslinking on sieving characteristics of chitosan/poly(acrylonitrile) composite nanofiltration membranes. Separation and Purification Technology 21(1–2): 27–38. doi: 10.1016/s1383-5866(00)00188-x

11. Zhou GJ, Wang G, Xu JJ, Chen HY (2002) Reagentless chemiluminescence biosensor for determination of hydrogen peroxide based on the immobilization of horseradish peroxidase on biocompatible chitosan membrane. Sensors and Actuators B 81(2): 334–339. doi: 10.1016/s0925-4005(01)00978-9

12. Xu Y, Du Y (2003) Effect of molecular structure of chitosan on protein delivery properties of chitosan nanoparticles. International Journal of Pharmaceuticals 250(1): 215–226. doi: 10.1016/s0378-5173(02)00548-3

13. Alsorra IA, Betigeri SS, Zhang H, Evans BA, Neau SH (2002) Molecular weight and degree of deacetylation effects on lipase-loaded chitosan bead characteristics. Biomaterials 23(17): 3637–3644. doi: 10.1016/s0142-9612(02)00096-0

14. Onda M, Ariga K, Kunitake T (1999) Activity and stability of glucose oxidase in molecular films assembled alternately with polyions. Journal of Bioscience and Bioengineering 87(1): 69–75. doi: 10.1016/s1389-1723(99)80010-3

15. Bindhu LV, Abraham ET (2003) Immobilization of horseradish peroxidase on chitosan for use in nonaqueous media. Journal of Applied Polymer Science 88(6): 1456–1464. doi: 10.1002/app.11815

16. Wang G, Xu JJ, Chen HY, Lu ZH (2003) Amperometric hydrogen peroxide biosensor with sol-gel/chitosan network-like film as immobilization matrix. Biosensors and Bioelectronics 18(4): 335–343. doi: 10.1016/s0956-5663(02)00152-5

17. Shu FR, Wilson GS (1976) Rotating ring-disk enzyme electrode for surface catalysis studies. Analytical Chemistry 48(12): 1679–1686. doi: 10.1021/ac50006a014

18. Chen C, Jiang Y, Kan J (2006) A noninterference polypyrrole glucose biosensor. Biosensors and Bioelectronics 22(5): 639–643. doi: 10.1016/j.bios.2006.01.023

19. Spagna G, Andreani F, Salatelli E, Romagnoli D, Pifferi PG (1997) Immobilization of α-L-arabinofuranosidase on chitin and chitosan. Process Biochemistry 33(1): 57–62. doi: 10.1016/s0032-9592(97)00067-8

20. .Bhatia RB, Brinker CJ, Gupta AK, Singh AK (2000) Aqueous sol-gel process for protein encapsulation. Chemistry of Materials 12(8): 2434–2441. doi: 10.1021/cm000260f

21. Yang YM, Wang JW, Tan RX (2004) Immobilization of glucose oxidase on chitosan-SiO$_2$ gel. Enzyme and Microbial Technology 34(2): 126–131. doi: 10.1016/j.enzmictec.2003.09.007

22. Xu JJ, Chen HY (2000) Amperometric glucose sensor based on glucose oxidase immobilized in electrochemically generated poly(ethacridine). AnalyticaChimicaActa 423(1): 101–106. doi: 10.1016/s0003-2670(00)01098-9

23. Karnes HT, Shiu G, Shah VP (1991) Validation of bioanalytical methods. Pharmaceutical Research 8(4): 421–426. doi: 10.1023/a:1015882607690

24. Zhu Y, Shen W, Dong X, Shi J (2005) Immobilization of hemoglobin on stable mesoporousmultilamellar silica vesicles and their activity and stability. Journal of Materials Research 20(10): 2682–2690. doi: 10.1557/jmr.2005.0360

25. Li J, Chia LS, Goh NK, Tan SN, Ge H (1997) Mediated amperometric glucose sensor modified by the sol-gel method. Sensors and Actuators B 40(2–3): 135–141. doi: 10.1016/s0925-4005(97)80252-3

26. Pan D, Chen J, Nie L, Tao W, Yao S (2004) Amperometric glucose biosensor based on immobilization of glucose oxidase in electropolymerized0-aminophenol film at Prussian blue-modified platinum electrode. Electrochimica Acta 49(5): 795–801. doi: 10.1016/j.electacta.2003.09.033

27. Shankaran DR, Kato NUT (2003) A metal dispersed sol-gel biocompositeamperometric glucose biosensor. Biosensors and Bioelectronics 18(5–6): 721–728. doi: 10.1016/s0956-5663(03)00005-8

28. Lim SH, Wei J, Lin J, Li Q, You JK (2005) A glucose biosensor based on electrodeposition of palladium nanoparticles and glucose oxidase onto Nafion-solubilized carbon nanotube electrode. Biosensors and Bioelectronics 20(11): 2341–2346. doi: 10.1016/j.bios.2004.08.005

29. Yang M, Yang Y, Liu B, Shen G, Yu R (2004) Amperometric glucose biosensor based on chitosan with improved selectivity and stability. Sensors and Actuators B 101(3): 269–276. doi: 10.1016/j.snb.2004.01.003

30. Wu BY, Hou SH, Yin F, Li J, Zhao ZX, et al. (2007) Amperometric glucose biosensor based on layer-by-layer assembly of multilayer films composed of chitosan, gold nanoparticles and glucose oxidase modified pt electrode. Biosensors and Bioelectronics 22(6): 838–844. doi: 10.1016/j.bios.2006.03.009

31. Xu JJ, Zhang XQ, Yu ZH, Fang HQ, Chen HY (2001) A stable glucose biosensor prepared by co-immobilizing glucose oxidase into poly(p-chlorophenol) at a platinum electrode. Fresenius Journal of Analytical Chemistry 369(6): 486–490. doi: 10.1007/s002160000690

32. Pekel N, Salih B, Güven O (2003) Activity studies of glucose oxidase immobilized onto poly(N-vinylimidazole) and metal ion-chelated poly(N-vinylimidazole) hydrogels. Journal of Molecular Catalysis B: Enzymatic 21(4): 273–382. doi: 10.1016/s1381-1177(02)00232-1

33. Magalhães JMCS, Machado AASC (1998) Urea potentiometric biosensor based on urease immobilized on chitosan membranes. Talanta 47(1):

183–191. doi: 10.1016/s0039-9140(98)00066-6

34. Bergmeyer HU, Bernt E (1974) Determination with Glucose Oxidase and Peroxidase.*In: H.U. Bergmeyer, (ed). Methods of Enzymatic Analysis,* New York and London: VerlagChemieWeinheim Academic Press, Inc. 1205–1215 p.

35. Segel IH (1976) Biochemical calculations. Canada: John Wiley & Sons, Inc. 235 p.

36. Orphan DD, Aslan M, Aktay G, Ergun E, Yesilada E, et al. (2003) Evaluation of hepatoprotective effect of *Gentianaolivieri*herbs on sub-acute administration and isolation active principle. Life Sciences 72(20): 2273–2283. doi: 10.1016/s0024-3205(03)00117-6

Chapter 6

UTILIZATION OF ENZYME-IMMOBILIZED MESOPOROUS SILICA NANOCONTAINERS (IBN-4) IN PRODRUG-ACTIVATED CANCER THERANOSTICS

Bau-Yen Hung [1], Yaswanth Kuthati [1], Ranjith Kumar Kankala [1], Shravan-kumar Kankala [2], Jin-Pei Deng [3], Chen-Lun Liu [1] and Chia-Hung Lee [1]

[1]Department of Life Science and Institute of Biotechnology, National Dong Hwa University, Hualien-974, Taiwan

[2]Department of Chemistry, Kakatiya University, Telangana State-506009, India

[3]Department of Chemistry, Tamkang University, New Taipei City 251, Taiwan

ABSTRACT

To develop a carrier for use in enzyme prodrug therapy, Horseradish peroxidase (HRP) was immobilized onto mesoporous silica nanoparticles (IBN-4: Institute of Bioengineering and Nanotechnology), where the nanoparticle surfaces were functionalized with 3-aminopropyltrimethoxysilane and further conjugated with glutaraldehyde. Consequently, the enzymes could be stabilized in nanochannels through the formation of covalent imine bonds. This strategy was used to protect HRP from immune exclusion, degradation and denaturation under biological conditions. Furthermore, immobilization of HRP in the nanochannels of IBN-4 nanomaterials exhibited good functional stability upon repetitive use and long-term storage (60 days) at 4 °C. The generation of functionalized and HRP-immobilized nanomaterials was further verified using various characterization techniques. The possibility of using HRP-encapsulated IBN-4 materials in prodrug cancer therapy was also demonstrated by evaluating their ability to convert a prodrug (indole-3-acetic acid (IAA)) into cytotoxic radicals, which triggered tumor cell apoptosis in human colon carcinoma (HT-29 cell line) cells. A lactate dehydrogenase (LDH) assay revealed that cells could be exposed to the IBN-4 nanocomposites without damaging their membranes, confirming apoptotic cell death. In summary, we demonstrated the potential of utilizing large porous mesoporous silica nanomaterials (IBN-4) as enzyme carriers for prodrug therapy.

INTRODUCTION

The delivery of fragile drugs to target sites at precise times and locations with high drug activity in a reproducible manner is an outstanding challenge. Various delivery approaches have been developed to create effective therapeutic methods to target cancerous sites using a variety of versatile drug formulations. The magic of bioactivation is that the therapeutic method acts by specifically activating the prodrug into active drug molecules upon reaching the targeting site. The prodrug approach was developed by Albert in 1958 for the purposes of increasing therapeutic efficacy and reducing cytotoxic side effects through the activation of toxic drugs at targeted sites. A prodrug is defined as a nontoxic precursor of an active drug that can be transformed into the active drug molecule by enzymatic catalysis and consequently deliver therapeutic effects to targeted sites. Enzyme prodrug therapy (EPT) is a novel therapeutic approach where prodrug-activating enzymes are initially delivered into the cancer cells using various targeting approaches; this enzyme delivery is followed by treatment with a nontoxic prodrug that is specifically activated into an anticancer drug through the enzymatic activity in the targeted cells. The concentration of the activated drug can be high in the tumor sites, thereby reducing systemic toxicity in normal tissues [1,2,3,4,5]. Hence, this therapeutic method can be applied to improve tumor targeting while reducing systemic toxicity. For the successful use of EPT in clinical applications, the catalytic enzymes in normal tissues that activate the prodrugs must be present at lower concentrations than in cancer tissues [6,7]. In recent years, EPT has been examined as a potential strategy for treating devastating diseases such as cancer [8,9,10]. The development of an efficient enzymatic delivery vehicle that can carry and activate prodrug molecules and kill cancer cells is extremely important. Many approaches have been developed to deliver various enzymes for the activation of anticancer prodrugs, such as antibodies, viruses, lectins, spores and liposomes [11,12,13,14,15,16,17,18,19]. Despite some of the advantages associated with these approaches, the sole use of antibodies and lectins is hindered by their poor bioavailability [20], whereas the scope of possible pathogenicity hampers the use of microbial vectors. However, the use of immunoliposomes has been shown to enhance enzymatic delivery compared to the sole use of antibody-enzyme conjugates [21,22], which has gained considerable attention with regards to nanoparticle-based drug delivery systems. Motivated by the advantages of immunoliposomes observed in drug delivery, many researchers have used liposomal, polymeric and other organic nanoparticle-based systems for prodrug therapy [23,24,25,26,27]. Nevertheless, liposomes and other organic nanoparticles have also been shown to have certain drawbacks, such as low encapsulation efficiencies, rapid leakage of water-soluble molecules, susceptibility to microbial attack in the presence of blood components and poor storage stabilities; these drawbacks may limit the application of liposomes in prodrug therapy [28].

In the last decade, exceptional developments have been made with respect to inorganic nanoparticles in the field of drug delivery. Researchers have developed various prodrugs for targeted delivery using a wide range of inorganic nanocomposites [17,29,30,31,32]. Many inorganic nanomaterials such as gold [33], silver, magnetic-Fe/Fe$_3$O$_4$ [34], gold-coated super paramagnetic iron oxide [35], carbon nanotubes, mesoporous silica [36,37], and layered double hydroxide (LDH) nanoparticles [38,39] have gained much attention due to their novel multifaceted characteristics that include low cytotoxicities, high loading capacities, large surface to volume ratios, good stabilities, rich functionalities, biocompatibilities, and targeting abilities; consequently, these types of inorganic nanomaterials are considered to be ideal enzyme carriers for various biological applications.

Many breakthroughs have been made for the loading of bioactive molecules such as enzymes placed in the tunable pores of mesoporous silica and great improvements in cellular uptake efficiencies [40,41]. Very recently, phosphonate@mesoporous silica nanoparticles (MSNs) were synthesized for the intracellular delivery of fluorescein-labeled bovine serum albumin (BSA) cargo using the folate receptor pathway; these MSNs were shown to have potential for use in targeted drug delivery for the treatment of cancer [42]. Sun et al. reported the development of gold nanoparticle-capped MSNs loaded with luciferase to retain enzyme bioactivity through intracellular-controlled catalysis and their use in tumor imaging. This provides a unique platform for monitoring metastasis by means of bioluminescence and using intracellular ATP and GSH levels as indicators [43]. Mou et al. proposed a new strategy for the intracellular delivery of superoxide dismutase, where multi-functional MSN nanocontainers were used to enhance the transmembrane permeability of the enzyme. This therapy was successful in delivering protein treatment for ROS-mediated diseases [44]. Interestingly, a variety of enzymes were immobilized in the mesoporous silica spheres in high loads, with safeguards to maintain stability [45,46,47,48], and the enzymes were shown to retain their bioactivities [41,49,50].

For the successful exploitation of EPT in treating cancer, it is a crucial to maintain therapeutic levels of the enzymes for catalytic activity at the tumor sites; simultaneously, the prodrug must be able to cross the cell membranes for successful prodrug activation. To meet these criteria, there is a need to employ safe delivery vehicles to protect the enzymes from biological stress and to retain their catalytic activity. Furthermore, the carriers must be able to selectively transport the enzymes to the desired sites after crossing the cellular membranes to maintain therapeutic levels of the prodrug and the enzyme even after prolonged circulation times. These outstanding and largely unmet challenges are best addressed by MSNs (Figure 1). The immobilization of enzymes in MSNs for use in EPT is one of the most promising developments in cancer therapy. MSNs have many advantages including high biocompatibility, easy biodegradation and metabolism, controllable particle and pore sizes [51,52,53,54,55,56], large surface areas [57], and an abundance of sites available for surface functionalization, which render them the most preferable enzyme delivery vehicle to date [58]. Earlier studies have shown that immobilization of enzymes in mesoporous silica are less susceptible to damage caused by changes in pH and temperature and interactions with organic solvents [46,59,60,61,62]. Enzymes can be safely transported to target sites by encapsulation in the nanochannels of mesoporous silica; the scope of these vehicles with regards to targeting tumor cells can be expanded by attaching various types of ligands (e.g., antibodies, peptides, aptamers, and vitamins) to the surface of the MSNs [63,64].

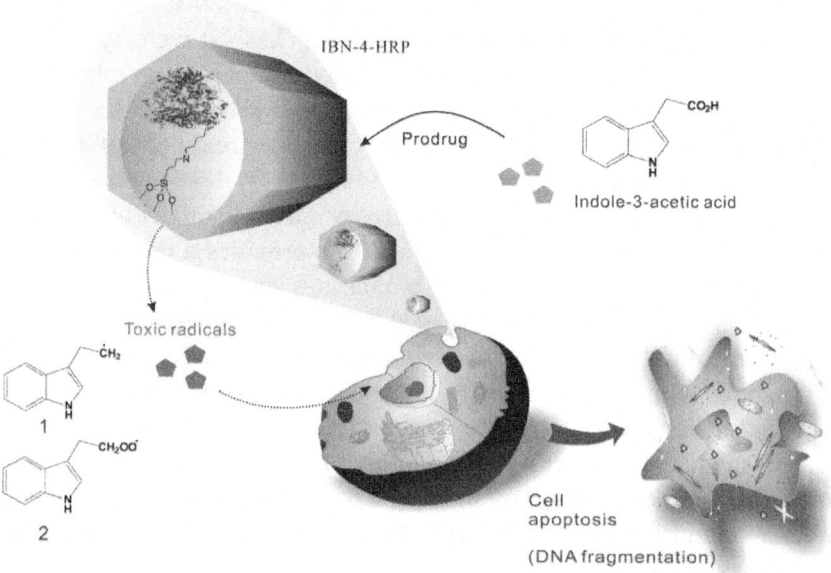

Figure 1: Schematic representation of enzyme prodrug therapy using IBN-4-HRP nanocomposites in the presence of indole-3-acetic acid and the resultant cell apoptosis (**1**. Skatolyl radical and **2**. Peroxyl radical).

RESULTS AND DISCUSSION

A schematic illustration for the synthesis of enzyme-immobilized MSNs is shown in Figure 1. The purpose of our study was to immobilize HRP onto IBN-4 nanoparticles via covalent linkages to facilitate the activation of the prodrug IAA (indole acetic acid) in the tumor microenvironment. Pristine IBN-4 nanoparticles were synthesized using a template (Pluronic P123) and an FC-4 nanoparticle overgrowth protectant (fluorocarbon surfactant). Hydrolysis and condensation of tetraethoxysilane (TEOS) were performed under dilute acidic conditions. The surfactant was then removed by oxidation with ammonium perchlorate and a 10 M HNO_3 solution; the resulting sample was labelled "IBN-4-extracted". After oxidation under acidic conditions, the surfaces had large amounts of silanol groups, which were easily functionalized with 3-aminopropyltrimethoxysilane (APTS) to afford the amine-modified mesoporous silica surfaces (IBN-4-NH_2). Subsequent conjugation of glutaraldehyde (GA) (IBN-4-NH-GA) allowed the aldehyde to serve as an anchor for the successful immobilization of enzymes on the surface of the silica particles through the formation of imine bonds (IBN-4-HRP). The reaction between IAA and HRP proceeded through a complex mechanism where IAA was activated by HRP and produced free radicals (Figure 1), such as indolyl, skatolyl, and peroxyl radicals, and reactive oxygen species (ROS), such as $\cdot O_2^-$ and H_2O_2. These reactive molecules caused morphological changes of the cells and initiated damage by inducing apoptotic-signaling cascades [65]. Thus, the proposed IAA/HRP combination may have enhanced cellular oxidative stress and may have led to cell death. Nanoparticle size and enzyme stability play a key role in the interaction and activation of prodrug molecules in cells during EPT. First, nanoparticles within a range of <300 nm were shown to have good dispersion, escape uptake by the reticuloendothelial system (RES) and enhance the EPR (Enhanced permeability and retention) effect [66,67,68]. However, recent research has shown that nanoparticles less than 100 nm in size could result in severe cytotoxicity due to nonspecific cellular interactions [69,70,71]. Therefore, in the current investigations, we employed IBN-4 nanoparticles that had a mean diameter that was greater than 100 nm, as shown in Figure 2. It was evident from the Transmission electron microscopic (TEM) images that the particles exhibited an average particle diameter of approximately 150–300 nm with aspect ratios (length to the width dimension ratio of uniformly sized nanoparticles) in the range of approximately 4 to 6. The morphology and particle size of IBN-4 did not change after the amine modification process (Figure 2c) and subsequent conjugation of HRP (Figure 2d). In addition, the DLS measurements (particle diameter and zeta potential) were recorded and are illustrated in Table 1. The particle sizes of all of the samples were nearly in accord with observations from the TEM images; however, a small amount of aggregation was observed in the case of IBN-4-HRP, which resulted in aggregates sizes of ~390 nm. Previous reports suggest that particles with an average particle size of <300 nm should be favorable due to ease of internalization considering the EPR effect [66,56]. The zeta values of pristine IBN-4 and IBN-4 extracted were −0.1 and −10.9 mV, respectively. IBN-4-HRP exhibited a negative shift in surface charge (*i.e.*, +17.7 mV) compared to IBN-4-NH$_2$ (*i.e.*, +26.3 mV). The positive zeta potential of enzyme-loaded IBN-4 nanoparticles facilitated cellular internalization through negatively charged cell membranes during biological investigations.

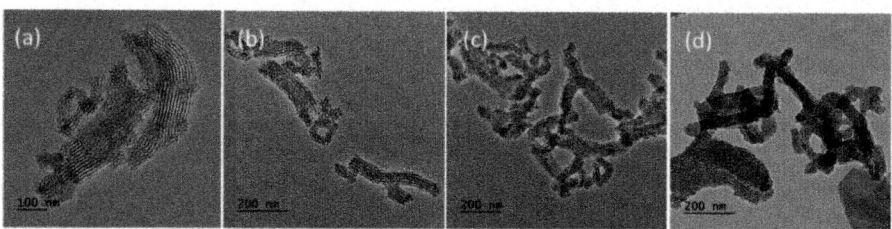

Figure 2: Transmission electron microscopic images of IBN-4 nanoparticles: (**a**) As-synthesized IBN-4, (**b**) IBN-4-extracted, (**c**) IBN-4-NH$_2$, and (**d**) IBN-4-HRP.

Table 1: Physical properties of the mesoporous materials

Sample	BET Surface Area (m²/g)	Pore Size (nm)	Pore Volume (cm³/g)	Particle Size * (nm)	Zeta Potential * (mV)
Pristine IBN-4	345	7.9	0.75	271 (±3.0)	−0.1 (±0.1)
IBN-4 extracted	812	9.3	1.71	254 (±7.6)	−10.9 (±0.4)
IBN-4-NH$_2$	357	6.0	0.98	296 (±6.4)	+26.3 (±0.7)
IBN-4-HRP	296	6.5	1.1	394 (±9.5)	+17.7 (±0.2)

The mesoporous characteristics of IBN-4 nanoparticles were determined using N$_2$-adsorption/desorption isotherms and BET theory with an ASAP 2020 Micromeritics surface area analyzer at −196 °C (Figure 3A, Table 1). The pore size distributions of pristine IBN-4, IBN-4 nanoparticles after surfactant removal (IBN-4-extracted), IBN-4 nanoparticles modified with APTS groups (IBN-4-NH$_2$) and glutaraldehyde conjugation nanoparticles for immobilization of HRP (IBN-4-HRP) were determined from adsorption branches of the isotherms using the Barrett–Joyner–Halenda (BJH) method (inset ofFigure 3A). There was a significant increase in the specific surface areas and pore volumes of the nanoparticles after surfactant removal from as-synthesized IBN-4 (345 m²·g⁻¹, 0.75 cm³·g⁻¹) (Figure 3A(a)) to 812 m²·g⁻¹ and 1.71 cm³·g⁻¹, respectively (Figure 3A(b)). Later, the final BET surface areas of the IBN-4-extracted sample decreased from 812 to 357 m²·g⁻¹ and the pore volumes decreased from 1.71 cm³·g⁻¹ to 0.98 cm³·g⁻¹ after amine modification (Figure 3A(c)). These changes were sufficient for the conjugation of enzymes in the mesoporous channels. Furthermore, the immobilization of the HRP enzyme (Figure 3A(d)) through a glutaraldehyde anchor decreased the surface area to 296 m²·g⁻¹, but the pore volume was increased to 1.1 cm³·g⁻¹ due to the contributions of enzyme molecules to the porosity of the particles. In addition, the area and intensity of the pore size distribution near 6.0 nm were further decreased, which showed that parts of the nanochannels were now occupied by enzymes.

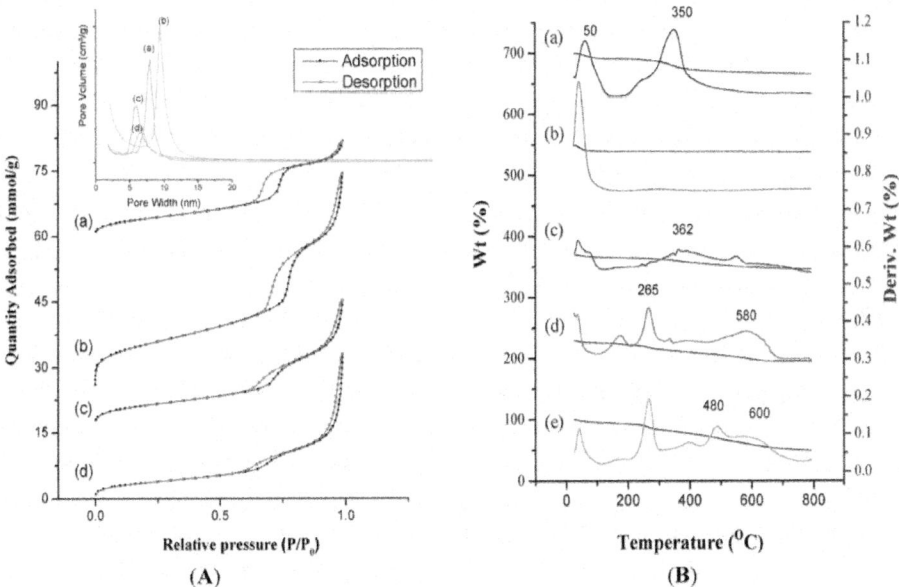

Figure. (A) Nitrogen adsorption–desorption isotherms of (a) as-synthesized IBN-4, (b) IBN-4 extracted, (c) IBN-4-NH₂ and (d) IBN-4-HRP. Corresponding pore size distribution plots are shown in the inset figure. (B) Thermogravimetric analysis curves of (a) as-synthesized IBN-4; (b) IBN-4 extracted, (c) IBN-4-NH₂, (d) IBN-4-NH-GA and (e) IBN-4-HRP.

Thermogravimetric (TGA) curves of all of the samples were recorded and the successive weight loss events were analyzed, as illustrated in Figure 3B. The primary weight loss (*i.e.*, 5%–8%) below 100 °C was attributed to loss of physisorbed moisture of the sample (Figure 3B(a)–(e)). The weight loss observed at 350 °C was due to surfactant decomposition in the as-synthesized sample (Figure 3B(a)) and the same observation was made for the IBN-4 sample extracted with ammonium perchlorate (Figure 3B(b)), which revealed the complete extraction of the surfactant. Similarly, a broad decomposition band centered at 362 °C was due to weight loss (*i.e.*, ~16%) from decomposition of the surface functionalized APTS. Furthermore, conjugation of glutaraldehyde to the surface amine resulted in decomposition at 180, 265 (*i.e.*, ~12%) and 580 °C. This confirmed the effective conjugation of the glutaraldehyde spacer for immobilization of HRP. The final events in the HRP-immobilized samples (IBN-4-HRP (Figure 3B(e)) resulted in weight loss in the range of 400–500 °C, demonstrating the effective immobilization of the enzymes (HRP weight loss alone *i.e.*, ~14%) that completely occupied the active aldehyde groups. It was detrimental for any of the aldehyde groups to remain free because glutaraldehyde exists as a low molecular weight polymer when it is used as spacer for protein immobilization. Thus, researchers should be quite cautious when using glutaraldehyde as a spacer. The amount of HRP weight loss from IBN-4 was approximately the same as the loading percentage obtained from ultraviolet-visible (UV-Vis) spectroscopy using a Bio-Rad enzyme quantification assay (IBN-4-HRP (13.13%)).

Fourier transform infrared spectroscopy (FT-IR) was performed to characterize the chemical modifications of the IBN-4 samples and their post-synthetic conjugations. The spectra of the as-synthesized and surfactant extracted IBN-4 samples are displayed in Figure 4. The bands at 1080 and 460 cm^{-1} confirms the Si-O-Si framework in all of the samples. Additionally, the typical bands observed at approximately 2900–3200 cm^{-1} were attributed to the asymmetric stretching vibrations of C-H in the aliphatic moieties of the surfactant. These bands vanished after oxidative extraction (Figure 4b), in contrast to those observed in Figure 4a, which confirmed that the surfactant was completely removed. However, the peaks of the Si–O–Si framework remained, which confirmed that the framework was not destroyed under oxidative reaction conditions. The C–H stretching bands at 2973 cm^{-1} and 2933 cm^{-1} appeared in the spectra of the amine functionalized IBN-4 samples (Figure 4c). The glutaraldehyde-conjugated IBN-4 displayed C=O (an aldehyde) stretching modes at or near 1667 cm^{-1} (Figure 4d). The spectrum of the HRP-loaded nanocarrier resulted in amide bands at 1452 and 1550 cm^{-1}. In addition, the band at 1550 cm^{-1} represented the NH bending vibrations in the plane; the intensity of the peak at 1452 was much lower than before [72]. Thus, the FT-IR spectra indicated the sequential modifications of IBN-4 were successful. The amino groups on the surface functionalized IBN-4 samples were quantitatively analyzed using UV-Vis spectrometry after treating the samples with ninhydrin to yield Ruhemann's purple coloration that displayed an absorption band at

580 nm upon reaction with primary amines. Unmodified IBN-4 nanoparticles (Figure 5b) did not show any prominent peaks at 580 nm due to the absence of amine groups. However, IBN-4-NH₂ samples exhibited a characteristic band at 580 nm (Figure 5c) along with the production of Ruhemann's purple, as shown in the respective inset sample images. This phenomenon indicated that large amounts of amino groups were conjugated to the surfaces of the IBN-4 samples. Glutaraldehyde-conjugated IBN-4-NH₂ nanomaterials (Figure 5d) showed very low absorption values for the active amine groups as confirmed by the lower absorption peaks at 580 nm; this observation could be explained by the reaction of the two terminal amino groups with glutaraldehyde moieties and their subsequent transformations into secondary amines.

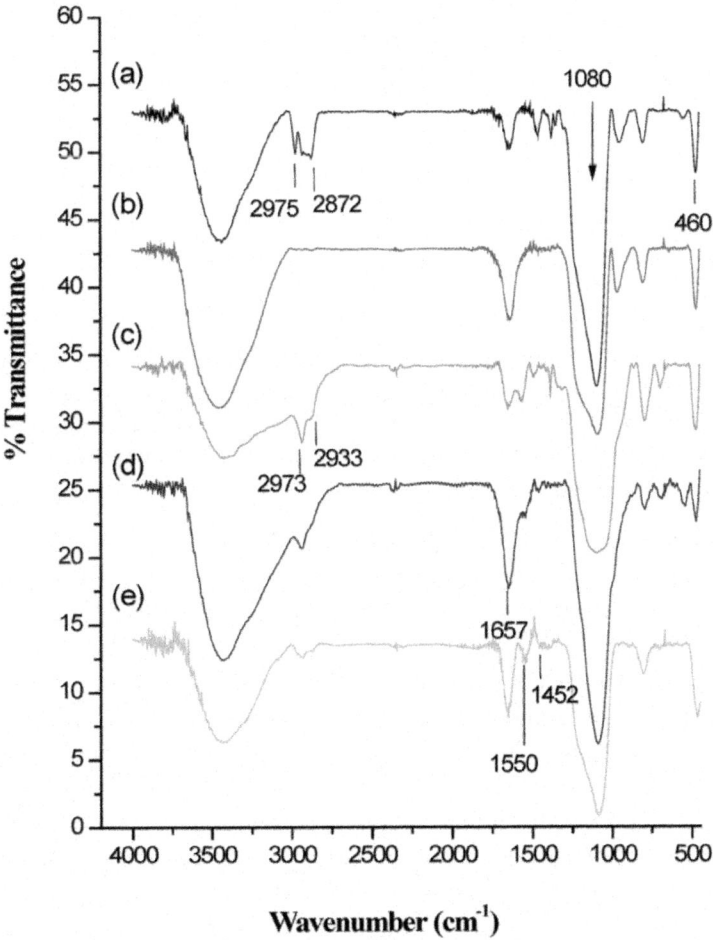

Figre 4.: Fourier transform infrared spectroscopy (FT-IR) spectra of the **(a)** as-syn-

thesized IBN-4 nanoparticles, (**b**) IBN-4 nanoparticles after surfactant removal (IBN-4-extracted), (**c**) IBN-4 nanoparticles modified with APTS groups (IBN-4-NH$_2$), (**d**) IBN-4-NH$_2$ nanoparticles modified with glutaraldehyde groups (IBN-4-NH-GA) and (**e**) IBN-4-NH-GA nanoparticles immobilized with HRP (IBN-4-HR**2.1. Detection of Primary Amine Groups**

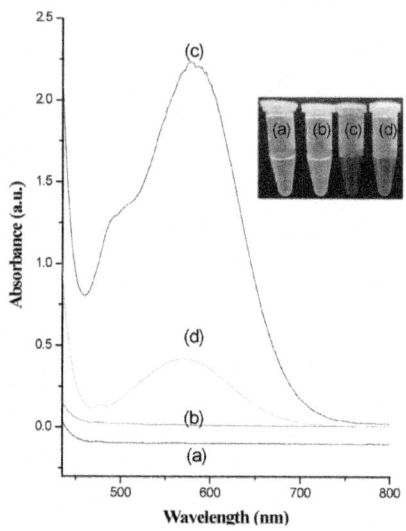

Figur5: . Ultraviolet–visible (UV-Vis) spectra and white-light sample images (inset) of (**a**) ninhydrin alone, (**b**) IBN-4 nanoparticles after surfactant removal, (**c**) IBN-4 nanoparticles modified with APTS groups (IBN-4-NH$_2$) and (**d**) IBN-4 nanoparticles modified with glutaraldehyde groups (IBN-4-NH-G**2.2. Properties of Immobilized HRP**

The specific activities per unit weight of enzymes for the HRP-immobilized nanoparticles were solely dependent on the method employed to immobilize the HRP molecules. We loaded HRP molecules using a typical two-step method (IBN-4-HRP) and utilized a one-pot synthetic approach (IBN-4-HRP (one pot)) to compare the resulting enzyme activities. The efficacies of delivery and the anti-tumor activities of the two methods were successfully compared. It was evident that enzyme immobilization onto IBN-4 nanomaterials using a two-step process (IBN-4-HRP (4144 units/g)) resulted in significantly higher yields than when the one pot approach was used (IBN-4-HRP (one pot)) (953.60 units/g)). The low enzymatic activities of the nanoparticles immobilized using the one-pot approach were attributed to the reaction between the amino groups of HRP and glutaraldehyde molecules, which formed undesirable enzyme complexes; these complexes may have caused the active sites to denature through saturation of lysine residues. These undesirable enzyme-bound complexes could be removed by washing steps, which limited enzyme loading in the functionalized IBN-4 nanoparticles. Consequently, this led to lower loading amounts with the one-pot synthetic approach (IBN-4-HRP (one pot *i.e.*, 9.87%) compared to IBN-4-HRP (two-step process *i.e.*, 13.13%)), as observed by UV-Vis spectroscopy with a Bio-Rad enzyme quantification assay. The long-term stability (stored at 4 °C) and the activity of HRP immobilized IBN-4 nanoparticles were also evaluated. Interestingly, over 90% of the original peroxidase activity of HRP was retained after two months. These results revealed that the stability of the HRP entrapped in IBN-4 increased without losing biological activ

IAA-Dependent Cytotoxicity of HRP Encapsulated IBN-4

The cytotoxic effects of using HRP-loaded IBN-4 in EPT were explored using an MTT (3-(4,5-dimethylthiazol-2-yl)-2,5-diphenyltetrazolium bromide) assay with a HT-29 (Human colon carcinoma) cell line (Figure 6), and the morphological changes were simultaneously observed using bright field microscopy (Figure 7). Previously, it was reported that the uptake of silica nanoparticles was fairly rapid and was observed after 2–3 h, with toxicity generally observed within 24 h [73].

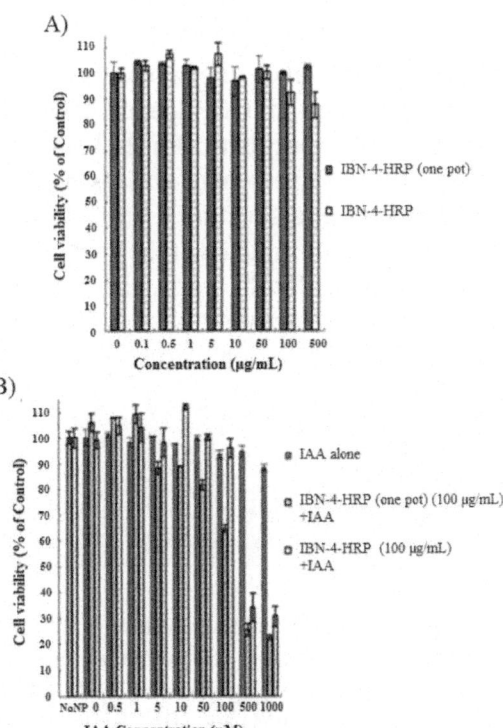

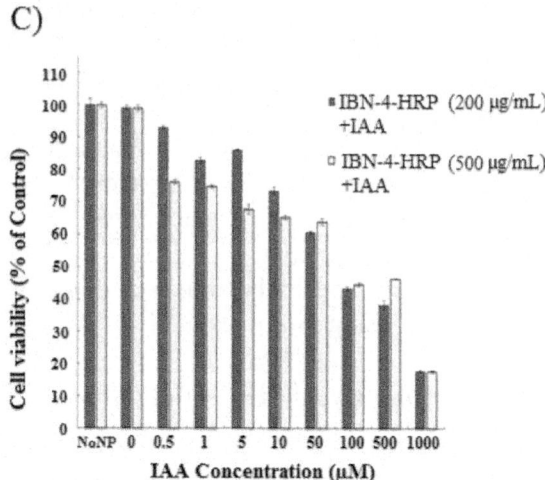

Figre 6:. The viability of Human colon carcinoma (HT-29) cells: (**A**) treatment with IBN-4 loaded with Horseradish peroxidase (HRP) prepared via two synthetic routes; (**B**) treatment of IBN-4-HRP synthesized using two different routes in the presence or absence of indole-3-acetic acid (IAA) (IAA expressed in µM); and (**C**) treatment of various concentrations (200 and 500 µg/mL) of IBN-4-HRP synthesized in the presence or absence of IAA (IAA expressed in µM).

To exclude the possible influence of enzyme-loaded nanoparticles on cell viability, various concentrations of HRP-loaded IBN-4 materials prepared using two synthetic routes were incubated for 24 h and the cell viability was evaluated. The tested concentration range was as high as 500 µg/mL, and the enzyme-loaded IBN-4 materials had no ostensible adverse effects on cell viability (Figure 6A). This demonstrated that the enzyme-loaded nanoparticles had no substantial cytotoxicity. Furthermore, the incubation of HT-29 cells with IAA alone did not show any signs of toxicity and, surprisingly, 90% of the cells were viable even at a concentration of 1000 µM (Figure 6B).

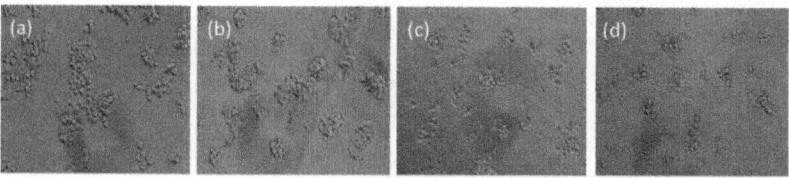

Fiure 7.: Bright field images of HT-29 cells without or with treatment of HRP-loaded IBN-4 nanoparticles and IAA: (**a**) Control (cells without treatment), (**b**) treatment with IAA (500 µM), (**c**) treatment with IAA (500 µM) and IBN-4-HRP (one-pot) (100 µg/mL) and (**d**) treatment with IAA (500 µM) and IBN-4-HRP (100 µg/mL).

However, when IAA was combined with enzyme-loaded IBN-4, it became toxic to HT-29 cells (Figure 6B). Cell growth inhibition was observed to be 35% in IBN-4-HRP (one pot) (100 µg/mL) at 100 µM, whereas 70% of cell death was observed when the concentration of IAA was increased to 500 µM. A very similar dose-dependent cytotoxic pattern was observed in cells treated with IBN-4-HRP in combination with IAA. Nevertheless, the activity of the enzymes immobilized with the one-pot approach showed significantly higher activity than the particles immobilized with the typical two-step approach at lower concentrations; at higher concentrations, both samples exhibited nearly identical cytotoxic effects. This contrast in enzyme activity and cytotoxicity with respect to different synthetic approaches may have been due to the enzyme release characteristics during the prodrug activation process. Similarly, when treatment was continued at higher concentrations of IBN-4-HRP (200 and 500 µg/mL) (Figure 6C) in the presence of IAA (0–1000 µM), increases in the nanoparticle–enzyme complex concentration did not result in further changes in activity. This demonstrated that the enzyme prodrug therapy exclusively depends on the concentration of the prodrug and that the respective concentration (from 100 to 500 µg/mL) is suitable for activation of the prodrug.

The morphological changes of the HT-29 cells incubated with enzyme-loaded IBN-4 nanoparticles and IAA were also studied using bright field microscopy (Figure 7). HT-29 cells treated with enzyme-loaded IBN-4 nanomaterials (100 µg/mL) and IAA (500 µM) exhibited nuclear condensation and substantially shrunken morphologies (Figure 7c,d), suggesting the induction of apoptosis. No morphological changes were observed in cells treated with either IAA alone (Figure 7b) or with the untreated control (Figure 7a). These results were in accordance with observations from the cytotoxicity study and indicated that the enzyme-loaded IBN-4 materials have the potential to replace free-HRP in enzyme-prodrug cancer therapy.

We used a lactate dehydrogenase (LDH) assay to measure the activity of the LDH enzyme that was released through cell membrane damage. For this experiment, maximum release of the positive control was obtained by treating the cells with 0.5% triton X-100 (Figure 8). Compared with the control experiment (CTL), treatment with IAA, (1.6 mM) enzyme-loaded IBN-4 materials, or a combination of the two did not result in a profound increase in the release of LDH even at doses as high as 100 µg/mL. From the results obtained, it was concluded that enzyme-loaded IBN-4 generated with either of the routes was highly biocompatible and retained membrane integrity by decreasing the amount of unexpected necrotic cell death. The majority of the anti-cancer drugs induce cell death through apoptotic or necrotic pathways. Induction of cell death by the apoptotic pathway confers many advantages over the necrotic pathway because the leakage of enzymes that results from the necrosis pathway usually causes an inflammatory response, which results in poor prognosis in cancer therapy. Furthermore, we analyzed the morphological changes of the HT-29 cells incubated with HRP-loaded IBN-4 and IAA by bright field microscopy (Figure 9). After treating the cells with IBN-4-HRP and IAA (1.6 mM), the morphologies of the cells became round and shrunk, which is a typical phenomenon observed for cell apoptosis. In comparison with the groups treated with IAA or IBN-4-HRP alone, the cell morphologies rarely had round shapes or shrunk, which demonstrates the low cytotoxicities of the individual treatment methods.

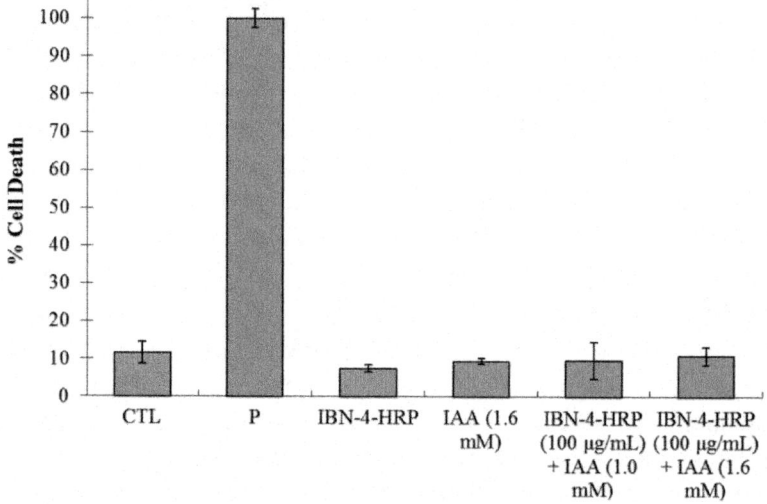

Figure 8.: Cytotoxic estimates of IBN-4 nanocomposites using an lactate dehydrogenase (LDH) leakage assay. Control experiment (CTL), P represents positive control (treated with Triton x-100) with various treatments using combinations of IAA and IBN-4-HRP.

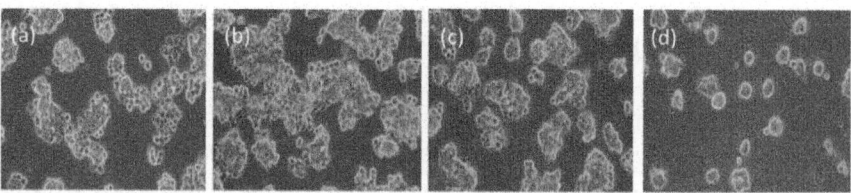

Figure 9.: Bright field images of HT-29 cells with or without treatment of IBN-4-HRP and IAA: (**a**) HT-29 cells without treatment; (**b**) treatment with IAA (1.6 mM); (**c**) treatment with IBN-4-HRP alone; and (**d**) treatment with IBN-4-HRP and IAA (1.6 mM).

Previous reports have suggested that the combination of IAA and HRP can induce apoptosis by activating p38 mitogen-activated protein (MAP) kinase and c-Jun N-terminal kinase (JNK). Furthermore, this combination was shown to activate caspase-8 and caspase-9 and to induce DNA damage [74]. To elucidate the mechanism of cell death, the DNA damage was investigated using a comet assay. Comet assays have been considered to be a very good technique for determining DNA damage and has been shown to have many advantages over existing methods that measure DNA fragmentation. The results of the comet assay showed significant DNA damage, which was indicated by the tail moment (Figure 10d) after treatment with enzyme loaded-IBN-4 materials (IBN-4-HRP) in the presence of IAA. Furthermore, the bright field images revealed profoundly shrunken cell morphologies when enzyme-loaded nanomaterials (IBN-4-HRP) were incubated along with IAA (500 μM) (Figure 10c), suggesting the induction of apoptosis. No morphological changes were observed in the control group (Figure 10a).

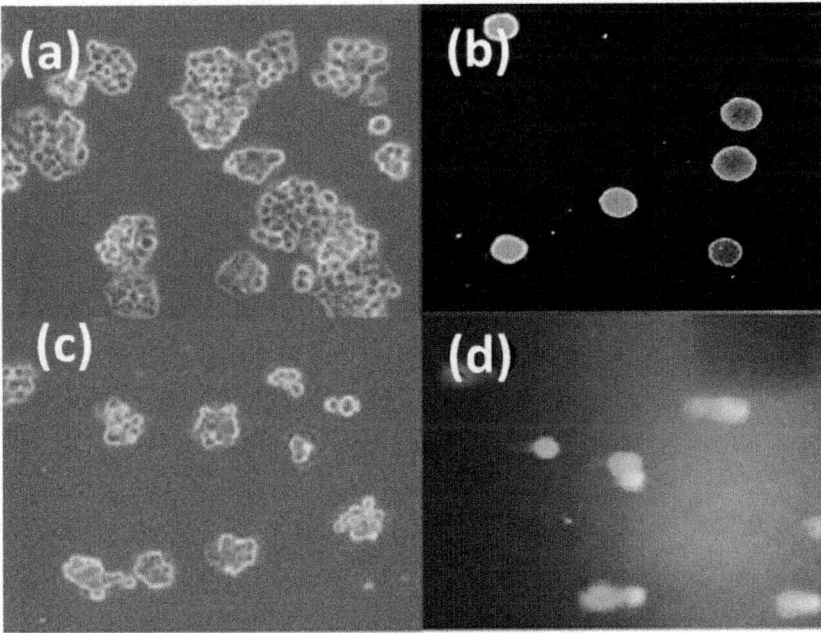

Figure 10. Microphotographs of bright field and fluorescent microscopic views of HT-29 cells: (**a**) Bright field view of untreated control cells (undamaged cells); (**b**) fluorescent microscopic view of untreated control cells (undamaged cells); (**c**) bright field view of cells treated with IAA (1 mM) and IBN-4-HRP (100 μg/mL) (shrunken morphologies); and (**d**) fluorescent microscopic view of cells IAA (1 mM) and IBN-4-HRP (100 μg/mL) (damaged cells with elongated ls).

EXPERIMENTAL ON

Materials

Pluronic P123, tetraethyl orthosilicate (TEOS), ammonium perchlorate (AP), (3-aminopropyl)trimethoxysilane (APTS), 3-(4,5-dimethylthiazol-2-yl)-2,5-diphenyltetrazolium bromide (MTT), potassium bromide (KBr) (FT-IR grade), glutaraldehyde (GA), horseradish peroxidase (HRP) and 3-indole acetic acid (IAA) were purchased from Sigma-Aldrich® (St. Louis, MO, USA). All other reagents and chemicals were analytical grade mats.

Characterization and Instruments

TEM images were captured on a Hitachi H-7100 (Hitachi High Technologies Corporation, Tokyo, Japan) instrument operating at 100 kV. Samples were prepared by dispersing them on carbon-coated copper (Cu) grids and drying them at room temperature. Fluorescence images were captured on an Olympus microscope hybridized with cooled color CCD of an Olympus DP73 (Olympus Corporation, Center Valley, PA, USA). The centrifuge used for nanomaterial syntheses and cell culturing processes at appropriate temperatures and rpm was an Hermle Z 36 HK (HERMLE Labortechnik GmbH, Wehingen, Germany) and a swing rotor Kubota KN-70 (Kubota Corporation, Tokyo, Japan), respectively. Thermogravimetric analyses were performed using a TGA Q50 V20, 13 Build 39 (Universal V4.5A TA Instruments, Lukens drive, New Castle, PA, USA). Dynamic light scattering (DLS) measurements to determine particle sizes and zeta potentials were performed using a Nano-HT Zetasizer 90 (Malvern Instruments Ltd., Worcestershire, UK). Ultraviolet-visible (UV-Vis) spectra were recorded on a Genequant-1300 series spectrophotometer (GE Healthcare Biosciences, Pittsburgh, PA, USA). MTT absorbance was recorded using an EnSpire Multi-label Plate Reader (Perkin Elmer, Santa Clara, CA).

Synthesis of IBN-4 Type Mesoporous Silica Nanoparticles

IBN-4 type mesoporous silica nanoparticles were prepared by following a previously published method [75], where 1 g of pluronic acid P123 and 2.8 g of FC-4 ($C_3F_7O(CFCF_3CF_2O)_2CFCF_3CONH(CH_2)_3N^+(C_2H_5)_2CH_3I^-$) were dissolved in 160 mL of 0.02 M HCl at 45 °C. After complete dissolution of the surfactant, TEOS (4 g) was added and the reaction mixture was stirred at room temperature for 20 h. The pristine IBN-4 particles were collected by centrifugation at 12,000 rpm for 17 min, and the copolymer templates were further removed by ammonium perchlorate (AP) [76]. To create oxidative conditions, approximately 0.2 g of the IBN-4 solids was treated with 0.4 g of AP in 20 mL of a 10 M HNO_3 solution under 100 °C for 12 h. The solid products were washed with dd-H_2O and collected by centrifugation at 12,000 rpm for n.

Synthesis of Amino-Functionalized IBN-4

Post-synthetic modification was carried out by adding 0.2 g of the IBN-4 solids in 30 mL of toluene and stirring the reaction mixture. A 0.3 mL aliquot of a pure APTS solution was added to the mixture and allowed to stir at 100 °C for 20 h. The amino-functionalized particles were collected by centrifugation at 12,000 rpm for 17 min and were washed twice using acetone and el.

Detection of Primary Amines with a Ninhydrin Test

The procedure employed methods following a previously published report on the detection of amine-functionalized silica material [77]. Typically, 50 mg of the samples was mixed with 5 mL of ethanolic ninhydrin solution (0.2 M). The solution was stirred at room temperature for 25 min. Subsequently, the supernatant was collected to and centrifuged for 10 min. The absorbance of the supernatant was monitored at 580 nm by UV-Vis spectry.

Covalent Immobilization of HRP onto IBN-4 Materials (IBN-4-HRP)

Enzymes were immobilized onto IBN-4 nanomaterials following a previously published method [78]. Initially, 0.15 g of amine-functionalized particles was dispersed in 22.5 mL of a 0.1 M Na_2HPO_4-NaH_2PO_4 buffer solution (pH 8.0) through ultra-sonication. A 50% glutaraldehyde (GA) solution was added to the particles dispersed in buffer to reach a final volume of 0.1% (v/v). The reaction mixture was continuously stirred for one hour at room temperature and the GA-modified particles were collected by centrifugation. Later, the HRP buffer solution was prepared by dissolving 22.5 mg of HRP in 15 mL of a 10 mM Na_2HPO_4-NaH_2PO_4 buffer solution (pH 8.0). Finally, 0.15 g of the GA-modified IBN-4 particles was redispersed in the HRP buffer solution by ultra-sonication. The reaction mixture was then stirred at 10 °C for 2 h and collected by centrifugation. The unbound enzymes were removed by washing them once with a 10 mM Na_2HPO_4-NaH_2PO_4 buffer solution. The enzyme-bound particles were redispersed in 15 mL of a 10 mM Na_2HPO_4-NaH_2PO_4 buffer solution (pH-8.0) and stored aC.

Covalent Immobilization of HRP into IBN-4 Materials by One pot Process (IBN-4-HRP (One Pot))

The direct method of HRP immobilization onto IBN-4 nanomaterials followed a previously published method [79]. At first, 10 mg of amino-functionalized particles was dispersed in 5 mL of a phosphate buffer saline (PBS) solution (pH 7.4) by ultrasonic dispersion and was followed by the addition of 5 mL of a 2.5% glutaraldehyde solution. The reaction mixture was stirred for 2 h at room temperature and then 5 mg of HRP was added. The reaction mixture was allowed to stir at 4 °C for 24 h and the particles were recovered by centrifugation. The unbound enzymes were removed by washing them with PBS. Finally, the enzyme-conjugated particles were redispersed in 1 mL of pH 7.4 PBS and stored °C.

HRP Enzymatic Assay

The peroxidase activity of HRP was measured colorimetrically. The principle behind this measurement involves the peroxidation of pyrogallol to purpurogallin in the presence of HRP. A reaction mixture (3.00 mL) was prepared to contain the following final concentrations: 100 mM potassium phosphate buffer (pH-6.0), 0.5% (*w/w*) hydrogen peroxide, 5.0% (*w/v*) pyrogallol, and IBN-4-HRP (1 mg/mL). The reaction was allowed to continue for 5 min while the contents were mixed properly. The absorbance was measured at 420 nm with a UV-Vis spectrophotometer through calibration for 10 min at 30 °C. The absorbance of a blank sample was obtained using water. One unit/g of the IBN-4-HRP formed 1.0 mg of purpurogallin from pyrogallol at specific pH and temperatures. All the activity measurements were performed in trite.

Cytotoxicity Studies

Cell Culture: a HT-29 (colon carcinoma) cell line was cultured in Roswell Park Memorial Institute (RPMI)-1640 medium supplemented with 10% (*v/v*) fetal bovine serum (FBS). Cultures were maintained in a humidified incubator at 37 °C with 5% CO_2 (carbon dioxide). Cell viability studies of the prepared particles were evaluated *in vitro* using a standard MTT (3-(4,5-dimethylthiazol-2-yl)-2,5-diphenyltetrazolium bromide) colorimetric assay. HT-29 cells were seeded in 96-well plates at a density of 1×10^4 cells per well. After letting the cells adhere overnight, the cells were serum-starved in RPMI-1640 medium. The cells were incubated for 24 h and then treated with various concentrations (0.5–1000 µM) of 3-indoleacetic acid and enzyme-conjugated nanoparticles (0.1 mg/mL) and were incubated for another 24 h. Subsequently, 50 µL of an MTT solution (1 mg of MTT in 1 mL of PBS) was added and incubation was continued for another 4 h. The relative percentages of metabolically active cells relative to the untreated control cells were then determined based on the mitochondrial conversion of MTT to formazan. Finally, the medium was pipetted out and 150 µL of DMSO was added to wells to dissolve the crystals; then, the absorbances of the individual wells were measured with a Microplate Reader atm.

LDH Release Assay

HT-29 cells (1×10^5 cells/well) were seeded in a 12-well plate. The cells were grown in 10% FBS-supplemented RPMI-1640. The cultured cells were treated with various sample combinations using IAA and IBN-4-HRP and were incubated for another 24 h. The levels of cytosolic LDH (lactate dehydrogenase) leakage were assessed to measure the extent of cellular membrane damage using a Sigma Tox-7 Kit (Sigma-Aldrich®, St Louis, MO, USA) following the manufacturer's instructions. This kit spectrophotometrically determines LDH activity by measuring the intensity of the reduced formazan at 490 nm, which is directly proportional to the LDH activity. These measurements were performed with an ELISA reader. The results presented represent mean values from triplicate measurements. The results are given as fractions of LDH release compared to the positive controls, which consisted of 0.5% Triton X-100 (yielding 100% LDH r).

Comet Assay

The DNA comet assay was performed using a Trevigen's Comet Assay® kit (Trevigen, Inc. Gaithersburg, MD, USA) according to the manufacturer's instructions. HT-29 cells were seeded at a density of 1×10^5 cells/well in a 6-well plate and various treatments were performed. The plates were placed in an incubator for 20 h. After harvesting, the cells were pooled in a 1% low melting point agarose gel at a ratio of 1:10 (v/v). The gel was run in accordance with the manufacturers' instructions for 20 min at 21 volts and 350 mA. The gel was then washed with water and ethanol to reanneal the DNA; eventually, the smear was stained with SYBR Green. DNA comets were considered to be markers of genotoxic effects. The images were captured using Olympus fluorescence image ayses.

CONCLUSIONS

To improve the enzymatic activity and stability of nanoparticles for use in therapeutic applications, the fragile enzyme molecules required immobilization on solid supports to prevent degradation and unfolding of the proteins. In this study, we immobilized an HRP enzyme onto nanochannels of mesoporous silica nanoparticles to demonstrate their ability to activate an anticancer prodrug (indole-3-acetic acid). The immobilization of HRP and activity experiments indicated that the confined spaces in mesoporous silica could maintain HRP microenvironments and therapeutic activities. The large pores of the mesoporous silica (IBN-4) nanocontainers were suitable for immobilizing HRP enzymes through covalent bonds. The conditions employed for enzyme loading using a typical approach via two chemical steps generated more favorable critical parameters for enzyme orientation and specific activities than the one-pot approach. It is expected that IBN-4 materials with large surface areas, pore sizes, and high biocompatibility can be used to enhance the stability of HRP enzymes and increase the diffusion rates of catalytic substrates, which can be used in EPT for cancer therapeutic applications.

ACKNOWLEDGMENTS

We are very thankful to the Ministry of Science and Technology, Taiwan for research grants (NSC 101-2113-M-259-003-MY2 and MOST 103-2113-M-259-005-MY2).

AUTHOR CONTRIBUTIONS

Bau-Yen Hung and Chia-Hung Lee conceived and designed the experiments; Bau-Yen Hung performed the experiments; Yaswanth Kuthati, Ranjith Kumar Kankala and Chia-Hung Lee analyzed the data and compiled the manuscript; Jin-Pei Deng performed the TEM experiments; and Shravankumar Kankala and Chen-Lun Liu assisted in data interf interest.

REFERENCES

1.   Weyel, D.; Sedlacek, H.H.; Muller, R.; Brusselbach, S. Secreted human beta-glucuronidase: A novel tool for gene-directed enzyme prodrug therapy. *Gene Ther.* **2000**, f] [PubMed]

2.   Springer, C.J.; Niculescu-Duvaz, I.I. Antibody-directed enzyme prodrug therapy (ADEPT): A review. *Adv. Drug Deliv. Rev.* **1997**, 2r] [PubMed]

3.   Springer, C.J.; Niculescu-Duvaz, I. Prodrug-activating systems in suicide gene therapy. *J. Clin. Investig.* **2000**, *105*,f] [PubMed]

4.   Connors, T.A. The choice of prodrugs for gene directed enzyme prodrug therapy of cancer. *Gene Ther.* **1995**, r] [PubMed]

5.   Hamstra, D.A.; Rehemtulla, A. Toward an enzyme/prodrug strategy for cancer gene therapy: Endogenous activation of carboxypeptidase

a mutants by the PACE/Furin family of propeptidases. *Hum. Gene Ther.* **1999**, *1*f] [PubMed]

6. Rigg, A.; Sikora, K. Genetic prodrug activation therapy. *Mol. Med. Today* **1997**, [CrossRef]

7. Rainov, N.G.; Dobberstein, K.U.; Sena-Esteves, M.; Herrlinger, U.; Kramm, C.M.; Philpot, R.M.; Hilton, J.; Chiocca, E.A.; Breakefield, X.O. New prodrug activation gene therapy for cancer using cytochrome P450 4B1 and 2-aminoanthracene/4-ipomeanol. *Hum. Gene Ther.* **1998**, *9*,f] [PubMed]

8. Wright, R.C.; Khakhar, A.; Eshleman, J.R.; Ostermeier, M. Advancements in the development of HIF-1□-activated protein switches for use in enzyme prodrug therapy. *PLoS Of*] [PubMed]

9. Zhang, J.; Kale, V.; Chen, M. Gene-directed enzyme prodrug therapy. *AAPS J.* **2014**, *17*f] [PubMed]

10. Kim, E.J.; Bhuniya, S.; Lee, H.; Kim, H.M.; Cheong, C.; Maiti, S.; Hong, K.S.; Kim, J.S. An activatable prodrug for the treatment of metastatic tumors. *J. Am. Chem. Soc.* **2014**, *136*, f] [PubMed]

11. Bagshawe, K.D. Antibody-directed enzyme prodrug therapy (ADEPT) for cancer. *Expert Rev. Anticancer Ther.* **2006**, *6*, f] [PubMed]

12. Mauger, A.B.; Burke, P.J.; Somani, H.H.; Friedlos, F.; Knox, R.J. Self-immolative prodrugs: Candidates for antibody-directed enzyme prodrug therapy in conjunction with a nitroreductase enzyme. *J. Med. Chem.* **1994**, *37*,f] [PubMed]

13. Napier, M.P.; Sharma, S.K.; Springer, C.J.; Bagshawe, K.D.; Green, A.J.; Martin, J.; Stribbling, S.M.; Cushen, N.; O'Malley, D.; Begent, R.H. Antibody-directed enzyme prodrug therapy: Efficacy and mechanism of action in colorectal carcinoma. *Clin. Cancer Res.* **2000**, r] [PubMed]

14. Palmer, D.H.; Mautner, V.; Mirza, D.; Oliff, S.; Gerritsen, W.; van der Sijp, J.R.; Hubscher, S.; Reynolds, G.; Bonney, S.; Rajaratnam, R.; *et al.* Virus-directed enzyme prodrug therapy: Intratumoral administration of a replication-deficient adenovirus encoding nitroreductase to patients with resectable liver cancer. *J. Clin. Oncol.* **2004**, *22*,f] [PubMed]

15. Tychopoulos, M.; Corcos, L.; Genne, P.; Beaune, P.; de Waziers, I. A virus-directed enzyme prodrug therapy (VDEPT) strategy for lung cancer using a CYP2B6/NADPH-cytochrome P450 reductase fusion protein. *Cancer Gene Ther.* **2005**,*1*f] [PubMed]

16. Garnier, P.; Wang, X.T.; Robinson, M.A.; van Kasteren, S.; Perkins, A.C.; Frier, M.; Fairbanks, A.J.; Davis, B.G. Lectin-directed enzyme activated

prodrug therapy (LEAPT): Synthesis and evaluation of rhamnose-capped prodrugs. *J. Drug Target.* **2010**, *1*f] [PubMed]

17. Dhar, S.; Liu, Z.; Thomale, J.; Dai, H.; Lippard, S.J. Targeted single-wall carbon nanotube-mediated Pt(IV) prodrug delivery using folate as a homing device. *J. Am. Chem. Soc.* **2008**, *130*, 1f] [PubMed]

18. Fonseca, M.J.; Jagtenberg, J.C.; Haisma, H.J.; Storm, G. Liposome-mediated targeting of enzymes to cancer cells for site-specific activation of prodrugs: Comparison with the corresponding antibody-enzyme conjugate. *Pharm. Res.***2003**, *2*f] [PubMed]

19. Vingerhoeds, M.H.; Haisma, H.J.; Belliot, S.O.; Smit, R.H.; Crommelin, D.J.; Storm, G. Immunoliposomes as enzyme-carriers (immuno-enzymosomes) for antibody-directed enzyme prodrug therapy (ADEPT): Optimization of prodrug activating capacity. *Pharm. Res.* **1996**, *1*f] [PubMed]

20. Chames, P.; van Regenmortel, M.; Weiss, E.; Baty, D. Therapeutic antibodies: Successes, limitations and hopes for the future. *Br. J. Pharmacol.* **2009**, *15*f] [PubMed]

21. Koning, G.A.; Morselt, H.W.; Velinova, M.J.; Donga, J.; Gorter, A.; Allen, T.M.; Zalipsky, S.; Kamps, J.A.; Scherphof, G.L. Selective transfer of a lipophilic prodrug of 5-fluorodeoxyuridine from immunoliposomes to colon cancer cells.*Biochim. Biophys. Acta* **1999**, *142* [CrossRef]

22. Vingerhoeds, M.H.; Haisma, H.J.; van Muijen, M.; van de Rijt, R.B.; Crommelin, D.J.; Storm, G. A new application for liposomes in cancer therapy. Immunoliposomes bearing enzymes (immuno-enzymosomes) for site-specific activation of prodrugs. *FEBS Lett.* **1993**, *33* [CrossRef]

23. Eshita, Y.; Higashihara, J.; Onishi, M.; Mizuno, M.; Yoshida, J.; Takasaki, T.; Kubota, N.; Onishi, Y. Mechanism of introduction of exogenous genes into cultured cells using DEAE-Dextran-MMA graft copolymer as non-viral gene carrier. *Molecules* **2009**, *14*,f] [PubMed]

24. Dhar, S.; Kolishetti, N.; Lippard, S.J.; Farokhzad, O.C. Targeted delivery of a cisplatin prodrug for safer and more effective prostate cancer therapy *in vivo*. *Proc. Natl. Acad. Sci. USA* **2011**, *108*,f] [PubMed]

25. Zhao, Y.; Duan, S.; Zeng, X.; Liu, C.; Davies, N.M.; Li, B.; Forrest, M.L. Prodrug strategy for psma-targeted delivery of TGX-221 to prostate cancer cells. *Mol. Pharm.* **2012**, *9*,f] [PubMed]

26. Jensen, S.S.; Andresen, T.L.; Davidsen, J.; Hoyrup, P.; Shnyder, S.D.; Bibby, M.C.; Gill, J.H.; Jorgensen, K. Secretory phospholipase A2 as a tumor-specific trigger for targeted delivery of a novel class of liposomal prodrug anticancer etherlipids. *Mol. Cancer Ther.* **2004**, *3*,r] [PubMed]

27. Zhu, S.; Lansakara, P.D.; Li, X.; Cui, Z. Lysosomal delivery of a lipophilic gemcitabine prodrug using novel acid-sensitive micelles improved its antitumor activity. *Bioconjug. Chem.* **2012**, *2*f] [PubMed]

28. Drbohlavova, J.; Chomoucka, J.; Adam, V.; Ryvolova, M.; Eckschlager, T.; Hubalek, J.; Kizek, R. Nanocarriers for anticancer drugs—New trends in nanomedicine. *Curr. Drug Metab.* **2013**, *14*f] [PubMed]

29. Pietersen, L.K.; Govender, P.; Kruger, H.G.; Maguire, G.E.; Govender, T. Enzymatic activation of a peptide functionalised gold nanoparticle system for prodrug delivery. *J. Nanosci. Nanotechnol.* **2011**, *11*,f] [PubMed]

30. Dhar, S.; Daniel, W.L.; Giljohann, D.A.; Mirkin, C.A.; Lippard, S.J. Polyvalent oligonucleotide gold nanoparticle conjugates as delivery vehicles for platinum(IV) warheads. *J. Am. Chem. Soc.* **2009**, *131*, 1f] [PubMed]

31. Fan, J.; Fang, G.; Wang, X.; Zeng, F.; Xiang, Y.; Wu, S. Targeted anticancer prodrug with mesoporous silica nanoparticles as vehicles. *Nanotechnolog*f] [PubMed]

32. Kuo, Y.-M.; Kuthati, Y.; Kankala, R.K.; Wei, P.-R.; Weng, C.-F.; Liu, C.-L.; Sung, P.-J.; Mou, C.-Y.; Lee, C.-H. Layered double hydroxide nanoparticles to enhance organ-specific targeting and the anti-proliferative effect of cisplatin. *J. Mater. Chem. B* **2015**, *3*, [CrossRef]

33. Gwenin, V.V.; Gwenin, C.D.; Kalaji, M. Colloidal gold modified with a genetically engineered nitroreductase: Toward a novel enzyme delivery system for cancer prodrug therapy. *Langmuir* **2011**, *27*, 1f] [PubMed]

34. Wang, H.; Shrestha, T.B.; Basel, M.T.; Dani, R.K.; Seo, G.M.; Balivada, S.; Pyle, M.M.; Prock, H.; Koper, O.B.; Thapa, P.S.; *et al.* Magnetic-Fe/Fe$_3$O$_4$-nanoparticle-bound SN38 as carboxylesterase-cleavable prodrug for the delivery to tumors within monocytes/macrophages. *Beilstein J. Nanotechnol.* **2012**, f] [PubMed]

35. Cude, M.P.; Gwenin, C.D. Development of gold coated superparamagnetic iron oxide nanoparticles for nitroreductase delivery. *ECS Trans.* **2011**,le Scholar]

36. Chiu, Y.-R.; Ho, W.-J.; Chao, J.-S.; Yuan, C.-J. Enzyme-encapsulated silica nanoparticle for cancer chemotherapy. *J. Nanopart. Res.* **2012**le Scholar]

37. Kankala, R.K.; Kuthati, Y.; Liu, C.-L.; Mou, C.-Y.; Lee, C.-H. Killing cancer cells by delivering a nanoreactor for inhibition of catalase and catalytically enhancing intracellular levels of ROS. *RSC Adv.* **2015**, *5*, 8 [CrossRef]

38. Kankala, R.K.; Kuthati, Y.; Liu, C.-L.; Lee, C.-H. Hierarchical coated metal hydroxide nanoconstructs as potential controlled release carriers of photosensitizer for skin melanoma. *RSC Adv.* **2015**, *5*, 4 [CrossRef]

39. Kuthati, Y.; Kankala, R.K.; Lee, C.-H. Layered double hydroxide nanoparticles for biomedical applications: Current status and recent prospects. *Appl. Clay Sci.* **2015**, *112–11* [CrossRef]

40. Giri, S.; Trewyn, B.G.; Lin, V.S.Y. Mesoporous silica nanomaterial-based biotechnological and biomedical delivery systems. *Nanomedicine* **2007**,f] [PubMed]

41. Ling, D.; Gao, L.; Wang, J.; Shokouhimehr, M.; Liu, J.; Yu, Y.; Hackett, M.J.; So, P.-K.; Zheng, B.; Yao, Z.; *et al.* A general strategy for site-directed enzyme immobilization by using NiO nanoparticle decorated mesoporous silica. *Chem. Eur. J.* **2014**, *20*ef] [PubMed]

42. Maddala, S.P.; Mastroianni, G.; Velluto, D.; Sullivan, A.C. Intracellular delivery of BSA by phosphonate@silica nanoparticles. *J. Mater. Chem. B* **2015**, *3*] [CrossRef]

43. Sun, X.; Zhao, Y.; Lin, V.S.Y.; Slowing, I.I.; Trewyn, B.G. Luciferase and luciferin co-immobilized mesoporous silica nanoparticle materials for intracellular biocatalysis. *J. Am. Chem. Soc.* **2011**, *133*, 1ef] [PubMed]

44. Chen, Y.-P.; Chen, C.-T.; Hung, Y.; Chou, C.-M.; Liu, T.-P.; Liang, M.-R.; Chen, C.-T.; Mou, C.-Y. A new strategy for intracellular delivery of enzyme using mesoporous silica nanoparticles: Superoxide dismutase. *J. Am. Chem. Soc.* **2013**,*135*ef] [PubMed]

45. Wang, Y.; Caruso, F. Mesoporous silica spheres as supports for enzyme immobilization and encapsulation. *Chem. Mater.* **2005**,r] [CrossRef]

46. Lee, C.H.; Lang, J.; Yen, C.W.; Shih, P.C.; Lin, T.S.; Mou, C.Y. Enhancing stability and oxidation activity of cytochrome c by immobilization in the nanochannels of mesoporous aluminosilicates. *J. Phys. Chem. B* **2005**, *109*sRef] [PubMed]

47. Méndez, J.; Morales Cruz, M.; Delgado, Y.; Figueroa, C.M.; Orellano, E.A.; Morales, M.; Monteagudo, A.; Griebenow, K. Delivery of chemically glycosylated cytochrome c immobilized in mesoporous silica nanoparticles induces apoptosis in HeLa cancer cells. *Mol. Pharm.* **2014**sRef] [PubMed]

48. Méndez, J.; Monteagudo, A.; Griebenow, K. Stimulus-responsive controlled release system by covalent immobilization of an enzyme into mesoporous silica nanoparticles. *Bioconj. Chem.* **201**ssRef] [PubMed]

49. Slowing, I.I.; Trewyn, B.G.; Lin, V.S.Y. Mesoporous silica nanoparticles

for intracellular delivery of membrane-impermeable proteins. *J. Am. Chem. Soc.* **2007**, ssRef] [PubMed]

50. Park, H.S.; Kim, C.W.; Lee, H.J.; Choi, J.H.; Lee, S.G.; Yun, Y.P.; Kwon, I.C.; Lee, S.J.; Jeong, S.Y.; Lee, S.C. A mesoporous silica nanoparticle with charge-convertible pore walls for efficient intracellular protein delivery.*Nanotechn*ssRef] [PubMed]

51. Yamada, H.; Urata, C.; Ujiie, H.; Yamauchi, Y.; Kuroda, K. Preparation of aqueous colloidal mesostructured and mesoporous silica nanoparticles with controlled particle size in a very wide range from 20 nm to 700 nm. *Nanoscale***2013**ssRef] [PubMed]

52. Yu, M.; Zhou, L.; Zhang, J.; Yuan, P.; Thorn, P.; Gu, W.; Yu, C. A simple approach to prepare monodisperse mesoporous silica nanospheres with adjustable sizes. *J. Colloid Interface Sci.* **20**ssRef] [PubMed]

53. Cheng, S.H.; Liao, W.N.; Chen, L.M.; Lee, C.H. pH-controllable release using functionalized mesoporous silica nanoparticles as an oral drug delivery system. *J. Mater. Chem.* **2011**,lar] [CrossRef]

54. Lee, C.H.; Lo, L.W.; Mou, C.Y.; Yang, C.S. Synthesis and characterization of positive-charge functionalized mesoporous silica nanoparticles for oral drug delivery of an anti-inflammatory drug. *Adv. Funct. Mater.* **2008**,lar] [CrossRef]

55. Lee, C.-H.; Cheng, S.-H.; Huang, I.P.; Souris, J.S.; Yang, C.-S.; Mou, C.-Y.; Lo, L.-W. Intracellular pH-responsive mesoporous silica nanoparticles for the controlled release of anticancer chemotherapeutics. *Angew. Chem.* **2010**, lar] [CrossRef]

56. Kuthati, Y.; Kankala, R.K.; Lin, S.-X.; Weng, C.-F.; Lee, C.-H. pH-Triggered controllable release of silver-indole-3 acetic acid complexes from mesoporous silica nanoparticles (IBN-4) for effectively killing malignant bacteria. *Mol. Pharm.***2015**,ssRef] [PubMed]

57. Colilla, M.; Balas, F.; Manzano, M.; Vallet-Regí, M. Novel method to synthesize ordered mesoporous silica with high surface areas. *Solid State Sci.* **200**lar] [CrossRef]

58. Kuthati, Y.; Sung, P.J.; Weng, C.F.; Mou, C.Y.; Lee, C.H. Functionalization of mesoporous silica nanoparticles for targeting, biocompatibility, combined cancer therapies and theragnosis. *J. Nanosci. Nanotechnol.* **2013**,ssRef] [PubMed]

59. Takahashi, H.; Li, B.; Sasaki, T.; Miyazaki, C.; Kajino, T.; Inagaki, S. Catalytic activity in organic solvents and stability of immobilized enzymes depend on the pore size and surface characteristics of mesoporous silica. *Chem. Mater.* **2000**lar] [CrossRef]

60. Díaz, J.F.; Balkus, K.J., Jr. Enzyme immobilization in MCM-41 molecular sieve. *J. Mol. Catal. B* 1olar] [CrossRef]

61. Takahashi, H.; Li, B.; Sasaki, T.; Miyazaki, C.; Kajino, T.; Inagaki, S. Immobilized enzymes in ordered mesoporous silica materials and improvement of their stability and catalytic activity in an organic solvent. *Microporous Mesoporous Mater.* **2001**, olar] [CrossRef]

62. Lee, C.H.; Wong, S.T.; Lin, T.S.; Mou, C.Y. Characterization and biomimetic study of a hydroxo-bridged dinuclear phenanthroline cupric complex encapsulated in mesoporous silica: Models for catechol oxidase. *J. Phys. Chem. B* **20**ossRef] [PubMed]

63. Siefker, J.; Karande, P.; Coppens, M.O. Packaging biological cargoes in mesoporous materials: Opportunities for drug delivery. *Expert Opin. Drug Deliv.* **2014**ossRef] [PubMed]

64. Marina, M.-C.; Montserrat, C.; Maria, V.-R. Smart mesoporous nanomaterials for antitumour therapy. *Nanomaterials***201**[Google Scholar]

65. Kim, D.-S.; Jeon, S.-E.; Park, K.-C. Oxidation of indole-3-acetic acid by horseradish peroxidase induces apoptosis in G361 human melanoma cells. *Cell. Signal.* olar] [CrossRef]

66. Slowing, I.I.; Vivero-Escoto, J.L.; Wu, C.-W.; Lin, V.S.Y. Mesoporous silica nanoparticles as controlled release drug delivery and gene transfection carriers. *Adv. Drug Deliv. Rev.* **2008**ossRef] [PubMed]

67. Gu, J.; Huang, K.; Zhu, X.; Li, Y.; Wei, J.; Zhao, W.; Liu, C.; Shi, J. Sub-150 nm mesoporous silica nanoparticles with tunable pore sizes and well-ordered mesostructure for protein encapsulation. *J. Colloid Interface Sci.* **201**ossRef] [PubMed]

68. Schmalenberg, K.E.; Frauchiger, L.; Nikkhouy-Albers, L.; Uhrich, K.E. Cytotoxicity of a unimolecular polymeric micelle and its degradation products. *Biomacromolecules* **20**ossRef] [PubMed]

69. Slowing, I.; Trewyn, B.G.; Lin, V.S.Y. Effect of surface functionalization of MCM-41-type mesoporous silica nanoparticles on the endocytosis by human cancer cells. *J. Am. Chem. Soc.* **2006**, *1*ossRef] [PubMed]

70. Yu, K.; Grabinski, C.; Schrand, A.; Murdock, R.; Wang, W.; Gu, B.; Schlager, J.; Hussain, S. Toxicity of amorphous silica nanoparticles in mouse keratinocytes. *J. Nanopart. Res.* olar] [CrossRef]

71. Lu, F.; Wu, S.; Hung, Y.; Mou, C. Size effect on cell uptake in well-suspended, uniform mesoporous silica nanoparticles. *Small* **200**ossRef] [PubMed]

72. Smeller, L.; Meersman, F.; Fidy, J.; Heremans, K. High-pressure FTIR study of the stability of horseradish peroxidase. Effect of heme substitution, ligand binding, Ca^{++} removal, and reduction of the disulfide bonds. *Biochemistry* **20**ossRef] [PubMed]

73. Lison, D.; Thomassen, L.C.J.; Rabolli, V.; Gonzalez, L.; Napierska, D.; Seo, J.W.; Kirsch-Volders, M.; Hoet, P.; Kirschhock, C.E.A.; Martens, J.A. Nominal and effective dosimetry of silica nanoparticles in cytotoxicity assays.*Toxicol. Sci.* **200**ossRef] [PubMed]

74. Jeong, Y.M.; Oh, M.H.; Kim, S.Y.; Li, H.; Yun, H.Y.; Baek, K.J.; Kwon, N.S.; Kim, W.Y.; Kim, D.S. Indole-3-acetic acid/horseradish peroxidase induces apoptosis in TCCSUP human urinary bladder carcinoma cells. *Pharmazie* **2**cholar] [PubMed]

75. Han, Y.; Ying, J.Y. Generalized fluorocarbon-surfactant-mediated synthesis of nanoparticles with various mesoporous structures. *Angew. Chem. Int. Ed.* **20**ossRef] [PubMed]

76. Yang, R.; Gao, D.; Huang, H.; Huang, B.; Cai, H. Mesoporous silicas prepared by ammonium perchlorate oxidation and theirs application in the selective adsorption of high explosives. *Microporous Mesoporous Mater.* **2**olar] [CrossRef]

77. Taylor, I.; Howard, A.G. Measurement of primary amine groups on surface-modified silica and their role in metal binding. *Anal. Chim. Acta* **1**olar] [CrossRef]

78. Jung, D.; Streb, C.; Hartmann, M. Covalent anchoring of chloroperoxidase and glucose oxidase on the mesoporous molecular sieve SBA-15. *Int. J. Mol. Sci.* **20**ossRef] [PubMed]

79. Kim, M.I.; Kim, J.; Lee, J.; Jia, H.; Na, H.B.; Youn, J.K.; Kwak, J.H.; Dohnalkova, A.; Grate, J.W.; Wang, P.; *et al.* Crosslinked enzyme aggregates in hierarchically-ordered mesoporous silica: A simple and effective method for enzyme stabilization. *Biotechnol. Bioeng.* **2007**, *96*, 210–218.

Chapter 7

CONSTRUCTING BIOPOLYMER-INORGANIC NANOCOMPOSITE THROUGH A BIOMIMETIC MINERALIZATION PROCESS FOR ENZYME IMMOBILIZATION

Jian Li[1], Jun Ma[1], Tao Jiang[1], Yanhuan Wang[1], Xuemei Wen[2] and Guozhu Li[3]

[1]School of Material Science and Chemical Engineering, Tianjin University of Science and Technology, Tianjin 300457, China

[2]Tianjin Synthetic Material Research Institute, Tianjin 300220, China

[3]Key Laboratory for Green Chemical Technology of Ministry of Education, School of Chemical Engineering and Technology, Tianjin University, Tianjin 300072, China

ABSTRACT

Inspired by biosilicification, biomimetic polymer-silica nanocomposite has aroused a lot of interest from the viewpoints of both scientific research and technological applications. In this study, a novel dual functional polymer, NH_2-Alginate, is synthesized through an oxidation-amination-reduction process. The "catalysis function" ensures the as-prepared NH_2-Alginate inducing biomimetic mineralization of silica from low concentration precursor (Na_2SiO_3), and the "template function" cause microscopic phase separation in aqueous solution. The diameter of resultant NH_2-Alginate micelles in aqueous solution distributed from 100 nm to 1.5 μm, and is influenced by the synthetic process of NH_2-Alginate. The size and morphology of obtained NH_2-Alginate/silica nanocomposite are correlated with the micelles. NH_2-Alginate/silica nanocomposite was subsequently utilized to immobilize β-Glucuronidase (GUS). The harsh condition tolerance and long-term storage stability of the immobilized GUS are notably improved due to the buffering effect of NH_2-Alginate and cage effect of silica matrix.

INTRODUCTION

Recent years, polymer-silica nanocomposite have aroused a lot of interest from the viewpoints of both scientific research and technological applications [1,2,3]. By conventional chemical methods, the preparation of silica-based materials often involved extreme temperature, pressure, and pH [4]. In contrast, biosilicification processes in nature can overcome these disadvantages [5,6]. Silicatein [7] and silaffin [8] identified in marine sponges and diatoms, respectively, can *in vitro* induce the formation of silica from precursor under ambient conditions. Therefore, various synthetic polymers (polypeptides, polyamines) as well as biopolymers (proteins, polysaccharides) have been used as inducers to mimic the biosilicification processes occurring in living organisms [9,10,11,12]. The obtained polymer-silica nanocomposite is widely used in areas of bioreactors, biosensors, and bio-deliveries [12,13,14].

To mimic the biosilicification processes occurring in aqueous solution, the polymers should provide two functions in general. One is catalysis function, and the other is template function [15]. "Catalysis function" means that the polymer can accelerate the precipitation of silica from a precursor with low concentration. For either synthetic polymers or biopolymers, it has been proved that the catalysis function is related to their charge at neutral pH [10,16]. Previous research showed that proteins with isoelectric point (pI) > 7.0 (lysozyme, protamine, *etc.*) can induce the precipitation of silica. However, no precipitation was observed for those proteins with pI < 7.0. The total charge and cationic residues ($-NH_2$, $-SH$) of the basic proteins play important roles for accelerating silica precipitation [17].

Template function is related to "microscopic phase separation" [18], one kind of self-assembly process which is generated by polymers when they contact with precursor [19]. Poulsen [20] proved that long-chain polyamines (LCPA) could undergo microscopic phase separation to form a microemulsion template, which significantly influenced the microstructure of polymer-silica nanocomposite. However, positively charged LCPA was incapable of precipitating silica individually unless polyanions or multivalent anions were added [20]. Brunner [21] explained the aggregation and phase separation of LCPA were formed by electrostatic interactions between the positively charged LCPA molecules and the negatively charged phosphate ions. Other multivalent anions [21,22] such as pyrophosphate, sulfate, or polyanion natSil-2 [20] were also capable of inducing the precipitation of silica from silicate-containing polyamine solutions.

In nature, few biopolymers but natSil-1A [23], which is isolated from diatoms, can serve above dual functions at the same time. Kroger [20] reported that natSil-1A could form the microscopic phase separation as template and induce silica-precipitating as catalyst, even in the absence of polyanions or multivalent anions. It can be explained that all the serine residues are phosphorylated and all the lysine residues are either methylated or covalently linked with polyamines. This structure ensures a highly zwitterionic polypeptide with the capability for self-assembly, which leads the microscopic phase separation driven by ionic interactions [23,24].

Inspired bynatSil-1A, we synthesized a novel dual functional polymer with zwitterionic structure. Specifically, alginate was chosen as the bulk polymer due to its long chain and abundant carboxyl group. Spermine was grafted onto alginate molecules through the oxidation-amination-reduction to obtain an aminated alginate (NH_2-Alginate). Therefore, both the catalyzing function and templating function are incorporated into one polymer. Herein, sodium silicate is used as the precursor and mixed with NH_2-Alginate. The feasibility of the nanocomposite for enzyme immobilization is demonstrated by using β-Glucuronidase (GUS) as the model enzyme.

RESULTS AND DISCUSSION

The Catalyzing Function of NH_2-Alginate

XPS spectra of NH_2-Alginate and NH_2-Alginate/silica nanocomposite are presented in Figure 1. A significant increase in nitrogen content demonstrates that the amine groups are successfully grafted onto the backbone of alginate. For the NH_2-Alginate/silica, an intensive peak of element Si appears and demonstrates the successful biomimetic mineralization of silica.

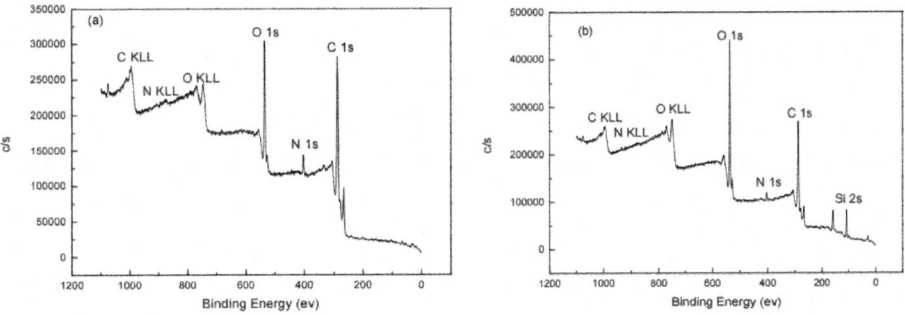

Figure 1: XPS spectra of NH_2-Alginate (**a**) and NH_2-Alginate/silica (**b**).

To investigate the catalysis function of NH_2-Alginate in silica precipitating, as shown in Figure 2a, the effect of pH value on the amount of NH_2-Alginate/silica nanocomposite products is determined. It was reported that cationic proteins can attract anionic inorganic precursors through electrostatic and hydrogen bonding interactions, and then promoted hydrolysis and condensation of the precursors [16]. Similarly, NH$_2$-Alginateis regarded as an acid-base catalyst with protonated amine group, which can induced the precursor hydrolysis and condensation. However, the amount of NH$_2$-Alginate/silica nanocomposite sharply decreases if pH value is greater than 7.0. When pH value reached 9.0, little precipitate is observed due to decreasing positive charge on the surface of NH$_2$-Alginate, which consequently weakens its electrostatic interaction with the precursor and then inhibits the catalysis function of NH$_2$-Alginatein alkaline conditions. ^{29}Si NMR is conducted to analyze the chemical structure and condensation degree of the mineralized silica catalyzed by NH$_2$-Alginate. As shown inFigure 2b, the two main peaks near -100 ppm and -110.5 ppm are attributed to Q^3 [Si(OSi)$_3$(OH)] and Q^4 [Si(OSi)$_4$] with relative percentages of 30.2% and 69.8% respectively, which suggests the formation of well-condensed silica under the catalysis of NH$_2$-Alginate. These results are in good agreement with the previous report that the mineralized silica using amine-containing polymers as catalysis and template exhibited a Q^4/Q^3 ratio from 1.0 to 2.5 [25,26,27,28,29].

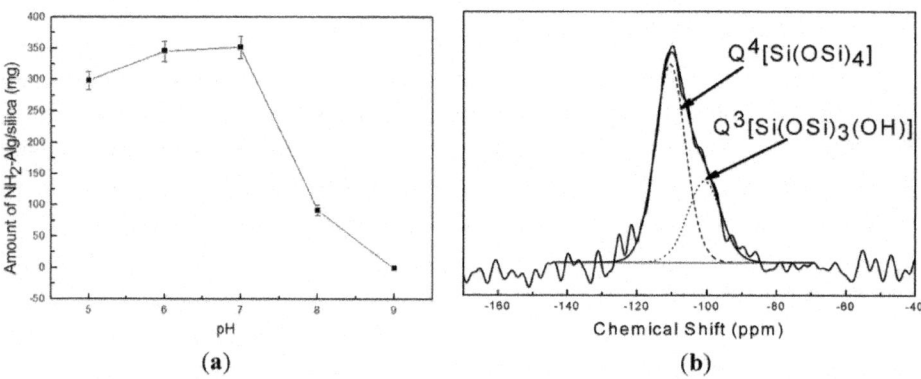

Figure 2: (a) The amount of NH$_2$-Alginate/silica as a function of pH value; (b) NMR spectra of NH$_2$-Alginate/silica.

The Template Function of NH$_2$-Alginate

The average diameters of different NH$_2$-Alginate micelles are characterized by DLS and the results are summarized inTable 1. Changing the ratio of [IO$_4^-$]/ [alginate unit] to 1.0, the diameter decreases with the increasing molecular weight (M_w) of amine agent, and the micelles diameter of 1#, 2#, 3# NH$_2$-Alginate are 216.1 nm (1,2-ethylenediamine), 173.6 nm (Diethylenetriamine) and 145.0 nm (Spermine) respectively. A similar trend is also observed in lower ratio of 0.25 (4#, 5#, 6#) or 0.1 (7#, 8#, 9#). During the conjugation process, the number of amine group kept constant, consequentely, the molar concentration of amine agent with high M_w is lower. Therefore, the conjugating point on the backbone of alginate is less, which results in nonhomogeneous distribution of amine and forms small micelle template. When using the same amine agent, the micelles diameters decrease with the increasing of $[IO_4^-]/[alginate\ unit]$ ratio (7# > 4# > 1#). Previous results showed that excess potassium periodate might induce further oxidation and break the main chain of alginate [30,31]. Therefore, NH_2-Alginate with lower M_w leads to smaller micelles diameters.

The morphology of nanocomposite is presented in Figure 3. The size of resultant 3# (Figure 3a), 6# (Figure 3b), 8# (Figure 3c), 9# (Figure 3d) NH_2-Alginate/silica nanocomposite is nearly 150 nm, 300 nm, 1000 nm, and 500 nm respectively, which is to the diameter of their related NH_2-Alginate micelles. Previous researches showed that the silica-precipitating polymers could pre-self-assemble into different "shapes" or exhibit different sizes in solution, and then be used as templates to direct the final silica morphology [19,32]. This research shows that the micelles could be formed via microscopic phase separation and then displayed the template function for producing silica-based nanocomposite. The higher oxidation ratio and dense amine group may lead to small diameters of NH_2-Alginate micelles and resultant nanocomposite (3#). In contrast, large Mw and long chain polymer could self-assemble and induce hexagonal [33] or square nanocomposite (8#). As 9# NH_2-Alginate, nonhomogeneous distribution of amine may lead amorphous nanocomposite product, and the TEM images show that 9# NH_2-Alginate/silica nanocomposite consists of nanoparticles with a diameter of 50 nm.

Table 1: The average diameter of different NH_2-Alginate micelles

No.	$[IO_4^-]/[alginate\ unit]$	Amination Agent	Diameter (nm)
1#	1	1,2-ethylenediamine	216.1
2#	1	Diethylenetriamine	173.6
3#	1	Spermine	145.0
4#	0.25	1,2-ethylenediamine	404.9
5#	0.25	Diethylenetriamine	344.5
6#	0.25	Spermine	331.9
7#	0.1	1,2-ethylenediamine	1535
8#	0.1	Diethylenetriamine	1160
9#	0.1	Spermine	598.7

It had been proved that multivalent anions were widely used for leading "microscopic phase separation", while protein or polyamine was incapable of precipitating silica if replacing the multivalent ions with monovalent ions [21]. In this study, the effect of Cl⁻ on the formation of NH_2-Alginate/silica nanocomposite is determined. About 350 mg of nanocomposite is obtained in deionized water, and nearly 200 mg is obtained upon preparing in NaCl solution. The zwitterionic structure of NH_2-Alginate (polyamine moieties and carboxy groups) causes the self-assembly process via electrostatic interactions even in NaCl solution. However, the amine groups of NH_2-Alginateare attracted by Cl^- ions, consequently, the electrostatic interaction is partly shielded and the microscopic phase separation is hindered. Therefore, the amount of NH_2-Alginate/silica nanocomposite is decreased to some extent.

Enzyme Activity Assay

Because excess potassium periodate might induce further oxidation and break the main chain of alginate in the synthesis of NH_2-Alginate [30,31], it might induce loosing networks and a high release rate of enzyme if NH_2-alginate individually formed hydrogel and encapsulated GUS. However, when it was used in biomimetic mineralization, the resultant silica presented dense network and confinement effect for the enzyme. In every enzymatic conversion process we measured the releasing GUS in the solution by the micro-Bradford method, but there was no GUS release detected. It is a significant advantage of biopolymer-inorganic nanocomposite.

The enzymatic conversion of baicalin follows the Michaelis-Menten kinetics (Figure 4). The corresponding Michaelis constant (K_m) and the maximum reaction rate (V_{max}) are calculated according to the Lineweaver-Burk plots and presented in Table 2. Michaelis constant (K_m) is an indicator for evaluating the binding ability between the substrate and enzymes, and usually increases after immobilization. However, in this study, the K_m for the immobilized GUS is increased slightly, indicating a well-preserved binding ability between GUS and the substrates [34]. The mesopores in the nanocomposite ensure the free motion of enzyme molecules, and the sufficient water in the nanocomposite enables enzyme molecules to catalyze in a nature-like microenvironment. While the maximum reaction rate (V_{max}) decreases due to the increased diffusion resistance for substrate/product molecules [35].

The storage stabilities of free and immobilized GUS are compared in Figure 5. It should be noticed that the storage stability was obviously enhanced after immobilization. During the first six days, the activity of immobilized enzyme decreased synchronously with the free one (from 100% to ~75%). From 7th to 22nd day, the relative activity of free GUS decreased to 7%, while 65% activity of GUS immobilized in NH_2-Alginate/silica nanocomposite was retained. This result is tentatively explained by the unfolding process, which is the main reason for enzyme deactivation during storage. It is supposed that in the initial storage time NH_2-Alginate/silica nanocomposite may allow the unfolding to some extent like it happened in solution, since GUS is immobilized in a nature-like microenvironment.

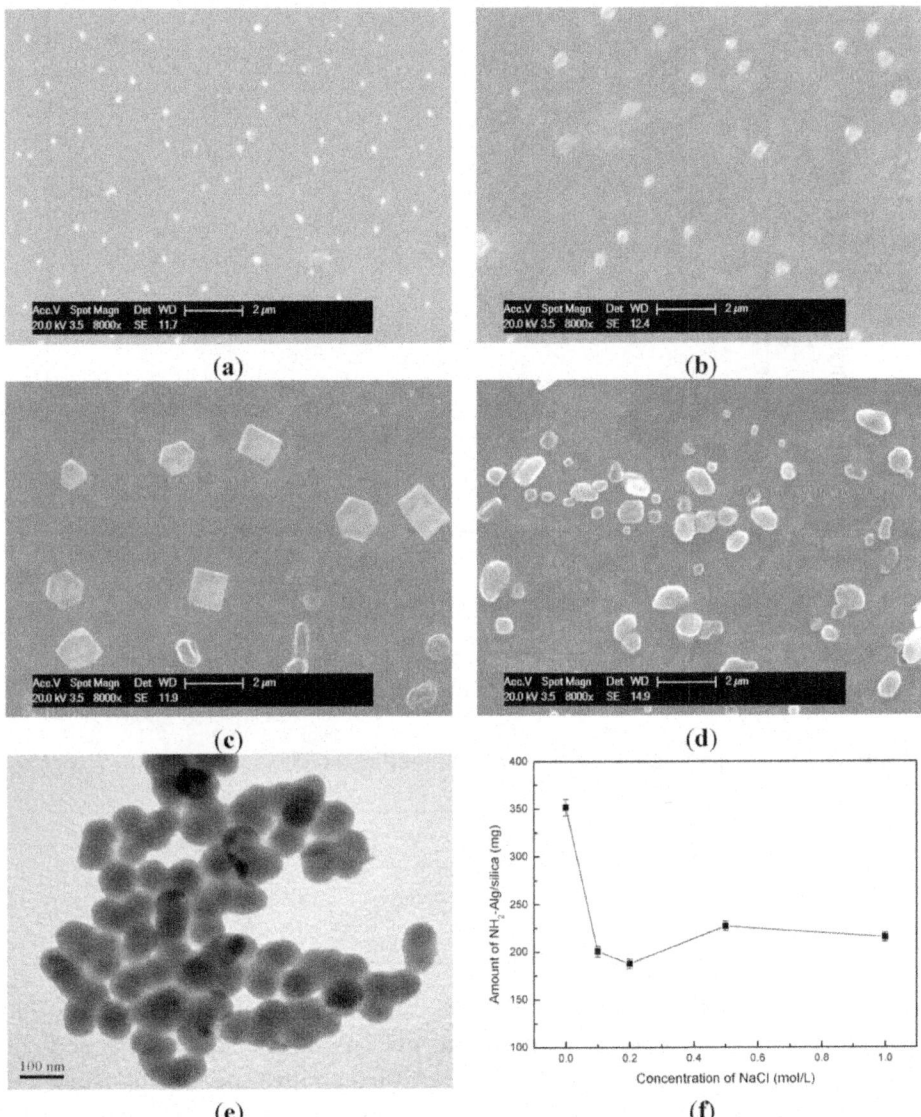

Figure 3: (**a–d**) SEM image of NH$_2$-Alginate/silica; (**e**) TEM image of NH$_2$-Alginate/silica; (**f**) the amount of NH$_2$-Alginate/silica as a function of NaCl concentration.

However, after a certain period of storage (six days in this case), the further unfolding of immobilized enzymes will be inhibited by the confinement of mesopores [35,36]. As a consequence, the decreased rate of activity for immobilized GUS become lower than its free form.

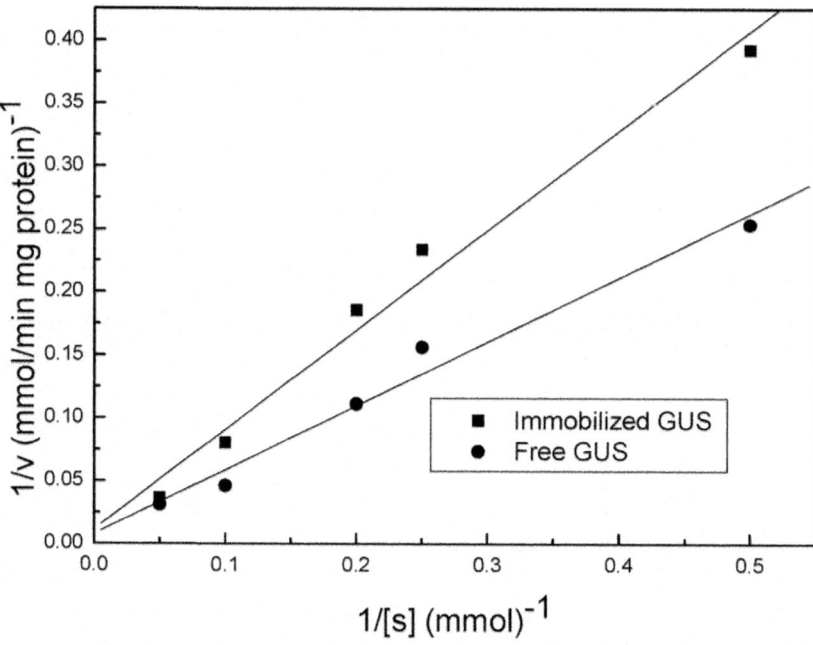

Figure 4: Typical lineweaver-Burk plots for free and immobilized GUS.

Table 2: Kinetic parameters for free and immobilized GUS

GUS	K_m (mM)	V_{max} (μmol/min· mg GUS)
free	67.52	135.29
immobilized	69.33	91.57

Figure 6 shows the relative activity for free and immobilized GUS with the variation of pH values. The optimal pH value remained unchanged (pH 7.0) after immobilization in the NH_2-Alginate/silica nanocomposite. The immobilized GUS keeps higher relative activity than its free form under extreme acidic conditions till pH 3.0 (85% *vs.* 60%) and alkaline conditions till pH 9.0 (40% *vs.* 0%), respectively. In general, the changes in pH would affect the charges carried by different amino acid residues of protein. Enzymes will undergo reversible or irreversible conformational changes and lose their activity under extreme pH conditions.

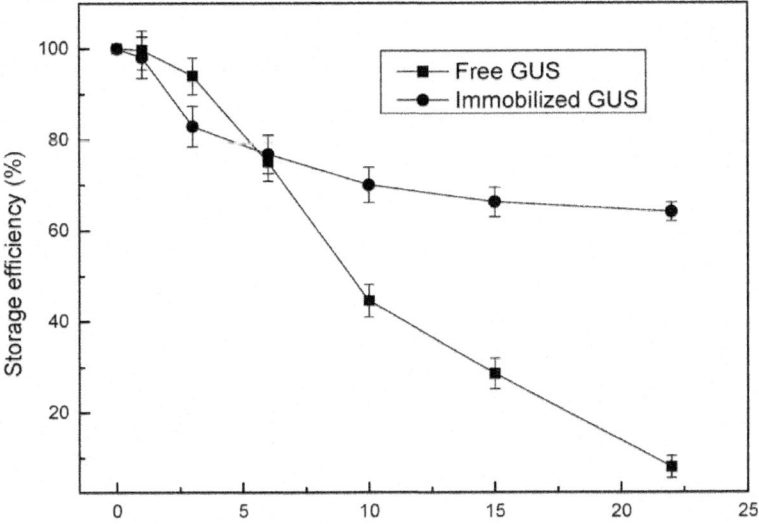

Figure 5: Storage stability of free and immobilized GUS.

Two kinds of functional groups, $-NH_2$ and $-COOH$ are distributed along the backbone. The functional groups can form abundant $-COO^-/-COOH$, $=NH/=NH_2^+$ and $-NH_2/-NH_3^+$ pairs, so it can tune the local pH value to some extent in case the bulk pH changed (the so-called buffering effect). In acidic medium, $=NH$, $-NH_2$, and $-COO^-$ can attract and consume the H^+ ions, thus preventing H^+ from diffusing into the inner polymer matrix and contacting the enzymes. On the contrary, in alkaline medium, $=NH_2^+$, $-NH_3^+$, and $-COOH$ will release H^+ ions, regulating the content of H^+ inside the NH_2-Alginate/silica nanocomposite. Thus, the zwitterionic structure of polymer provides above properties called "buffering effect", which was also presented in our previous research about another type of NH_2-alginate [37]. By this way, the pH change in the microenvironment is envisaged to be smaller than that happening in the bulk solution [38,39]. Additionally, several studies also had reported that the immobilized enzymes exhibited higher relative activity than their free form in extreme acidic conditions and alkaline conditions, and mainly attributed that to "confinement effect" of enzymes in mesoporous silica, which leads to the restricted conformational changes under extreme pH conditions [40,41,42]. Therefore, the above two effects act synergistically so that the extreme pH tolerance was enhanced.

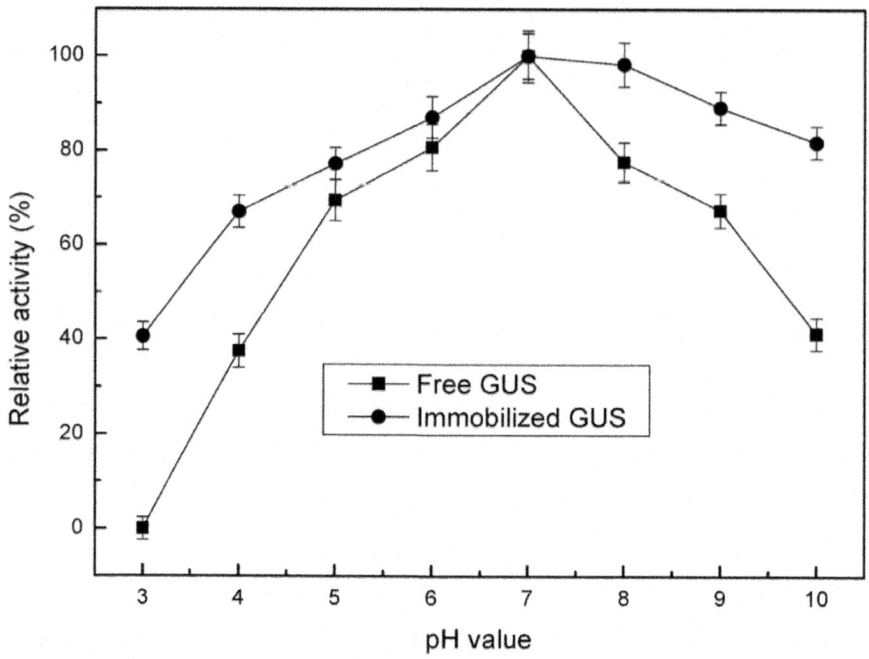

Figure 6: Effect of pH value on the activity of free and immobilized GUS.

EXPERIMENTAL SECTION

Materials

GUS (EC 3.2.1.31) from *Escherichia coli* (type IX-A, lyophilized powder, 1,000,000–5,000,000 units/g protein) and Sodium alginate were obtained from Sigma Chemical Co. (St. Louis, MO, USA). Spermine, sodium periodate and sodium borohydride were obtained from Fluka Chemie (Buchs, Switzerland). All of other solvents and reagents were of analytical grade.

Synthesis of NH$_2$-Alginate

Preparing NH$_2$-alginate is similar with the process of dextran bioconjugation, which was described in previous research by Domb's group [43]. The scheme is presented in Figure 7.

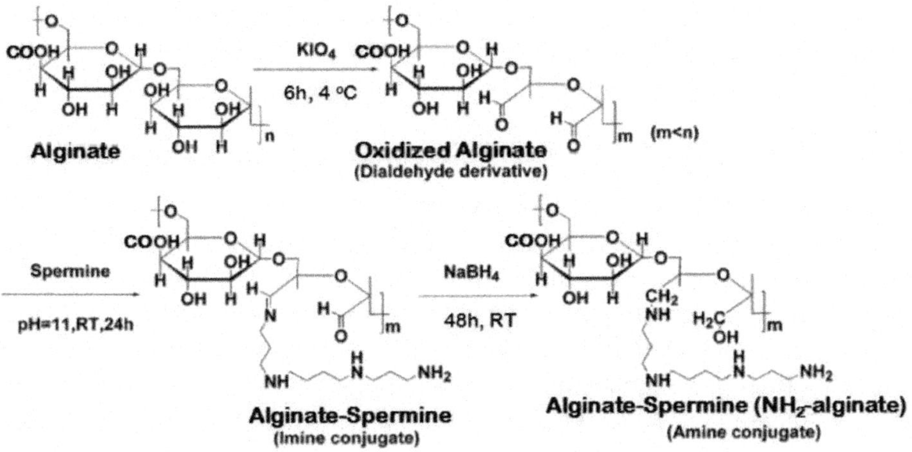

Figure 7: Illustration of NH_2-Alginate synthesis process.

Oxidation

Alginate (10 g, 50.5 mmol of units) was dissolved in 200 mL of double deionized water (DDW). Potassium periodate at a mole ratio (IO_4^-/alginate units) from 0.1 to 1.0 was separately added to this solution and the mixture was vigorously stirred in the dark at 4 °C until a clear yellow solution was obtained (12 h). The unreacted periodate (IO_4^-) was removed by adding ethylene glycol, followed by extensive dialysis against DDW (12,000 cut-off cellulose tubing) for 2 days and at 4 °C. Purified oxidized alginate were freeze-dried to obtain a white powder.

Amination

Amination agent containing 63.125 mmol amine group was dissolved in 50 mL of borate buffer (0.1 M, pH 11). A solution of oxidized alginate in 100 mL of DDW was slowly added during 5 h (sage metering pump) into the aminnation agent solution. The mixture was gently stirred at room temperature for 24 h and dialyzed against DDW at 4 °C, applying 3500 cutoff cellulose tubing (Membrane Filtration Products, Inc., San Antonio, TX, USA). The imine-based conjugate was obtained by following lyophilization.

Reduction

The amine based conjugates (reduced) were obtained after reducing the imine conjugates with excess 1 g $NaBH_4$ in water at room temperature for 24 h. The reduction was repeated with additional portion of $NaBH_4$ (1 g) and stirring for 24 h at the same conditions. Then the resulting light-yellow solution was dialyzed against DDW at 4 °C for 2 days followed by freeze-drying to obtain amined alginate (NH_2-Alginate).

Preparation of NH_2-Alginate/Silica Nanocomposite and Encapsulation of GUS

Nanocomposite preparation: NH_2-Alginate was suspended in DDW or NaCl solution (10 mg/mL). A freshly prepared 100 mmol/L sodium silicate solution was obtained by dissolving sodium silicate in water followed by acidification to a specific pH with HCl. The NH_2-Alginate solution (20 mL) was then mixed with 30 mL sodium silicate solution, and allowed to react for 10 min. The resultant precipitates were collected by centrifugation, rinsed twice with deionized water to remove unreacted agent, and lyophilized to dryness.

GUS encapsulation: The GUS could be dissolved in NH_2-Alginate solution and immobilized in a co-pericipitating process. A mixture of NH_2-Alginate (10 mg/mL) and GUS (0.05 mg/mL) was prepared by dissolving them in Tris-HCl buffer solution (pH 7.0, 30 mmol/L). Freshly prepared 100 mmol/L sodium silicate solution was obtained by dissolving sodium silicate in water followed by anacidification to pH 7.0 with HCl. 20 mL of NH_2-Alginate/GUS solution was then added to 30 mL of sodium silicate solution. After 10 minutes, the resultant precipitates were centrifuged and rinsed twice with deionized water to remove residual GUS, NH_2-Alginate and silicate. The supernatant was collected and the GUS content was determined by the Bradford method using a UV spectrophotometer (U-2800, Hitachi, Tokyo, Japan). The encapsulation efficiency was calculated according to Equation (1):

$$\text{Encapsulation efficiency (\%)} = \left(1 - \frac{C_1 V_1}{C_0 V_0}\right) \times 100$$

(1)

where C_0 (mg·mL^{-1}) and V_0 (mL) were the concentration and volume of introduced GUS solution, respectively; $C_0 V_0$ (mg) was the introduced GUS amount in the immobilization medium; C_1 (mg·mL^{-1}) and V_1 (mL) were the GUS concentration and supernatant volume when preparing GUS-containing nanocomposite, respectively; $C_1 V_1$ (mg) was the amount of GUS leaked out during the preparation process of GUS-containing nanocomposite.

Characterizations

The surface properties of microcapsules were characterized by X-ray photoelectron spectroscopy (XPS) in a Perkin-Elmer PHI 1600 ESCA system with a monochromatic Mg $K\alpha$ source and a charge neutralizer.

Solid-state ^{29}Si MAS NMR spectra of the NH$_2$-Alginate/silica nanocomposite were recorded on an Infinity Plus-300 MHz spectrometer (Varian, CA, USA) with resonance frequencies of 59.63 MHz for ^{29}Si. The magnetic field was 7.05 T, and the spin rate of the sample was spun at 3 kHz.

For Scanning electron microscopy (SEM) analysis, samples were prepared by applying a drop of the particle suspension to a glass slide and then drying overnight. After that, the samples were sputtered with gold. Measurements were conducted using Philips XL30 ESEM and Hitachi S-4800 instrument (Hitachi, Tokyo, Japan) at an operation voltage of 20.0 keV and 0.7 keV.

Transmission electron microscope (TEM) observation was performed on a JEM-100CXII instrument (JEOL, Tokyo, Japan).

The average size of NH$_2$-Alginate micelles was measured by BrookhavenInstruments BI200SM dynamic light scattering (DLS) system. All samples was dispersed in 0.05 M Tris-HCl buffer solution (pH 7.0).

Enzyme Activity Assay

Bioconversion of baicalin to baicalein is catalyzed by GUS, which had been described in our previous research [44]. Free or immobilized GUS was introduced into a beaker containing 20 mL of 0.09 mM baicalin and 0.1% w/v Na$_2$SO$_3$, both dissolved in 30 mM Tris-HCl buffer. The beaker was sealed and the reaction was performed under stirring. After 60 min, 100 µL of the reacting solution was sampled and analyzed by HPLC (HP1100, Agilent, CA, USA). The enzyme activities and stabilities were studied and compared by measuring the amount of baicalein produced, and each data point was replicated three times. The optimum conditions for GUS activity were determined by testing the enzyme activity under a range of pH values (3–10). The activity of GUS was calculated based on the amount of baicalein produced and was expressed as relative activity compared with the activity at the optimum pH.

GUS immobilized in NH$_2$-Alginate/silica nanocomposite was stored at 4 °C for a certain period of time. The storage stability was compared in terms of storage efficiency which was defined as the ratio of immobilized enzyme activity after storage to their initial activity.

ACKNOWLEDGMENTS

The authors are thankful for the financial support from the National Natural Science Foundation of China (No. 21306139, No. 21306132), Natural Science Foundation of Tianjin (No. 12JCQNJC06000), and Lab Opening Foundation of Tianjin University of Science & Technology (No. 1103A207).

AUTHOR CONTRIBUTIONS

Jian Li performed the experiments and analysis, and contributed to a main part of manuscript writing. Jian Li, Jun Ma and Tao Jiang are equally contributed in conceiving and designing the original research. Yanhuan Wang, Xuemei Wen and Guozhu Li analyzed filtration testing, conducted data extraction, and commented on manuscript writing.

REFERENCES

1.   Lee, J.E.; Lee, N.; Kim, T.; Kim, J.; Hyeon, T. Multifunctional mesoporous silica nanocomposite nanoparticles for theranostic applications. *Acc. Chem. Res.* **2011**, *44*, 893–902.

2.   Ciriminna, R.; Fidalgo, A.; Pandarus, V.; Beland, F.; Ilharco, L.M.; Pagliaro, M. The sol-gel route to advanced silica-based materials and recent applications. *Chem. Rev.* **2013**, *113*, 6592–6620.

3.   Ab Rahman, I.; Padavettan, V. Synthesis of silica nanoparticles by sol-gel: Size-dependent properties, surface modification, and applications in silica-polymer nanocomposites—A review. *J. Nanomater.* **2012**, *2012*.

4.   Bounor-Legare, V.; Cassagnau, P. *In situ* synthesis of organic-inorganic hybrids or nanocomposites from sol-gel chemistry in molten polymers. *Prog. Polym. Sci.* **2014**, *39*, 1473–1497.

5.   Ruiz-Hitzky, E.; Darder, M.; Aranda, P.; Ariga, K. Advances in biomimetic and nanostructured biohybrid materials.*Adv. Mater.* **2010**, *22*, 323–336.

6.   Wang, S.J.; Cai, Q.W.; Du, M.X.; Cao, M.W.; Xu, H. Biomimetic mineralization of silica. *Prog. Chem.* **2015**, *27*, 229–241.

7.   Cha, J.N.; Shimizu, K.; Zhou, Y.; Christiansen, S.C.; Chmelka, B.F.; Stucky, G.D.; Morse, D.E. Silicatein filaments and subunits from a marine sponge direct the polymerization of silica and silicones *in vitro. Proc. Natl. Acad. Sci. USA***1999**, *96*, 361–365.

8.   Kroger, N.; Deutzmann, R.; Sumper, M. Polycationic peptides from diatom biosilica that direct silica nanosphere formation. *Science* **1999**, *286*, 1129–1132.

9.   Liu, X.L.; Zhu, P.X.; Gao, Y.F.; Jin, R.H. Polyamine-promoted growth of one-dimensional nanostructure-based silica and its feature in catalyst design. *Materials* **2012**, *5*, 1787–1799.

10.   Shiomi, T.; Tsunoda, T.; Kawai, A.; Mizukami, F.; Sakaguchi, K. Biomimetic synthesis of lysozyme-silica hybrid hollow particles using sonochemical treatment: Influence of pH and lysozyme concentration on morphology. *Chem. Mater.***2007**, *19*, 4486–4493.

11. Leng, B.X.; Chen, X.; Shao, Z.Z.; Ming, W.H. Biomimetic synthesis of silica with chitosan-mediated morphology. *Small*2008, *4*, 755–758.

12. Kawachi, Y.; Kugimiya, S.; Nakamura, H.; Kato, K. Enzyme encapsulation in silica gel prepared by polylysine and its catalytic activity. *Appl. Surf. Sci.* **2014**, *314*, 64–70.

13. Dolatabadi, J.E.N.; de la Guardia, M. Applications of diatoms and silica nanotechnology in biosensing, drug and gene delivery, and formation of complex metal nanostructures. *Trac. Trend. Anal. Chem.* **2011**, *30*, 1538–1548.

14. Patwardhan, S.V. Biomimetic and bioinspired silica: Recent developments and applications. *Chem. Commun.* **2011**, *47*, 7567–7582.

15. Wu, H.; Li, J.; Li, L.; Jiang, Y.; Jiang, Y.; Jiang, Z. Protamine-templated biomimetic hybrid capsules: Efficient and stable carrier for enzyme encapsulation. *Chem. Mater.* **2007**, *20*, 1041–1048.

16. Jiang, Y.; Yang, D.; Zhang, L.; Sun, Q.; Zhang, Y.; Li, J.; Jiang, Z. Biomimetic synthesis of titania nanoparticles induced by protamine. *Dalton Trans.* **2008**, *31*, 4165–4171.

17. Roth, K.M.; Zhou, Y.; Yang, W.J.; Morse, D.E. Bifunctional small molecules are biomimetic catalysts for silica synthesis at neutral pH. *J. Am. Chem. Soc.* **2005**, *127*, 325–330.

18. Sumper, M. A phase separation model for the nanopatterning of diatom biosilica. *Science* **2002**, *295*, 2430–2433.

19. Fernandes, F.M.; Coradin, T.; Aime, C. Self-assembly in biosilicification and biotemplated silica materials.*Nanomaterials* **2014**, *4*, 792–812.

20. Poulsen, N.; Sumper, M.; Kroger, N. Biosilica formation in diatoms: Characterization of native silaffin-2 and its role in silica morphogenesis. *Proc. Natl. Acad. Sci. USA* **2003**, *100*, 12075–12080.

21. Brunner, E.; Lutz, K.; Sumper, M. Biomimetic synthesis of silica nanospheres depends on the aggregation and phase separation of polyamines in aqueous solution. *Phys. Chem. Chem. Phys.* **2004**, *6*, 854–857.

22. Lutz, K.; Groger, C.; Sumper, M.; Brunner, E. Biomimetic silica formation: Analysis of the phosphate-induced self-assembly of polyamines. *Phys. Chem. Chem. Phys.* **2005**, *7*, 2812–2815.

23. Kroger, N.; Deutzmann, R.; Sumper, M. Silica-precipitating peptides from diatoms-the chemical structure of silaffin-1a from cylindrotheca fusiformis. *J. Biol. Chem.* **2001**, *276*, 26066–26070.

24. Kroger, N.; Lorenz, S.; Brunner, E.; Sumper, M. Self-assembly of highly phosphorylated silaffins and their function in biosilica morphogenesis. *Science* **2002**, *298*, 584–586.

25. Groger, C.; Lutz, K.; Brunner, E. NMR studies of biomineralisation. *Prog. Nucl. Magn. Reson. Spectrosc.* **2009**, *54*, 54–68.

26. Yuan, J.J.; Jin, R.H. Temporally and spatially controlled silicification for self-generating polymer@silica hybrid nanotube on substrates with tunable film nanostructure. *J. Mater. Chem.* **2012**, *22*, 5080–5088.

27. Yuan, J.J.; Zhu, P.X.; Fukazawa, N.; Jin, R.H. Synthesis of nanofiber-based silica networks mediated by organized poly(ethylene imine): Structure, properties, and mechanism. *Adv. Funct. Mater.* **2006**, *16*, 2205–2212.

28. Shiu, C.C.; Wang, S.A.; Chang, C.H.; Jan, J.S. Poly(L-glutamic acid)-decorated hybrid colloidal particles from complex particle-templated silica mineralization. *J. Phys. Chem. B* **2013**, *117*, 10007–10016.

29. Hu, J.J.; Hsieh, Y.H.; Jan, J.S. Polyelectrolyte complex-silica hybrid colloidal particles decorated with different polyelectrolytes. *J. Colloid Interface Sci.* **2015**, *438*, 94–101.

30. Kong, H.J.; Kaigler, D.; Kim, K.; Mooney, D.J. Controlling rigidity and degradation of alginate hydrogels via molecular weight distribution. *Biomacromolecules* **2004**, *5*, 1720–1727.

31. Boontheekul, T.; Kong, H.J.; Mooney, D.J. Controlling alginate gel degradation utilizing partial oxidation and bimodal molecular weight distribution. *Biomaterials* **2005**, *26*, 2455–2465.

32. Patwardhan, S.V.; Clarson, S.J.; Perry, C.C. On the role(s) of additives in bioinspired silicification. *Chem. Commun.***2005**, *9*, 1113–1121.

33. Tomczak, M.M.; Glawe, D.D.; Drummy, L.F.; Lawrence, C.G.; Stone, M.O.; Perry, C.C.; Pochan, D.J.; Deming, T.J.; Naik, R.R. Polypeptide-templated synthesis of hexagonal silica platelets. *J. Am. Chem. Soc.* **2005**, *127*, 12577–12582.

34. Chen, L.; Wei, B.; Zhang, X.T.; Li, C. Bifunctional graphene/gamma-Fe_2O_3 hybrid aerogels with double nanocrystalline networks for enzyme immobilization. *Small* **2013**, *9*, 2331–2340.

35. Li, J.; Jiang, Z.; Wu, H.; Long, L.; Jiang, Y.; Zhang, L. Improving the recycling and storage stability of enzyme by encapsulation in mesoporous $CaCO_3$-alginate composite gel. *Compos. Sci. Technol.* **2009**, *69*, 539–544.

36. Ravindra, R.; Shuang, Z.; Gies, H.; Winter, R. Protein encapsulation in mesoporous silicate: The effects of confinement on protein stability, hydration, and volumetric properties. *J. Am. Chem. Soc.* **2004**, *126*, 12224–12225.

37. Li, J.; Wu, H.; Liang, Y.; Jiang, Z.; Jiang, Y.; Zhang, L. Facile fabrication of organic-inorganic hybrid beads by aminated alginate enabled gelation and biomimetic mineralization. *J. Biomater. Sci. Polym. Ed.* **2013**, *24*, 119–134.

38. Song, X.K.; Wu, H.; Shi, J.F.; Wang, X.L.; Zhang, W.Y.; Ai, Q.H.; Jiang, Z.Y. Facile fabrication of organic-inorganic composite beads by gelatin induced biomimetic mineralization for yeast alcohol dehydrogenase encapsulation. *J. Mol. Catal. B. Enzym.* **2014**, *100*, 49–58.

39. Li, J.; Jiang, Z.; Wu, H.; Zhang, L.; Long, L.; Jiang, Y. Constructing inorganic shell onto LBL microcapsule through biomimetic mineralization: A novel and facile method for fabrication of microbioreactors. *Soft Matter* **2010**, *6*, 542–550.

40. Lai, J.K.; Chuang, T.H.; Jan, J.S.; Wang, S.S.S. Efficient and stable enzyme immobilization in a block copolypeptide vesicle-templated biomimetic silica support. *Colloid Surf. B* **2010**, *80*, 51–58.

41. Luckarift, H.R.; Spain, J.C.; Naik, R.R.; Stone, M.O. Enzyme immobilization in a biomimetic silica support. *Nat. Biotechnol.* **2004**, *22*, 211–213.

42. Naik, R.R.; Tomczak, M.M.; Luckarift, H.R.; Spain, J.C.; Stone, M.O. Entrapment of enzymes and nanoparticles using biomimetically synthesized silica. *Chem. Commun.* **2004**, *15*, 1684–1685.

43. Yudovin-Farber, I.; Azzam, T.; Metzer, E.; Taraboulos, A.; Domb, A.J. Cationic polysaccharides as antiprion agents. *J. Med. Chem.* **2005**, *48*, 1414–1420.

44. Zhang, Y.; Wu, H.; Li, L.; Li, J.; Jiang, Z.; Jiang, Y.; Chen, Y. Enzymatic conversion of baicalin into baicalein by β-glucuronidase encapsulated in biomimetic core-shell structured hybrid capsules. *J. Mol. Catal. B Enzym.* **2009**, *57*, 130–135.

Chapter 8

CHITIN-LIGNIN MATERIAL AS A NOVEL MATRIX FOR ENZYME IMMOBILIZATION

Jakub Zdarta [1], Łukasz Klapiszewski [1], Marcin Wysokowski [1], Małgorzata Norman [1], Agnieszka Kołodziejczak-Radzimska [1], Dariusz Moszyński [2], Hermann Ehrlich [3], Hieronim Maciejewski [4,5], Allison L. Stelling [6] and Teofil Jesionowski [1]

[1]Institute of Chemical Technology and Engineering, Faculty of Chemical Technology, Poznan University of Technology, Berdychowo 4, 60965 Poznan, Poland

[2]Institute of Inorganic Chemical Technology and Environmental Engineering, West Pomeranian University of Technology, Pulaskiego 10, 70322 Szczecin, Poland

[3]Institute of Experimental Physics, TU Bergakademie Freiberg, Leipziger Str. 23, 09599 Freiberg, Germany

[4]Adam Mickiewicz University in Poznan, Faculty of Chemistry, Umultowska 89b, 61614 Poznan, Poland

[5]Poznan Science and Technology Park, Adam Mickiewicz University Fundation, Rubież 46, 61612 Poznan, Poland

[6]Duke University, Center for Materials Genomics, Department of Mechanical Engineering and Materials Science,144 Hudson Hall, Durham, NC 27708, USA

ABSTRACT

Innovative materials were made via the combination of chitin and lignin, and the immobilization of lipase from *Aspergillus niger*. Analysis by techniques including FTIR, XPS and ^{13}C CP MAS NMR confirmed the effective immobilization of the enzyme on the surface of the composite support. The electrokinetic properties of the resulting systems were also determined. Results obtained from elemental analysis and by the Bradford method enabled the determination of optimum parameters for the immobilization process. Based on the hydrolysis reaction of para-nitrophenyl palmitate, a determination was made of the catalytic activity, thermal and pH stability, and reusability. The systems with immobilized enzymes were found to have a hydrolytic activity of 5.72 mU, and increased thermal and pH stability compared with the native lipase. The products were also shown to retain approximately 80% of their initial catalytic activity, even after 20 reaction cycles. The immobilization process, using a cheap, non-toxic matrix of natural origin, leads to systems with potential applications in wastewater remediation processes and in biosensors.

INTRODUCTION

Continuing technological progress means that scientists are constantly finding new solutions that make use of lignin and its derivatives. When suitably modified, lignin is a polarographically active material [1] capable of undergoing a variety of electrochemical reactions, in the course of both oxidation and reduction [2]. Consequently, in recent years it has found interesting applications in electrochemistry. One of these was the creation of a cheap and fully environmentally friendly cathode, developed by Milczarek and Inganäs [3]. The valuable properties of lignin and its particular structure had previously been exploited by Milczarek in the construction of electrochemical sensors and detectors, as described in [4,5,6,7]. Interesting work using lignin-based material to create an innovative, cheap battery was reported by Gnedenkov et al. [8,9,10]. Literature reports also indicate the possibility of using lignocellulose materials, including pure lignin, as a filler in a wide range of polymers, both in strongly polar (poly(ethylene terephthalate)—PET; poly(ethylene oxide)—PEO) [11,12] and in hydrophobic (polypropylene—PP) [13,14] polymer matrices. Studies have also been carried out using poly(vinyl chloride) [15]. The biopolymer may also serve as a potential cheap and easily available biosorbent for environmentally harmful metal ions [16,17,18,19,20]. As a sorbent, lignin may be obtained chiefly as a waste product of the paper industry, and subjected to chemical modification to increase the number of functional groups [21,22]. It has also been reported that lignin has multifunctional barrier properties, protecting against harmful UV radiation, as well as antibacterial properties [23]. There are also promising possibilities for the use of lignin in the pharmaceutical industry and in medicine.

Chitin is an aminopolysaccharide, built of a long polymer chain consisting of N-acetylglucosamine units connected by β-1,4-glycoside bonds [24]. Chitin is a natural polymer, obtained chiefly from the shells of marine invertebrates, including the marine sponges [25,26,27,28]. It is friendly to the natural environment, and it exhibits high chemical stability and high reactivity, and is also non-toxic, bioactive, biodegradable and biocompatible [29]. Because of these features it is used in many areas of biomedicine and biotechnology [30,31]. One of these fields is the immobilization of enzymes [32,33,34]. Krajewska [35] presents a wide-ranging review of the literature concerning the use of chitin as a support for many catalytic proteins. Enzymes were immobilized by cross-linking with chitin by glutaraldehyde to reduce the viscosity of fruit and vegetable juices [36]. Outside the food industry, mention might be made of the use of enzymes immobilized on chitin via physisorption [37] or with the formation of covalent bonds [38] to detect and remove phenols. One of the most industrially useful groups of enzymes are the lipases, which are hydrophobic enzymes. To take full advantage of their technical and economic possibilities, they are used in a form immobilized on chitin [39]. An important factor in the widespread use of chitin as a support is the universality of the forms in which it can be used. Available morphological forms include powder, flakes, beads, nanoscale whiskers and fibers [40].

The creation of a stable material with defined properties provides the possibility of combining the undoubted advantages of both precursors, such as the aforementioned biocompatibility and non-toxicity, in the process of enzyme immobilization. The presence of multiple reactive functional groups in the structure of both materials increases their affinity to biomolecules [41]. It should be noted that the fact that the matrix is made using relatively cheap waste materials has a positive impact with regard to the economic aspects of the immobilization process [42]. The systems so produced may have potential uses in many fields where there is a need for highly pure and non-toxic catalysts.

The aim of the present study was to use a chitin-lignin material as a novel matrix for immobilization by adsorption of the lipase from *Aspergillus niger*. This is work of an innovative aspect, because there are no reports in the literature concerning the use of this system in enzyme immobilization. The systems produced may find uses in the transesterification and hydrolysis of a wide range of compounds, as well as in the production of biosensors. The results of the analysis confirmed the effective immobilization of the lipase on the chitin-lignin support. A detailed analysis was also made of the effect of process parameters on the properties of the resulting systems, and it was shown that lipase immobilized on the composite offers greater thermal and chemical stability than the native enzyme.

RESULTS AND DISCUSSION

Physicochemical Evaluation

FTIR Spectroscopy

Figure 1 shows the FTIR spectra of the chitin–lignin material, lipase from *Aspergillus niger* (Figure 1a), and the products following enzyme immobilization (Figure 1b). The major bands are summarized in Table 1.

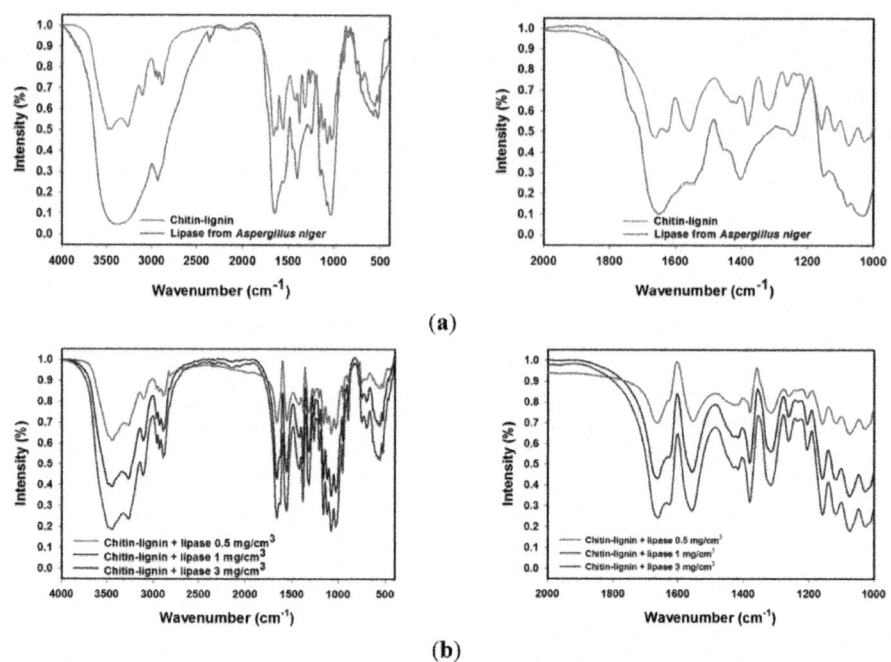

Figure 1: FTIR spectra of chitin-lignin composite and lipase (**a**) and selected products following 24 h of enzyme immobilization (**b**), in two different spectral range.

Table 1: Maximal vibrational wavenumbers (cm⁻¹) attributed to lipase from *Aspergillus niger*, chitin-lignin material, and products following immobilization

Lipase from *Aspergillus niger*	Chitin-Lignin Material	Products after Immobilization	Vibrational Assignment
3460	3444	3457	O-H stretching
3242	3257	3264	N-H stretching
-	3111	3112	C_{Ar}-H stretching
2931	2965, 2930, 2877	2966, 2935, 2879	CH_x stretching
-	1674	1676	C=O stretching
1647	1625	1639	amide I stretching
1546	1556	1552	amide II bending
1448	1432	1438	CH_2 bending
-	1420	1417	C_{Ar}-C_{Ar} stretching
1402	1388	1401	O-H stretching
-	1323	1329	C-O (syringyl unit) streching
1257	1268	1261	amide III bending
1151, 1073, 1037	1158, 1116, 1077, 1022	1162, 1113, 1081, 1027	C-O-C (ring), C-O stretching
-	953	957	CH_3 bending
-	903	905	β-1,4-glycosidic bonds
-	745	745	aromatic C-H(guaiacyl unit), bending
576	558	571	N-H bending
531	527	530	C-C scissoring

Analysis of the FTIR spectrum of the enzyme prior to immobilization shows the presence of a band in the range 3550–3200 cm^{-1} associated with stretching vibrations of O-H and N-H groups, and one at wavenumber 2931 cm^{-1} from stretching vibrations of C-H (CH$_3$ and CH$_2$). The most important signals in the spectrum of the native lipase are peaks at wavenumbers 1647 cm^{-1}, 1546 cm^{-1} and 1257 cm^{-1}, whose presence is characteristic of stretching vibrations of amide I, II and III bonds [43,44]. The FTIR spectrum of the enzyme also features a peak at wavenumber 1402 cm^{-1}, generated by stretching vibrations of O-H groups, and a low-intensity signal at 1448 cm^{-1} confirming the presence of bending vibrations of CH$_2$. The group of signals at 1151 cm^{-1}, 1073 cm^{-1} and 1037 cm^{-1} are associated with the presence of C-O-C bonds in the protein structure [45]. In addition, of note are two signals below 1000 cm^{-1}: at 576 cm^{-1} a band of N-H stretching vibrations, and at 531 cm^{-1} a band of scissor vibrations of the C-C bonds forming the skeleton of the enzyme structure [46].

Analysis of the spectrum of the chitin–lignin matrix confirms that the expected product was obtained. It also features a large number of bands, this being a result of the complex structure of the system. Attention is drawn to the bands with maxima at 3444 cm^{-1} and 3257 cm^{-1}, attributed to stretching vibrations of O-H and N-H groups. A peak with a maximum at 3111 cm^{-1} is associated with stretching vibrations of C$_{Ar}$-H groups present in the lignin structure [47]. A series of signals in the range 2970–2870 cm^{-1} confirms the presence of CH$_2$ and CH$_3$ groups in the structure of the composite, while the distinct band with a maximum at 1674 cm^{-1} comes from stretching vibrations of C=O bonds. Four signals between 1160 cm^{-1} and 1020 cm^{-1} can be attributed to stretching vibrations of C-O-C bonds in the glucose ring in chitin, as well as other C-O bonds in the material [48]. The interpretation of the carbon–oxygen bonds present in the system is supplemented by a peak at wavenumber 905 cm^{-1}, which is a consequence of the β-1,4-glycosidic bonds in chitin [49]. Note should also be taken of the signals originating from vibrations of amide I, II and III bonds. These are bands analogous to those present in the enzyme structure, but appearing at slightly different wavenumbers, respectively 1639 cm^{-1}, 1552 cm^{-1} and 1261 cm^{-1}, as a result of the different chemical environment of the bonds. Very significant bands, confirming the production of a chitin–lignin material, are present at 1420 cm^{-1}, 1329 cm^{-1} and 745 cm^{-1}, and originate from the stretching and bending vibrations of the aromatic structures present in lignin [50].

The FTIR spectra of the systems following immobilization carried out for 24 h using solutions of the enzyme in various concentrations are shown in Figure 1b. Analysis of the data obtained shows that the lipase was effectively immobilized on the matrix surface. In spite of the similarity of the bands present on the spectra of the support and the enzyme, an indication is provided by the presence of signals associated with vibrations of amide I, II and III bonds contained in the protein structure, at wavenumbers 1639 cm⁻¹, 1552 cm⁻¹ and 1261 cm⁻¹ respectively [51]. The intensity of these bands increases, and their absorption maxima are shifted, compared with the spectrum of the support. Analogous observations apply to the signals from stretching vibrations of O-H groups at wavenumber 3457 cm⁻¹, and from stretching vibrations of C=O bonds at 1676 cm⁻¹. The changes provide additional evidence confirming the immobilization, as well as indicating hydrogen bonding between the matrix and enzyme [52]. It is also interesting that as the concentration of the enzyme solution used for immobilization increases, particular bands in the product spectra become more intense. This provides indirect evidence that there is also an increase in the quantity of the enzyme deposited on the matrix surface.

^{13}C CP MAS NMR Spectroscopy

Figure 2 shows the ^{13}C CP MAS NMR spectra of the obtained chitin–lignin material, the native lipase, and the product following 24 h of immobilization of the enzyme from solution at a concentration of 3 mg/cm³.

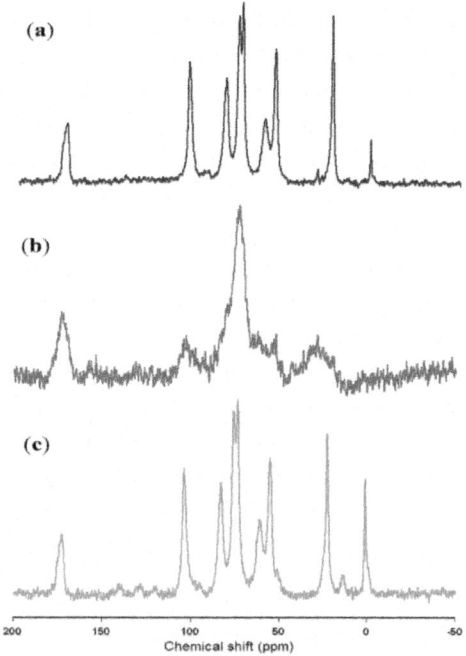

Figure 2: ^{13}C CP MAS NMR spectra of chitin-lignin (**a**); lipase (**b**) and chitin-lignin matrix with immobilized enzyme (**c**).

The ^{13}C CP MAS NMR spectrum of the chitin-lignin material shows the presence of signals characteristic of the precursors, which provides confirmation of the effective formation of the expected material. The signal at 22 ppm originates from the carbon in CH_3 in acetamide groups from chitin, while the entire group of peaks in the range 55–105 ppm is generated by carbon atoms in *N*-acetylglucosamine mers [53]. The distinct signal at 175 ppm originates from the carbonyl carbons in acetamide groups in the chitin structure [54]. The spectrum of the immobilized enzyme provides confirmation of the previous findings concerning the great similarity in structure of the lipase and the chitin; which is the chief component of the composite. The spectrum of the protein contains two clear signals, with maxima at 76 and 177 ppm, as well as several bands of much smaller intensity and wider range. The spectrum of the product formed after immobilization, in view of the similarity of the spectra of the precursors, does not show many changes. There is a different shape, particularly at the base, in the signals at 56 and 107 ppm. There is also a characteristic area between 115 and 145 ppm, where there appear signals which were not observed in the spectrum of the support, but which appear with low intensity in the spectrum of the native enzyme. Analysis of the ^{13}C CP MAS NMR spectra confirms the effectiveness of the immobilization process and the immobilization of the enzyme on the surface of the chitin-lignin matrix. In addition, in the case of the signals on the spectrum of the system after immobilization, there is seen to be a small shift in their maxima, which may suggest that the protein is attached to the support by way of the formation of hydrogen bonds.

Elemental Analysis

Table 2 contains the results of elemental analysis, describing the change in the content of such elements as nitrogen, carbon, hydrogen and sulfur in the immobilized enzyme preparations and in the matrix used.

Table 2: Elemental content of examined elements in the chitin-lignin matrix and in products following immobilization

Enzyme Solution Concentration (mg/cm³)	Immobilization Time	Elemental Content (%)			
		N	C	H	S
Chitin-lignin matrix		5.07	33.86	4.93	0.03
0.5	1 min	5.23	35.42	5.40	0.02
	2 h	5.58	37.17	5.67	0.01
	24 h	6.41	37.77	5.73	0.03
1.0	1 min	5.75	38.31	5.54	0.01
	2 h	5.96	38.77	5.78	0.03
	24 h	6.66	39.81	5.95	0.02
3.0	1 min	5.96	39.01	5.91	0.03
	2 h	6.03	39.30	6.05	0.02
	24 h	6.77	39.92	6.07	0.02

The initial matrix, prior to enzyme immobilization, has a carbon content of 33.86% and a hydrogen content of 4.93%. These elements are present in the structure of both lignin and chitin. Nitrogen, found in the elemental composition of the hybrid material with a content of 5.07%, is associated with the presence of N-acetylglucosamine groups in chitin. The presence of sulfur in the composite is explained by the use of sulfuric acid in the kraft process used to produce the lignin precursor.

The elemental analysis of systems resulting from the immobilization of lipase on the surface of the chitin-lignin matrix showed an increase in the contents of carbon, nitrogen and hydrogen, compared with the initial material. These changes are a result of the presence of those three elements in the structure of the enzyme, and confirm the effective immobilization of the protein on the surface of the support. The increase in the content of the analyzed components with higher initial concentration of protein solution and longer time of immobilization indicates that both of these parameters have a significant effect on the quantity of enzyme immobilized. The most distinct changes compared with the chitin–lignin material were observed for the system produced following a process lasting 24 h using a solution of concentration 3 mg/cm³, which may be taken as confirmation that the greatest quantity of protein was immobilized under such conditions.

XPS Analysis

The surface composition for samples of lipase, chitin–lignin material and the product following enzyme immobilization was examined with X-ray photoelectron spectroscopy. The surface of all samples is composed of carbon, oxygen and nitrogen. Some traces of calcium, potassium and sulfur were detected, but these are not considered in the quantitative calculations. The elemental surface compositions calculated from XPS data are given in Table 3.

Table 3: Elemental composition of the surface of samples

Sample Name	Atomic %			N/C Ratio	O/C Ratio
	C	O	N	H	S
Lipase	58.2	30.7	11.1	0.19	0.53
Chitin-lignin matrix	61.4	32.6	6.0	0.10	0.53
Chitin-lignin + lipase	62.5	30.0	7.5	0.12	0.48

The elemental composition of the lipase as reported by Tomizuka *et al.* and expressed as a C:O:N molar ratio is 61:25:14 [55]. These values are in good agreement with the ratio obtained in the present study for the surface of lipase, namely 58:31:11. Similar good agreement is obtained for the surface composition of the chitin-lignin matrix, which was reported previously [53]. The oxygen-carbon ratio close to 0.5 obtained for chitin-lignin, as well as the surface composition of the matrix, are very close to the values observed for nanocrystalline chitin [56]. Since the elemental composition of lignin differs significantly from the ratio observed here, it is concluded that the surface of the support matrix is composed mainly of chitin. The nitrogen-carbon ratio is almost twice as high for the lipase as for the chitin-lignin material. Therefore an increase in this parameter can be used as an indicator for successful enzyme immobilization, as reported previously [57]. Indeed the N/C ratio increases from 0.10 for the pure chitin-lignin matrix to 0.12 for the sample after immobilization. The elemental analysis of samples before and after immobilization, as described in Section 2.1.3., indicates an increase of approximately 20% in the nitrogen content after enzyme immobilization. This is corroborated by XPS data. This increase in nitrogen concentration following the immobilization process is taken as indirect evidence of successful lipase immobilization.

Evaluation of the chemical composition of the surface of the examined materials is based mainly on analysis of the XPS C 1s peak. The spectra have a relatively complex profile (Figure 3). Deconvolution of the experimental data was performed using a model consisting of four basic components of the C 1s transition: C_1–C_4. Component C_1, with a binding energy of 284.4 ± 0.1 eV, corresponds essentially to non-functionalized carbon atoms located in the aromatic rings expected to be in the lignin structure. Component C_2, with a binding energy of 284.8 eV, is attributed to all other non-functionalized sp^2 and sp^3carbon atoms, bonded either to other carbon or to hydrogen atoms. Component C_3, shifted by 1.4 ± 0.2 eV from component C_2 in the direction of increasing binding energies, is attributed to a set of groups with a carbon atom bonded to one atom of oxygen or nitrogen. These include the following functional groups which are presumed to be present in the studied materials:

C-O-C, C-OH, C-N-C, C-NH$_2$. Component C$_4$, shifted by 2.9 $\pm$ 0.2 eV from component C$_2$ in the direction of increasing binding energies, also corresponds to a set of functional groups: C=O, O-C-O, N-C-O and N-C=O. The binding energy interpretations given above are based on the energy shifts given in Appendix E [58]. A relative surface functional group composition obtained from decomposition of the C 1s signal is given in Table 4. The total C 1s peak intensity is taken as 100.

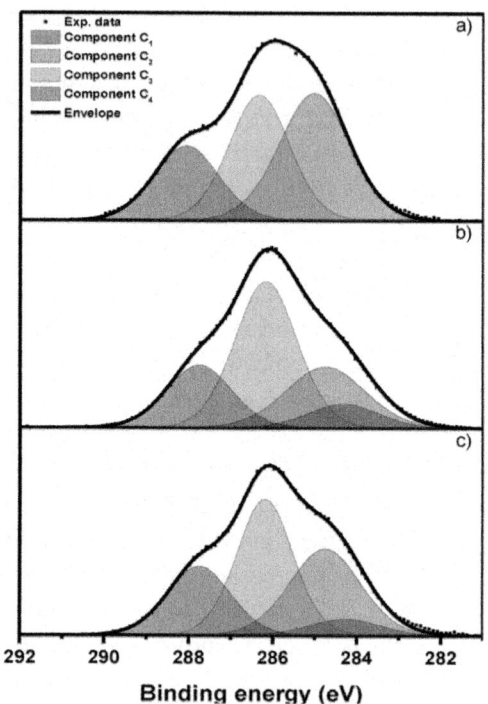

Figure 3: The XPS C 1s spectra for chitin-lignin (**a**); lipase (**b**); and the chitin-lignin + lipase product (**c**).

Table 4: Distribution of functional groups calculated on the basis of the deconvolution model of the XPS C 1s peak.

Sample Name	Total C 1s Peak Intensity (%)			
	C$_1$	C$_2$	C$_3$	C$_4$
Lipase	-	42	36	22
Chitin-lignin	9	25	46	20
Chitin-lignin + lipase	6	32	39	23

Since lipase contains a relatively small number of aromatic rings, originating from amino acids such as phenylalanine or tyrosine [55], the component C_1 is not considered in the deconvolution of the C 1s spectrum for that substance. Component C_2 prevails in the XPS signal, followed by C_3. The support material is a mixture of chitin and lignin. The expected component ratio for pure chitin is $C_2:C_3:C_4 = 25:50:25$ [59], while the ratio $(C_1 + C_2):C_3:C_4$ observed for lignin is 65:29:3 [60]. On the surface of the chitin–lignin matrix observed here, the contributions of components C_1 and C_2 are lower than would be given by a simple average for the mixture of chitin and lignin. Therefore, as suggested earlier, it is concluded that chitin prevails on the surface of the support. Comparison of the spectra of the chitin-lignin material and the product following enzyme immobilization indicates that C_1 diminishes slightly, while C_2 increases. Since C_2 is dominant in the XPS spectrum of lipase, we believe this to be an indication of successful enzyme immobilization.

Some additional evidence of the successful immobilization of lipase on the chitin-lignin matrix can be observed in the XPS O 1s spectra shown in Figure 4. The XPS O 1s transition observed for lipase is symmetric, with a maximum at binding energy 531.8 eV (dotted curve). In the case of the chitin-lignin matrix and the product of enzyme immobilization, the maximum of the O 1s peak is shifted in the direction of high binding energy to 532.4 eV. The structure of both chitin and lignin is dominated by C-OH groups, while in the case of the lipase a more equal ratio between hydroxyl and carboxyl groups is expected. The characteristic position of the O 1s peak for C-OH groups is approximately 532.5 eV, while its position for C=O groups is reported to be about 531.3 eV [61]. Accordingly, a shift in the XPS O 1s spectra is observed between the lipase and chitin-lignin. A small difference is also observed between the profile of the O 1s peak for chitin-lignin and for the chitin-lignin + lipase product. On the high-energy side of the spectrum the intensity of the O 1s peak obtained for the product following enzyme immobilization is slightly higher than the intensity of the peak obtained for the chitin-lignin support. The difference is small, but considering the relatively low quantity of immobilized lipase, it can be taken as confirmation of the increased concentration of C=O groups, which is an expected result of lipase being attached to the support.

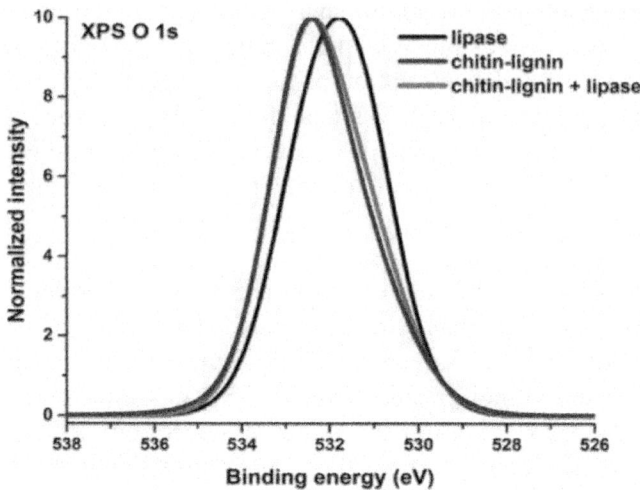

Figure 4: XPS O 1s spectra for lipase, chitin-lignin matrix and the product following enzyme immobilization.

XPS analysis provides no direct confirmation of lipase immobilization, since there is no apparent evidence of the formation of a new chemical environment. However, the formation of hydrogen bonds is not excluded. Moreover, the increase in the nitrogen-carbon ratio in combination with the subtle changes in the C 1s and O 1s component ratios can be considered an indication of successful immobilization of the enzyme.

Electrokinetic Characteristic

Studies of zeta potential and the effect of pH provide very valuable data about the electrokinetic properties of dispersed systems. Figure 5 shows the results obtained. Determination of the zeta potential of the biocomposite with and without immobilized enzyme provides indirect confirmation of the effectiveness of the suggested method of immobilization. The graph shows the values of the zeta potential obtained for selected samples following immobilization for 24 h.

The zeta potential of the chitin–lignin system is negative over the whole of the investigated pH range, and the isoelectric point is not attained. This results from the presence of specific functional groups (-COOH and -OH) on the surface of the component biopolymers. The electrokinetic potential of pure kraft lignin is even more negative; its value increased when the lignin was combined with chitin (due to the presence of surface NH_2 functional groups, which in an acidic environment can undergo protonation to NH_3^+) [53]. Lipase consists of several amino acids. The high percentage of acidic amino acids (Asp and Glu) gives the molecule a net negative charge, which is higher than the total for the positively charged residues (Arg, Lys, and His) [62]. That is why the isoelectric point of this protein is about 4 [63,64]. This value indicates that only at pH values below it will the surface charge (and indirectly zeta potential) be positive. The absolute value of zeta potential of chitin-lignin + lipase is smaller than this for matrix, especially in acidic condition, which can be explained by adsorption of lipase.

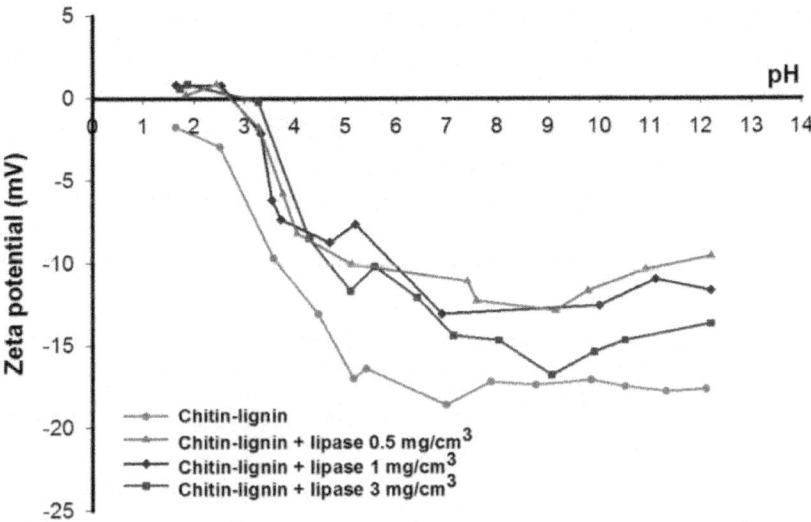

Figure 5: The zeta potential, as a function of pH, of the chitin-lignin material and selected products following immobilization.

Following immobilization of the enzyme on the surface of the support, as a result of interactions between the surface groups of the support and of the enzyme, the absolute values of the zeta potential decreased. This provides indirect evidence of the adsorptive nature of the attachment of the enzyme to the chitin–lignin support [65,66,67]. There was a decrease in the number of the free functional groups which are responsible for generating the charge. In addition, the chitin-lignin products upon addition of enzyme attain their isoelectric point (the pH at which the zeta potential is zero), which had previously not been observed. From the measured values of zeta potential it can be concluded that the quantity of immobilized enzyme influences its electrokinetic properties [68]. Nevertheless, irrespective of the quantity of adsorbed enzyme, the value of the isoelectric point is 2.7.

Quantity of Immobilized Enzyme

Based on the Bradford method [69] it was determined how the quantity of enzyme immobilized on the surface of the chitin-lignin support is affected by the concentration of the solution used in the immobilization process, and by the duration of the process. Table 5 contains detailed data on the quantity of biocatalyst adsorbed, depending on the concentration of the protein solution and the time of the process. The results are presented in terms of milligrams of enzyme per 1 gram of used matrix.

The results show that increasing the time of the immobilization process causes greater quantities of enzyme to be adsorbed. It should nonetheless be noted that the greatest increase in adsorbed protein occurs in the initial stages of the process. After the process time exceeds 4 h, the quantity of immobilized biocatalyst does not increase significantly, and the maximum change, depending on the concentration of the enzyme solution, is approximately 2 mg/g.

Table 5: Content of investigated elements in the chitin-lignin matrix and in the products following immobilization

Immobilization Time	Concentration of Enzyme Solution (mg/cm³)		
	0.5	1	3
	Amount of Immobilized Enzyme (mg/g)		
1 min	1.45	5.13	6.19
1 h	6.23	9.76	14.97
2 h	8.17	10.84	18.46
4 h	8.58	11.37	18.72
24 h	9.22	11.84	19.31
96 h	9.94	12.57	20.28

Another parameter having a significant effect on the quantity of protein in the products following immobilization is the concentration of the solution used. The results show that the greatest quantity of protein is adsorbed from the solution with a concentration of 3 mg/cm³. When identical times of immobilization are compared, this solution enables the adsorption of more than twice as much protein as when a solution of concentration 0.5 mg/cm³ is used.

The greatest quantity of the enzyme was adsorbed from the solution with a concentration of 3 mg/cm³ following a process lasting 96 h. However, the optimum time of the immobilization process is 4 h, enabling comparable quantities of protein to be immobilized in a much shorter time, which has a positive impact on the economics of the studied process.

Hydrolytic Activity

Determination of Hydrolytic Activity

The hydrolytic activity of the free and immobilized enzyme was assessed spectrophotometrically based on the hydrolysis reaction of para-nitrophenyl palmitate. Figure 6 shows the results for catalytic activity of preparations with immobilized lipase obtained using enzyme solutions with concentrations of 0.5, 1 and 3 mg/cm³, subjected to immobilization over different time intervals. The measurements were performed at 30 °C.

The systems with immobilized enzymes have lower catalytic activity than the native lipase, for which the activity is measured at 7.46 mU. Irrespective of the concentration of the protein solution, the greatest activity is found for the products formed after 4 h of immobilization. The results showed the enzyme solution with a concentration of 3 mg/cm³ to be optimum for immobilization on a chitin–lignin support. The resulting immobilized lipase has the highest activity of all of the systems investigated, equal to 5.76 mU. This sample was selected for further analysis to determine the stability of the resulting system depending on the conditions of the catalyzed reaction. The results show unambiguously that a greater quantity of immobilized enzyme does not lead directly to an increase in the system's catalytic activity. The products obtained following 96 h of immobilization, which have the greatest quantities of immobilized protein, exhibit a lower activity. This is caused by the accumulation of too great a quantity of the enzyme on the matrix surface, blocking the active sites on the biocatalyst and thus reducing its activity [70].

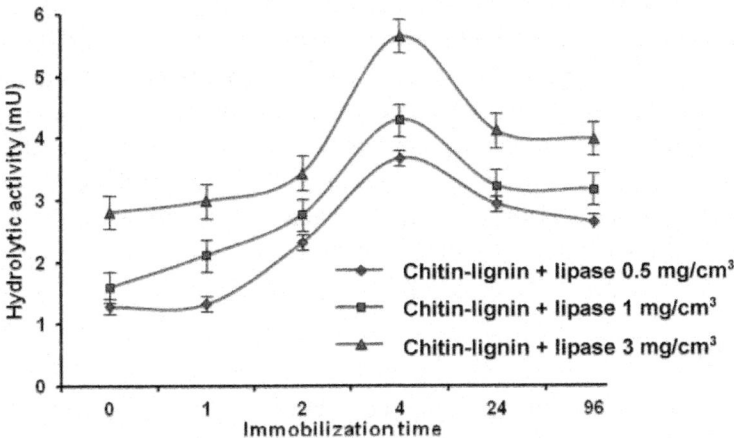

Figure 6: Graph showing changes in the catalytic activity of products depending on the time of immobilization and the concentration of the enzyme solution.

Thermal Stability

Thermal stability is one of the most important properties of immobilized enzymes. The thermal stability of the immobilized lipase was studied, in comparison with the native enzyme, over a temperature range of 10–80 °C. For this analysis, the system selected was one that underwent 4 h immobilization in the enzyme solution at a concentration of 3 mg/cm^3 in phosphate buffer at pH = 7. Figure 7 shows a comparison of the thermal stability of the native lipase with that of the lipase immobilized on a chitin-lignin matrix.

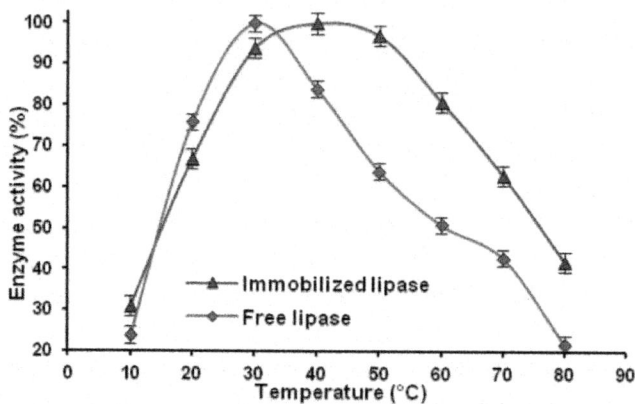

Figure 7: Graph of thermal stability of immobilized and native lipase in the temperature range 10–80 °C.

The native lipase attains its maximum hydrolytic activity at 30 °C, while that of the immobilized enzyme occurs at 40 °C. It should be noted, however, that the immobilized lipase retains more than 90% of its initial activity even at 50 °C, where the properties of the free enzyme are lost to a significant degree. These results show clearly that attaching the biocatalyst to a solid support has a positive effect on its resistance to denaturation at high temperature. This has been shown to be a result of an increase in the rigidity of the protein structure [71]. The thermal stability increased because the immobilization process could protect the tertiary structure of the peptide from conformational changes caused by the higher temperature [72].

pH Stability

The pH stability is an important characteristic of systems resulting from immobilization. The pH stability of the immobilized lipase, compared with that of the native enzyme, was studied over a pH range of 3 to 11 at 30 °C. Figure

8shows a comparison of the pH stability of the native lipase with that of the lipase immobilized on a chitin-lignin matrix.

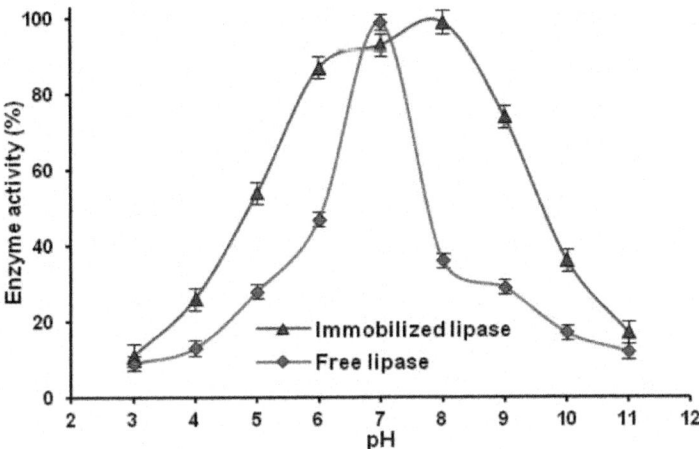

Figure 8: Graph showing changes in the catalytic active of immobilized and native lipase over the pH range 3–11.

The data above show that the pH has a large effect on the activity of the lipase in an aqueous environment. The activity of native lipase reaches a maximum at pH = 7, and small changes in pH cause a large decrease in hydrolytic activity, by as much as 50%. The immobilized lipase has its highest activity at pH = 8, which is characteristic of immobilized enzymes in this catalytic group [73]. The attachment of the enzyme to a solid support also causes it to retain more than 70% of its activity in the pH range 6–9. The improved stability of the immobilized enzyme compared with the native protein is probably a result of conformational changes taking place in the protein tertiary and quaternary structure following immobilization [74]. An increase in pH stability of the immobilized lipase is also connected with the changes in spatial orientation of secondary structure of the protein backbone, caused by the formation of hydrogen bonds between the enzyme and matrix [75].

Reusability

Figure 9 shows the reusability of the lipase immobilized on the chitin-lignin matrix over 20 cycles. In each cycle, the immobilized lipase was separated and washed with phosphate buffer, and the activity was calculated for p-NPP hydrolysis.

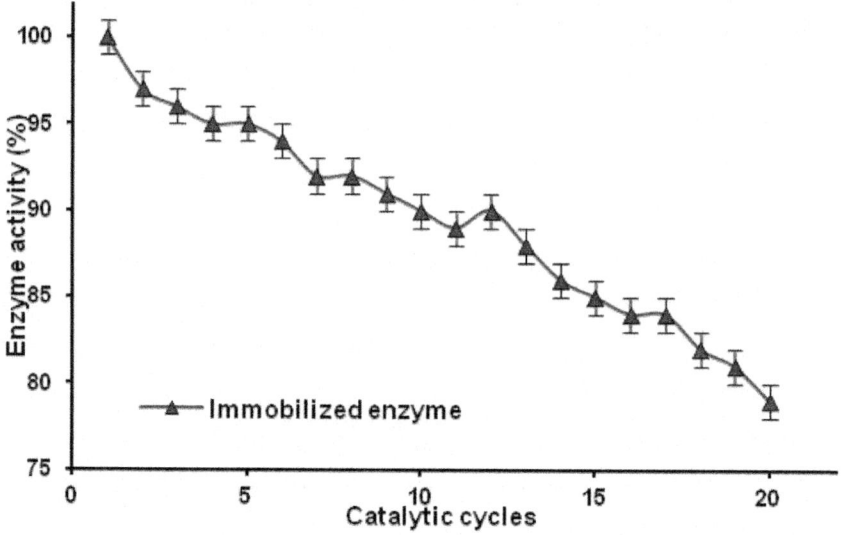

Figure 9: Changes in catalytic activity of immobilized lipase over 20 catalytic cycles.

The immobilized lipase was tested over 20 catalytic cycles, and was found to retain approximately 80% of its initial activity. The high reusability of products based on a chitin-lignin matrix may also lead to widespread use of this support in the immobilization of enzymes of other catalytic groups. Prolongation of the catalytic activity of these products may also lead to a significant reduction in the costs of carrying out reactions in real-life applications.

EXPERIMENTAL SECTION

Materials

The precursors, α-chitin powder from crab shells (technical grade) and kraft lignin (reagent grade), and 15% hydrogen peroxide as an oxidizing agent, were obtained from Sigma-Aldrich (Munich, Germany). Immobilization was carried out using commercial lipase from *Aspergillus niger* (Sigma-Aldrich, Munich, Germany) and phosphate buffer at pH = 7 (Amresco, Solon, OH, USA). The 85% phosphoric acid and 96% ethyl alcohol used in the Bradford method were obtained from Chempur (Piekary Śląskie, Poland). Coomassie Brilliant Blue G-250 (CBB G-250) was obtained from Sigma-Aldrich (Munich, Germany). The catalytic activity tests used para-nitrophenyl palmitate, Triton X-100 and gum arabic from Sigma-Aldrich (Munich, Germany) and 2-propanol from Chempur (Piekary Śląskie, Poland).

Preparation of Chitin-Lignin Material

The process of obtaining the chitin–lignin material (precursors ratio 1:1, *m/m*) began with the addition of 15 cm³ of 15% hydrogen peroxide to the lignin, according to the procedure reported in previously published work [53]. The mixture was subjected to intensive mixing at approximately 800 rpm for about 30 min using a high-speed stirrer (Eurostar Digital, IKA Werke GmbH, Staufen, Germany). Chitin was then added to the reactor, and mixing continued for 60 min. The resulting chitin-lignin material was filtered under reduced pressure and washed with distilled water. The product was then dried in a convectional dryer (Memmert, Munich, Germany) at approximately 105 °C for about 24 h.

Enzyme Immobilization

The process of immobilization of lipase from *Aspergillus niger* on the surface of the chitin-lignin composite was carried out using solutions of the enzyme at concentrations of 0.5, 1 and 3 mg/cm³ in a phosphate buffer at pH = 7, for times of 1 min and 1, 2, 4, 24 and 96 h. Quantities of 250 mg of the previously obtained matrix were placed in conical flasks, and 15 cm³ of the solution of the enzyme in the required concentration was added. The mixture was placed in a KS260 BASIC shaker (IKA Werke GmbH, Staufen, Germany), and shaken for the required length of time. Afterwards the precipitate was filtered under reduced pressure and left to dry at room temperature for 24 h.

Physicochemical Evaluation

The presence of the expected functional groups was confirmed by Fourier transform infrared (FTIR) spectroscopy, using a Vertex 70 spectrophotometer (Bruker, Karlsruhe, Germany). The materials were analyzed in the form of tablets, made by placing a mixture of anhydrous KBr (*ca.* 0.25 g) and 1.5 mg of the tested substance in a steel ring under a pressure of 10 MPa. The tests were performed at a resolution of 0.5 cm⁻¹ in the wavenumber range 4000–400 cm⁻¹.

^{13}C CP MAS NMR measurement was carried out on a DSX spectrometer (Bruker, Karlsruhe, Germany). For the determination of NMR spectra, a sample of approximately 100 mg was placed in a ZrO_2 rotator with diameter 4 mm, which enabled spinning of the sample. Centrifugation at the magic angle was performed at a spinning frequency of 8 kHz. The ^{13}C CP MAS NMR spectra were recorded at 100.63 MHz in a standard 4 mm MAS probe using single pulse excitation with high power proton decoupling (pulse repetition 10 s, spinning speed 8 kHz).

The elemental contents of the chitin-lignin hybrid material and the immobilized enzyme were determined using a Vario EL Cube instrument (Elementar Analysensysteme GmbH, Hanau, Germany), which is capable of registering the percentage content of carbon, hydrogen, nitrogen and sulfur in samples after high-temperature combustion. A properly weighed sample was placed in an 80-position autosampler and subjected to combustion. The decomposed sample was transferred in a stream of inert gas into an adsorption column. The results are given to $\pm0.01\%$, and each is obtained by averaging three measurements.

The X-ray photoelectron spectra were obtained using Al K (hv = 1486.6 eV) radiation with a Prevac system equipped with a Scienta SES 2002 (VG Scienta, Uppsala, Sweden) electron energy analyzer operating at constant transmission energy (E_p = 50 eV). The spectrometer was calibrated using the following photoemission lines (with reference to the Fermi level): EB Cu $2p_{3/2}$ = 932.8 eV, EB Ag $3d_{5/2}$ = 368.3 eV, EB Au $4f_{7/2}$ = 84.0 eV. The instrumental resolution, as evaluated by the full width at half maximum (FWHM) of the Ag $3d_{5/2}$ peak, was 1.0 eV. The samples were placed loose in a grooved molybdenum sample holder. The analysis chamber was evacuated during the experiments to better than $1\cdot10^{-9}$ mbar. Data processing involved background subtraction by means of an "S-type" integral profile and a curve-fitting procedure (a mixed Gaussian–Lorentzian function was employed) based on a least-squares method (CasaXPS software). The experimental errors were estimated to be ±0.1 eV for the photoelectron peaks of carbon and oxygen. Charging effects were corrected using the C 1s component attributed after deconvolution to aliphatic carbon bonds (component C_2) and determined at 284.8 eV. The reproducibility of the peak position thus obtained was ±0.1 eV. The surface composition of the samples was obtained on the basis of the peak area intensities of the C 1s, O 1s, and N 1s transitions using the sensitivity factor approach and assuming homogeneous distribution of elements in the surface layer.

The electrokinetic stability of the materials with immobilized enzyme was determined on the basis of zeta potential dependence on pH, using a Zetasizer Nano ZS (Malvern Instruments Ltd., Worcestershire, UK) equipped with an autotitrator. Measurements were made in a 0.001 M NaCl solution over the pH range 2–12, using 0.001 M NaCl solution.

The quantity of immobilized enzyme was determined by the Bradford method [69]. A solution of the Bradford reagent was prepared by dissolving 10 mg of Coomassie Brilliant Blue G-250 in 5 cm³ of 96% ethyl alcohol, 15 cm³ of 85% phosphoric acid and 80 cm³ of water. In a quartz cuvette, 4 cm³ of the Bradford reagent was mixed with 800 μL of the analyzed protein solution and 100 µL water, and the analysis was performed 10 min after the preparation of the mixture. Measurements were made at wavelength 595 nm, using a JASCO 650 spectrophotometer (Jasco, Tokyo, Japan).

Evaluation of Hydrolytic Activity

The activity of the immobilized lipase was measured by the method used in our previous work [76], with slight modifications. Spectrophotometric measurements were made for 2 min at wavelength 410 nm at 30 °C, based on the transesterification reaction of para-nitrophenyl palmitate (p-NPP) to para-nitrophenyl (p-NP). Hydrolytic activity was measured in 1 cm³ quartz cuvettes containing 5 mg of immobilized lipase with 2.7 cm³ of substrate solution containing 10 mM phosphate buffer, 10 mM of p-NPP solution in 2-propanol, 0.44% mass fraction of Triton X-100 and 0.11% mass fraction of gum arabic. One mUnit of immobilized enzyme activity was defined as the release of 1 µmoL of p-NP per minute.

Thermal Stability

The thermal stability of the immobilized and native lipase was determined over a temperature range of 10–80 °C. Hydrolytic activity was calculated as described in Section 3.5.

pH Stability

The pH stability of the immobilized and native lipase was determined by incubating the substrate solution at different pH values (3, 5, 7, 9, 11) to compare the activity of the free and immobilized lipase. Catalytic activity was calculated as described in Section 3.5.

Reusability

The reusability of the immobilized lipase was determined by testing over 20 cycles. Between each reaction step, the chitin-lignin matrix with the immobilized enzyme was separated from the substrate solution by centrifugation and washed with phosphate buffer. The hydrolytic activity was calculated as described in Section 3.5.

CONCLUSIONS

In this study, a chitin-lignin system was used as an innovative matrix in the process of immobilizing lipase from *Aspergillus niger*. Detailed characteristics of the obtained matrix, and confirmation of the effective immobilization of the enzyme, were obtained using such techniques as FTIR, XPS, ^{13}C CP MAS NMR and elemental analysis. It was shown that both the time of the process and the initial concentration of the protein solution have a significant effect on the properties of the products obtained. A determination was also made of the quantity of enzyme immobilized on the surface of the system, and of the catalytic activity of the system following lipase immobilization. It was found that the immobilized lipase exhibits lower activity than the free enzyme, but retains its catalytic properties for a greater number of reaction cycles. The enzyme bound to the chitin-lignin matrix also has greater thermal and chemical stability than the native protein. Measurement of the zeta potential enabled determination of the electrokinetic properties of the systems obtained. Detailed analysis of the FTIR spectra of the products of the immobilization process, and changes in the zeta potential and shifts in signal maxima on ^{13}C CP MAS NMR spectra, indicate that the enzyme is attached by way of physical adsorption, probably through the formation of hydrogen bonds.

ACKNOWLEDGMENTS

The study was financed within the Polish National Center of Science funds according to decision No. DEC-2013/09/B/ST8/00159.

AUTHOR CONTRIBUTIONS

J.Z.: Planning studies. Preparation of functional chitin-lignin biosorbent. Evaluation of enzyme immobilization efficiency. Results development. Ł.K.: Analysis of physicochemical properties of the materials obtained. Results development. M.W.: Analysis of structural properties of the materials obtained. Results development. D.M.: Implementation and description of the XPS analysis. M.N.: Implementation and description of the zeta potential analysis. A.K.-R.: Evaluation of enzyme immobilization efficiency. Results development. H.E.: Supervising manuscript with data interpretation. H.M.: Supervising manuscript with data interpretation. A.L.S: Supervising manuscript with data interpretation. T.J.: Coordination of all tasks in the paper. Planning studies. Results development.

REFERENCES

1. Evstigneyev, E.; Shevchenko, S.; Mayorova, H.; Platonow, A. Polarographically active structural fragments of lignin. II. Dimeric model compounds and lignins. *J. Wood Chem. Technol.* **2004**, *24*, 263–278.

2. Lund, H.; Baizer, M.M. *Organic Electrochemistry—An Introduction and Guide*; Marcel Dekker: New York, NY, USA, 1991.

3. Milczarek, G.; Inganäs, O. Renewable cathode materials from biopolymer/conjugated polymer interpenetrating networks. *Science* **2012**, *335*, 1468–1471.

4. Milczarek, G. Preparation and characterization of a lignin modified electrode. *Electroanalsia* **2007**, *19*, 1411–1414.

5. Milczarek, G. Preparation, characterization and electrocatalytic properties of an iodine/lignin modified gold electrode. *Electrochim. Acta* **2009**, *54*, 3199–3205.

6. Milczarek, G. Lignosulfonate-modified electrodes: Electrochemical properties and electrocatalysis of NADH oxidation. *Langmuir* **2009**, *25*, 10345–10353.

7. Milczarek, G.; Rębiś, T. Synthesis and electroanalytical performance of a composite material based on poly(3,4-ethylenedioxythiophene) doped with lignosulfonate. *Int. J. Electrochem.* **2012**, *130980*, 1–7.

8. Gnedenkov, S.V.; Opra, D.P.; Sinebryukhov, S.L.; Tsvetnikov, A.K.; Ustinov, A.Y.; Sergienko, V.I. Hydrolysis lignin-based organic electrode material for primary lithium batteries. *J. Solid State Electrochem.* **2014**, *17*, 2611–2621.

9. Gnedenkov, S.V.; Opra, D.P.; Sinebryukhov, S.L.; Tsvetnikov, A.K.; Ustinov, A.Y.; Sergienko, V.I. Hydrolysis lignin: Electrochemical properties of the organic cathode material for primary lithium battery. *J. Ind. Eng. Chem.* **2014**, *20*, 903–910.

10. Opra, D.P.; Gnedenkov, S.V.; Sinebryukhov, S.L.; Tsvetnikov, A.K.; Sergienko, V.I. Fabrication of battery cathode material based on hydrolytic lignin. *Solid State Phenom.* **2014**, *213*, 154–159.

11. Kadla, J.F.; Kubo, S. Lignin-based polymer blends: Analysis of intermolecular interactions in lignin-synthetic polymer blends. *Compos. A Appl. Sci. manuf.* **2004**, *35*, 395–400.

12. Canetti, M.; Bertini, F. Supermolecular structure and thermal properties of poly(ethylene terephthalate)/lignin composites. *Compos. Sci. Technol.* **2007**, *67*, 3151–3157.

13. Chen, F.; Dai, H.; Dong, X.; Yang, J.; Zhong, M. Physical properties of lignin-based polypropylene blends. *Polym. Compos.* **2011**, *32*, 1019–1025.

14. Borysiak, S. Fundamental studies on lignocellulose/polypropylene composites: Effects of wood treatment on the transcrystalline morphology and mechanical properties. *J. Appl. Polym. Sci.* **2013**, *127*, 1309–1322.

15. Gozdecki, C.; Wilczyński, A.; Kociszewski, M.; Zajchowski, S.

Mechanical properties of wood-polypropylene composites with industrial wood particles of different sizes. *Wood Fiber. Sci.* **2012**, *44*, 14–21.

16. Guo, X.; Zhang, S.; Shan, X. Adsorption of metal ions on lignin. *J. Hazard. Mater.* **2008**, *151*, 134–142.

17. Betancur, M.; Bonelli, P.R.; Velásquez, J.A.; Cukierman, A.L. Potentiality of lignin from the Kraft pulping process for removal of trace nickel from wastewater: Effect of demineralization. *Bioresour. Technol.* **2009**, *100*, 1130–1137.

18. Bulgariu, L.; Bulgariu, D.; Malutan, T.; Macoveanu, M. Adsorption of lead(II) ions from aqueous solution onto lignin.*Adsorp. Sci. Technol.* **2009**, *27*, 435–445.

19. Harmita, H.; Karthikeyan, K.G.; Pan, X.J. Copper and cadmium sorption onto kraft and organosolv lignins. *Bioresour. Technol.* **2009**, *100*, 6183–6191.

20. Ahmaruzzaman, M. Industrial wastes as low-cost potential adsorbents for the treatment of wastewater laden with heavy metals. *Adv. Colloid Interface Sci.* **2011**, *166*, 36–59.

21. Lei, Y.; Huizhen, Y. Modification of reed alkali lignin to adsorption of heavy metals. *Adv. Mater. Res.* **2013**, *622*, 1646–1650.

22. Ge, Y.; Li, Z.; Kong, Y.; Song, Q.; Wang, K. Heavy metal ions retention by bi-functionalized lignin: Synthesis, applications, and adsorption mechanisms. *J. Ind. Eng. Chem.* **2014**, *20*, 4429–4436.

23. Toh, K.; Yokoyama, H.; Takahashi, C.; Watanabe, T.; Noda, H. Effect of herb lignin on the growth of enterobacteria. *J. Gen. Appl. Microbiol.* **2007**, *53*, 201–205.

24. Ehrlich, H. Chitin and collagen as universal and alternative templates in biomineralization. *Int. Geol. Rev.* **2010**, *52*, 661–699.

25. Ehrlich, H.; Krautter, M.; Hanke, T.; Simon, P.; Knieb, C.; Heinemann, S.; Worch, H. First evidence of the presence of chitin in skeleton of marine sponges. Part II. Glass sponges. (Hexactinellida: Porifera). *J. Exp. Zool. B* **2007**, *308*, 473–478.

26. Ehrlich, H.; Maldonado, M.; Spindler, K.D.; Eckert, C.; Hanke, T.; Born, R.; Goebel, C.; Simon, P.; Heinemann, S.; Worch, H. First evidence of chitin as a component of the skeletal fibers of marine sponges. Part I. Verongidae (Demospongia: Porifera). *J. Exp. Zool. B* **2007**, *308*, 347–356.

27. Brunner, E.; Ehrlich, H.; Schupp, P.; Hedrich, R.; Hunoldt, S.; Kammer, M.; Machill, S.; Paasch, S.; Bazhenov, V.V.; Kurek, D.V.; *et al.* Chitin-

based scaffolds are an integral part of the skeleton of the marine demosponge *Ianthella basta.J. Struct. Biol.* **2009**, *168*, 539–547.

28. Ehrlich, H.; Ilan, M.; Maldonado, M.; Muricy, G.; Bavestrello, G.; Kljajic, Z.; Carballo, J.L.; Schiaparelli, S.; Ereskovsky, A.; Schupp, P.; *et al.* Three-dimensional chitin-based scaffolds from Verongida sponges (Demospongiae: Porifera). Part I. Isolation and identification of chitin. *Int. J. Biol. Macromol.* **2010**, *47*, 132–140.

29. Yang, T.C.; Zall, R.R. Absorption of metals by natural polymers generated from seafood processing wastes. *Ind. Eng. Chem. Prod. Res. Dev.* **1984**, *23*, 168–172.

30. Muzzarelli, R.A.A. Chitins and chitosans for the repair of wounded skin, nerve, cartilage and bone. *Carbohydr. Polym.* **2009**, *76*, 167–182.

31. Jayakumar, R.; Nair, A.; Sanoj Rejinold, N.; Maya, S.; Nair, S.V. Doxorubicin-loaded pH-responsive chitin nanogels for drug delivery to cancer cells. *Carbohydr. Polym.* **2012**, *87*, 2352–2356.

32. Liu, H.S.; Chen, W.H.; Lai, J.T. Immobilization of isoamylase on carboxymethyl-cellulose and chitin. *Appl. Biochem. Biotechnol.* **1997**, *66*, 57–67.

33. Chang, R.C.; Shaw, J.F. The immobilization of *Candida cylindracea* lipase on PVC, chitin and agarose. *Bot. Bull. Acad. Sin.* **1987**, *28*, 33–42.

34. Romo-Sanchez, S.; Arevalo-Villena, M.; Garcia Romero, E.; Ramirez, H.L.; Briones Perez, A. Immobilization of β-glucosidase and its application for enhancement of aroma precursors in muscat wine. *Food Bioprocess Technol.* **2014**, *7*, 1381–1392.

35. Krajewska, B. Application of chitin- and chitosan-based materials for enzyme immobilizations: A review. *Enzyme Microb. Technol.* **2004**, *35*, 126–139.

36. Vaillant, F.; Millan, A.; Millan, P.; Dormier, M.; Decloux, M.; Reynes, M. Co-immobilized pectinlyase and endocellulase on chitin and nylon supports. *Process Biochem.* **2000**, *35*, 989–996.

37. Batra, R.; Gupta, M.N. Non-covalent immobilization of potato (*Solanum tuberosum*) polyphenol oxidase on chitin. *Biotechnol. Appl. Biochem.* **1994**, *19*, 209–215.

38. Wang, G.; Xu, J.J.; Ye, L.H.; Zhu, J.J.; Chen, H.Y. Highly sensitive sensors based on the immobilization of tyrosinase in chitosan. *Bioelectrochemistry* **2002**, *57*, 33–38.

39. Gomes, F.M.; Pereira, E.B.; de Castro, H.F. Immobilization of lipase on chitin and its use in nonconventional biocatalysis. *Biomacromolecules* **2004**, *5*,

17–23.

40. Zeng, J.B.; He, Y.S.; Li, S.L.; Wang, Y.Z. Chitin whiskers: An overwiev. *Biomacromolecules* **2012**, *13*, 1–11.

41. Filipkowska, U. Desorption of reactive dyes from modified chitin. *Environ. Technol.* **2008**, *29*, 681–690.

42. Jesionowski, T.; Zdarta, J.; Krajewska, B. Enzymes immobilization by adsorption: A review. *Adsorption* **2014**, *20*, 801–821.

43. Wong, P.T.T.; Nong, R.K.; Caputo, T.A.; Godwin, T.A.; Rigas, B. Infrared spectroscopy of exfoliated human cervical cells: Evidence of extensive structural changes during carcinogenesis. *Proc. Natl. Acad. Sci. USA* **1991**, *88*, 10988–10992.

44. Dousseau, F.; Pezolet, M. Determination of the secondary structure content of proteins in aqueous solutions from their amide I and amide II infrared bands. Comparison between classical and partial least-squares methods. *Biochemistry* **1990**, *29*, 8771–8779.

45. Dong, L.; Ge, C.; Qin, P.; Chen, Y.; Xu, Q. Immobilization and catalytic properties of candida lipolytic lipase on surface of organic intercalated and modified MgAl-LDHs. *Sol. Sci.* **2014**, *31*, 8–15.

46. Cabrera-Padilla, R.Y.; Lisboa, M.C.; Pereira, M.M.; Figueiredo, R.T.; Franceschi, E.; Fricks, A.T.; Lima, A.S.; Silva, D.P.; Soares, C.M.F. Immobilization of *Candida rugosa* lipase onto an eco-friendly support in the presence of ionic liquid. *Bioprocess Biosyst. Eng.* **2014**.

47. Klapiszewski, Ł.; Zdarta, J.; Szatkowski, T.; Wysokowski, M.; Nowacka, M.; Szwarc-Rzepka, K.; Bartczak, P.; Siwińska-Stefańska, K.; Ehrlich, H.; Jesionowski, T. Silica/lignosulfonate hybrid materials: Preparation and characterization. *Cent. Eur. J. Chem.* **2014**, *12*, 719–735.

48. Lavall, R.L.; Assis, O.B.G.; Campana-Filho, S.P. β-Chitin from the pens of *Loligo* sp.: Extraction and characterization. *Bioresource Technol.* **2007**, *98*, 2465–2472.

49. Jang, M.K.; Kong, B.G.; Jeong, Y.I.; Lee, C.H.; Nah, J.W. Physicochemical characterization of α-chitin, β-chitin, and γ-chitin separated from natural resources. *J. Polym. Sci. A* **2004**, *42*, 3423–3432.

50. Klapiszewski, Ł.; Wysokowski, M.; Majchrzak, I.; Szatkowski, T.; Nowacka, M.; Siwińska-Stefańska, K.; Szwarc-Rzepka, K.; Bartczak, P.; Ehrlich, H.; Jesionowski, T. Preparation and characterization of multifunctional chitin/lignin materials. *J. Nanomater.* **2013**, 1–13.

51. Naidja, A.; Liu, C.; Huang, P.M. Formation of protein-birnessite complex: XRD, FTIR, and AFM analysis. *J. Colloid Interface Sci.* **2002**, *251*, 46–

56.

52. Portaccio, M.; Della Ventura, B.; Mita, D.G.; Manolova, N.; Stoilova, O.; Rashkov, I.; Lepore, M. FT-IR microscopy characterization of sol–gel layers prior and after glucose oxidase immobilization for biosensing applications. *J. Sol-Gel Sci. Technol.* **2011**, *57*, 204–211.

53. Wysokowski, M.; Klapiszewski, Ł.; Moszyński, D.; Bartczak, P.; Majchrzak, I.; Siwińska-Stefańska, K.; Bazhenov, V.V.; Jesionowski, T. Modification of chitin with kraft lignin and development of new biosorbents for removal of cadmium(II) and nickel(II) ions. *Mar. Drugs* **2014**, *12*, 2245–2268.

54. Cardenas, G.; Cabrera, G.; Taboada, E.; Miranda, S.P. Chitin characterization by SEM, FTIR, XRD, and ^{13}C cross polarization/mass angle spinning NMR. *J. Appl. Polym. Sci.* **2004**, *93*, 1876–1885.

55. Tomizuka, N.; Ota, Y.; Yamada, K. Lipase from *Candida cylindracea II*. Amino acid composition, carbohydrate component, and some physical properties. *Agric. Biol. Chem.* **1966**, *30*, 1090–1096.

56. Wang, B.; Li, J.; Zhang, J.; Li, H.; Chen, P.; Gu, Q.; Wang, Z. Thermo-mechanical properties of the composite made of poly (3-hydroxybutyrate-co-3-hydroxyvalerate) and acetylated chitin nanocrystals. *Carbohydr. Polym.* **2013**, *95*, 100–106.

57. Song, J.; Kahveci, D.; Chen, M.; Guo, Z.; Xie, E.; Xu, X.; Besenbacher, F.; Dong, M. Enhanced catalytic activity of lipase encapsulated in PCL nanofibers. *Langmuir* **2012**, *28*, 6157–6162.

58. Briggs, D.; Grant, J.T. *Surface Analysis by Auger and X-ray Photoelectron Spectroscopy*; IM Publications and SurfaceSpectra Limited: Charlton, UK, 2003.

59. Wang, J.; Wang, Z.; Li, J.; Wang, B.; Liu, J.; Chen, P.; Miao, M.; Gu, Q. Chitin nanocrystals grafted with poly(3-hydroxybutyrate-co-3-hydroxyvalerate) and their effects on thermal behavior of PHBV. *Carbohydr. Polym.* **2012**, *87*, 784–789.

60. De Lange, P.J.; Mahy, J.W.G. ToF-SIMS and XPS investigations of fibers, coatings and biomedical materials. *Fresenius' J. Anal. Chem.* **1995**, *353*, 487–493.

61. Rouxhet, P.G.; Genet, M.J. XPS analysis of bio-organic systems. *Surf. Interface Anal.* **2011**, *43*, 1453–1470.

62. Namboodiri, V.M.H.; Chattaopadhyaya, R. Purification and biochemical characterization of a novel thermostable lipase from *Aspergillus niger*. *Lipids* **2000**, *35*, 495–502.

63. Pokorny, D.; Cimerman, A.; Steiner, W. *Aspergillus niger* lipases: Induction, isolation and characterization of two lipases from a MZKI Al 16 strain. *J. Mol. Cat. B Enzym.* **1997**, *2*, 215–222.

64. Xiaoming, L.; Breddam, K. A novel carboxylesterase from *Aspergillus niger* and its hydrolysis of succinimide esters. *Carlsberg Res. Commun.* **1989**, *54*, 241–249.

65. Rezwan, K.; Studart, A.R.; Volrols, J.; Gauckler, L.J. Change of ζ potential of biocompatible colloidal oxide particles upon adsorption of bovine serum albumin and lysozyme. *J. Phys. Chem. B* **2005**, *109*, 14469–14474.

66. Rezwan, K.; Meier, L.P.; Rezwan, M.; Voros, J.; Textor, M.; Gauckler, L.J. Bovine serum albumin adsorption onto colloidal Al_2O_3 particles: A new model based on zeta potential and UV-Vis measurements. *Langmuir* **2004**, *20*, 10055–10061.

67. Bernsmann, F.; Frisch, B.; Ringwald, C.; Ball, V. Protein adsorption on dopamine–melanin films: Role of electrostatic interactions inferred from ζ-potential measurements *versus* chemisorption. *J. Colloid Interface Sci.* **2010**, *344*, 54–60.

68. Li, S.; Hu, J.; Liu, B. A study on the adsorption behavior of protein onto functional microspheres. *Chem. Technol. Biotechnol.* **2005**, *80*, 531–536.

69. Bradford, M.M. Rapid and sensitive method for the quantitation of microgram quantities of protein utilizing the principle of protein-dye binding. *Anal. Biochem.* **1976**, *72*, 248–254.

70. Sheldon, R.A.; van Pelt, S. Enzyme immobilisation: Why, what and how? *Chem. Soc. Rev.* **2013**, *42*, 6223–6235.

71. Abdel-Naby, M.A. Immobilization of Aspergillus niger NRC 107 xylanase and beta-xylosidase, and properties of the immobilzed enzymes. *Appl. Biochem. Biotechnol.* **1993**, *38*, 69–81.

72. Jia, J.; Hu, Y.; Liu, L.; Jiang, L.; Zou, B.; Huang, H. Enhancing catalytic performance of porcine pancreatic lipase by covalent modification using functional ionic liquids. *ACS Catal.* **2013**, *3*, 1976–1983.

73. Emregul, E.; Sungur, S.; Akbulut, U. Polyacrylamide-gelatine carrier system used for invertase immobilization. *Food Chem.* **2006**, *97*, 591–597.

74. Zhu, Y.T.; Ren, X.Y.; Liu, Y.M.; Wei, Y.; Qing, L.S.; Liao, X. Covalent immobilization of porcine pancreatic lipase on carboxyl-activated magnetic nanoparticles: Characterization and application for enzymatic inhibition assays. *Mater. Sci. Eng. C Mater Boil. Appl.* **2014**, *38*, 278–

285.

75. Melgosa, R.; Sanz, M.T.; Solaesa, A.G.; Bucio, S.L.; Beltran, S. Enzymatic activity and conformational and morphological studies of four commercial lipases treated with supercritical carbon dioxide. *J. Supercrit. Fluids* **2015**,*97*, 51–62.

76. Zdarta, J.; Sałek, K.; Kołodziejczak-Radzimska, A.; Siwińska-Stefańska, K.; Szwarc-Rzepka, K.; Norman, M.; Klapiszewski, Ł.; Bartczak, P.; Kaczorek, E.; Jesionowski, T. Immobilization of *Amano Lipase A* onto Stöber silica surface: Process characterization and kinetic studies. *Open Chem.* **2015**, *13*, 138–148.

Chapter 9

FROM PROTEIN ENGINEERING TO IMMOBILIZATION: PROMISING STRATEGIES FOR THE UPGRADE OF INDUSTRIAL ENZYMES

Raushan Kumar Singh, Manish Kumar Tiwari, Ranjitha Singh and Jung-Kul Lee

Department of Chemical Engineering, Konkuk University, 1 Hwayang-Dong, Gwangjin-Gu, Seoul 143-701, Korea

ABSTRACT

Enzymes found in nature have been exploited in industry due to their inherent catalytic properties in complex chemical processes under mild experimental and environmental conditions. The desired industrial goal is often difficult to achieve using the native form of the enzyme. Recent developments in protein engineering have revolutionized the development of commercially available enzymes into better industrial catalysts. Protein engineering aims at modifying the sequence of a protein, and hence its structure, to create enzymes with improved functional properties such as stability, specific activity, inhibition by reaction products, and selectivity towards non-natural substrates. Soluble enzymes are often immobilized onto solid insoluble supports to be reused in continuous processes and to facilitate the economic recovery of the enzyme after the reaction without any significant loss to its biochemical properties. Immobilization confers considerable stability towards temperature variations and organic solvents. Multipoint and multisubunit covalent attachments of enzymes on appropriately functionalized supports via linkers provide rigidity to the immobilized enzyme structure, ultimately resulting in improved enzyme stability. Protein engineering and immobilization techniques are sequential and compatible approaches for the improvement of enzyme properties. The present review highlights and summarizes various studies that have aimed to improve the biochemical properties of industrially significant enzymes.

INTRODUCTION

Biocatalysts are extensively used in the industrial production of bulk chemicals and pharmaceuticals, and over 300 processes have already been implemented [1]. In the vast majority of processes, native microbial enzymes of microbial with that exhibit the desired properties are used. Often, for a given biocatalytic process, the native enzyme does not meet the requirements for large-scale application, and its properties thus need to be optimized or modulated. Many industrial enzymes, such as lipases, with a wide range of substrate specificities, are utilized in many processes, often compromising the desired productivity. The role of protein engineering is to overcome the limitations of natural enzymes as biocatalysts and engineer process-specific biocatalysts. This includes optimizing the chemoselectivity, regioselectivity, and, especially, stereoselectivity of the biocatalyst, as well as process-related aspects, such as long-term stability at certain temperatures or pH-values and activity in the presence of high substrate concentrations to achieve maximal productivity. Some improvement in process efficiency can be achieved by modifying the chemical manufacturing process to suit the sensitivities of the biocatalyst (e.g., in terms of pH, temperature, and solvents). The other alternative is to use protein engineering methodologies to generate new biocatalysts to function under more ideal process conditions, followed by immobilization to establish more robust processes. Due to recent advances in protein engineering techniques, numerous examples of the optimization of certain enzyme traits (e.g., thermostability, tolerance towards organic solvents, enantioselectivity) have been reported. This has been achieved both by developing new screening systems and by advancements in the understanding of protein structure. Rational protein design by default variants with very few amino acid exchanges and simultaneous saturation mutagenesis allows the generation of synergistic effects of neighboring mutations. On the other hand, semi-rational approaches have recently been shown to be very efficient in cases where the key amino acids governing the property of interest are known [2]. The choice of method therefore is still a case-to-case decision, depending on the property of interest and existing structural and mechanistic knowledge, as well as on practical considerations, such as the availability of a high-throughput screening or selection system.

Enzymes are considered to be sensitive, unstable at elevated temperatures, and require an aqueous medium for function; these are features that are not ideal for a catalyst, and are undesirable in most syntheses. In many cases a simple way to avoid at least some of these drawbacks is to immobilize enzymes [3]. The immobilization of enzymes has proven particularly valuable and has been exploited over the last four decades to enhance enzyme properties such as activity, stability, and substrate specificity for their successful utilization in industrial processes (Figure 1). In spite of the long history and obvious advantages of enzyme immobilization [4], Straathof *et al.* (2002) estimated that only 20% of biocatalytic processes involve immobilized enzymes [5]. Initially, the main challenge was to find suitable immobilization methods to allow multiple uses of enzymes for the same reaction. With the advancement in immobilization techniques, the focus has shifted to the development of modulated enzymes with the desired properties for certain specific applications. Immobilization has its associated advantages (it allows for multiple, repetitive, or continuous use and has minimum reaction time, high stability, improved process control, multienzyme system, easy product separation, while it is less labor intensive and more cost effective, safe to use, and environmentally friendly) [6] and disadvantages (its lowered activity, conformational change of the enzyme, possibility of enzyme denaturation, changes in properties, mass transfer limitations, and lowered efficacy against insoluble substrates). This review covers different strategies of protein engineering and immobilization to modulate the properties of enzymes to suit industrial processes.

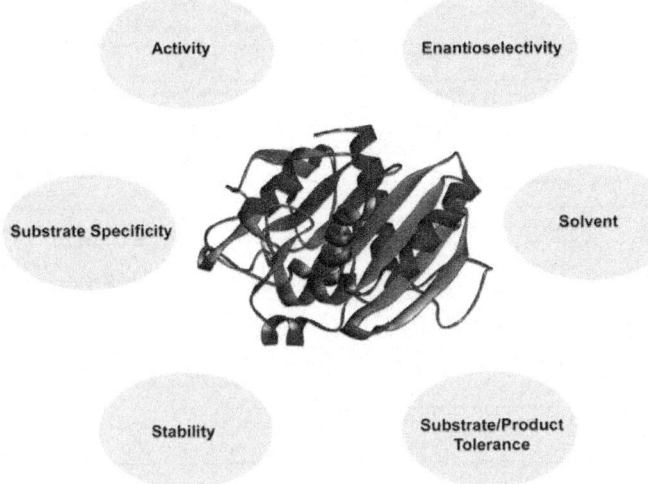

Figure 1: Evolvable enzyme properties for its successful utilization in industrial processes.

PROTEIN ENGINEERING TO UPGRADE INDUSTRIAL ENZYMES

In the past few decades, biocatalysts have been successfully exploited for the synthesis of various complex drug intermediates, speciality chemicals, and even commodity chemicals in the pharmaceutical, chemical, and food industries due to their inherent ability to catalyze reactions with high velocity and unmet specificity under a variety of conditions, as well as their potential as a greener alternative to chemical catalysts. The increasing interest in applying enzymes in industrial and household catalysis has spurred the development of protein engineering methodologies for novel biocatalysts with new or improved properties. Recent advances in recombinant DNA technology, high throughput technology, genomics, and proteomics have fueled the development of new biocatalysts and biocatalytic processes. Since the beginning of large-scale (recombinant) enzyme production for industrial applications, protein engineering has emerged as a powerful tool to improve enzyme properties. Enzymes with the desired properties such as enhanced activity, high thermostabilty, and specificity under industrial conditions can be obtained by optimizing process conditions and by protein engineering (Table 1).

Table 1: List of the enzymes engineered by protein engineering

Enzyme	Organism	Improved property	Method	Application	Reference
Hydantoinase	Arthrobacter sp.	Enantioselective hydantoinase and 5-fold more productivity	Saturation mutagenesis, screening	Production of L-Met (L-amino acids)	[7]
Cyclodextrin glucanotransferase	Bacillus stearothermophilus ET1	Modulation of cyclizing activity and thermostability	Site-directed mutagenesis	Bread industry	[8]
Lipase B	Candida antarctica	20-fold increase in half-life at 70 °C	epPCR	Resolution and desymmetrization of compound	[9]
Tagatose-1,6-Bisphosphate aldolase	E. coli	80-fold improvement in k_{cat}/K_m and 100-fold change in stereospecificity	DNA shuffling and screening	Efficient syntheses of complex stereoisomeric products	[10]
Xylose isomerase	Thermotoga neapolitana	High activity on glucose at low temperature and low pH	Random Mutagenesis and screening	Used in preparation of high fructose syrup	[11]
Amylosucrase	Neisseria polysaccharea	5-fold increased activity	Random mutagenesis, gene shuffling, and directed evolution	Synthesis or the modification of polysaccharides	[12]
Galactose oxidase	F. graminearum	3.4–4.4 fold greater V_{max}/K_m and increased specificity	epPCR and screening	Derivatization of guar gum	[13]
Fructose bisphosphate aldolase	E. coli	Increased thermostablity and stability to treatment with organic solvent	DNA shuffling	Use in organic synthesis	[14]
1,3-1,4-α-D-glucanase	Fibrobacter succinogenes	3–4-fold increase in the turnover rate (k)	PCR-based gene truncation	Beer industry	[15]
Lipase	P. aeruginosa	2-fold increase in amidase activity	Random mutagenesis and screening	Understanding lipase inability to hydrolyze amides	[16]
Protease BYA	Bacillus sp. Y	Specific activity 1.5-fold higher	Site-directed mutagenesis	Detergents products	[17]
p-Hydroxybenzoate hydroxylase	Pseudomonas fluorescens NBRC 14160	Activity, reaction specificity, and thermal stability	Combinatorial mutagenesis	Degrading various aromatic compounds in the environment	[18]
Endo-1,4-β-xylanase II	Trichoderma reesei	Increased alkali stability	Site-directed mutagenesis	Sulfate pulp bleaching	[19]
Xylose isomerase	Thermotoga neapolitana	2.3-fold increases in catalytic efficiency	Random mutagenesis	Production of high fructose corn syrup	[11]

Enzyme	Source	Improvement	Method	Application	Ref
α-Amylase	*Bacillus* sp. TS-25	10 °C enhancement in thermal stability	Directed evolution	Baking industry	[20]
Xylanase		Tm improved by 25 °C	Gene site-saturation mutagenesis	Degradation of hemicellulose	[21]
Fructosyl peptide oxidase	*Coniochaeta* sp	79.8-fold enhanced thermostability	Directed evolution and site-directed mutagenesis	Clinical diagnosis	[22]
Endo-β-1,4-xylanase	*Bacillus subtilis*	Acid stability	Rational protein engineering	Degradation of hemicellulose	[23]
Subtilase	*Bacillus* sp.	6-fold increase in caseinolytic activity at 15–25 °C	Directed evolution and site-directed mutagenesis	Detergent additives and food processing	[24]
CotA laccase	*B. subtilis*	120-fold more specific for ABTS	Directed evolution	Catalyze oxidation of polyphenols	[25]
Pyranose 2-oxidase	*Trametes multicolor*	Altered substrate selectivity for D-galactose, D-glucose	Semi-rational enzyme engineering approach	Food industry	[26]
Xylanase XT6	*Geobacillus stearothermophilus*	52-fold enhancement in thermostability; increased catalytic efficiency	Directed evolution and site-directed mutagenesis	Degradation of hemicellulose	[27]
Lipase	*Bacillus pumilus*	Thermostability and 4-fold increase in k_{cat}	Site-directed mutagenesis	Chemical, food, leather and detergent industries	[28]
Bgl-licMB	*Bacillus amyloliquefaciens* (Bgl) and *Clostridium thermocellum* (licMB)	2.7 and 20-fold higher k_{cat}/K_m than that of the parental Bgl and licMB, respectively	Splicing-by-overlap extension	Brewing and animal-feed industries	[29]
β-agarase AgaA	*Zobellia galactanivorans*	Catalytic activity and thermostability	Site-directed mutagenesis	Production of functional neo-agarooligosaccharides	[30]
Prolidase	*Pyrococcus horikoshii*	Thermostability	Random mutagenesis	Detoxification of organophosphorus nerve agents	[31]
Lipases	*Geobacillus* sp. NTU 03	79.4-fold increment in activity; 6.3–79-fold enhanced thermostability	Error-prone PCR and site-saturation mutagenesis	Transesterification	[32]
Xylanase	*Hypocrea jecorina*	Thermostability	Look-through mutagenesis (LTMTM) and combinatorial beneficial mutagenesis (CBMTM)	Degradation of hemicellulose	[33]
Amylase	*Bacillus* sp. US149	Thermostability	Site-directed mutagenesis	Bread industry	[34]
Cholesterol oxidase	*Brevibacterium* sp.	Thermostability and enzymatic activity	Site-directed mutagenesis	Detection and conversion of cholesterol	[35]
Lipase B	*Candida antarctica*	Enhancement of thermostability	Molecular dynamics (MD) simulation and site-directed mutagenesis	Detergent industries	[36]
Laccase	*Bacillus* HR03	3-fold improved k_{cat} and thermostability	Directed mutagenesis	Catalyze oxidation of polyphenols, and polyamines	[37]
D-psicose 3-epimerase	*Agrobacterium tumefaciens*	Thermostability	Random and site-directed mutagenesis	Industrial producer of D-psicose	[38]
1,3-1,4-β-D-glucanase	*Fibrobacter succinogenes*	Thermostability and specific activity	Rational mutagenesis	Widely used as a feed additive	[39]
α-Amylase	*Bacillus licheniformis*	Acid stability	Direct evolution	Starch hydrolysis	[40]
Alkaline amylase	*Alkalimonas amylolytica*	Oxidative stability	Site-directed mutagenesis	Detergent and textile industries	[41]
Endoglucanase	*Thermoascus aurantiacus*	4-fold increase in k_{cat} and 2.5-fold improvement in hydrolytic activity on cellulosic substrates	Site-directed mutagenesis	Bioethanol production	[42]
D-glucose 1-dehydrogenase isozymes	*Bacillus megaterium*	Substrate specificity	Site-directed mutagenesis	Measurements of blood glucose level	[43]
Glycerol dehydratase	*Klebsiella pneumoniae*	2-fold pH stability; enhanced specific activity	Rational design	Synthesis of 1,3-Propanediol	[44]

Cyclodextrin Glucanotransferase	*Bacillus sp.* G1	Enhancement of thermostability	Rational mutagenesis	Starch is converted into cyclodextrins	[45]
Cellobiose phosphorylase	*Clostridium thermocellum*	Enhancement of thermostability	Combined rational and random approaches	Phosphorolysis of cellobiose	[46]
Superoxide dismutase	*Potentilla atrosanguinea*	Thermostability	Site-directed mutagenesis	Scavenging of O_2^-	[47]
Endoglucanase Cel8A	*Clostridium thermocellum*	Thermostability	Consensus-guided mutagenesis	Conversion of cellulosic biomass to biofuels	[48]
Endo β-glucanase Egl499	*Bacillus subtilis* JA18	Increase in half life from 10 to 29 mins at 65 °C	Deletion of C-terminal region	Animal feed production	[49]
Pyranose 2-oxidase	*Trametes multicolor*	Increase half life from 7.7 min to 10 h (at 60 °C)	Designed triple mutant	Food industry	[50]
Xylanase XT6	*Geobacillus stearothermophilus*	52× increase in thermal stability, k_{cat} increase by 10 °C, catalytic efficiency increase by 90%	Directed evolution and site-directed mutagenesis	Biobleaching	[27]
Tyrosine phenol-lyase	*Symbiobacterium toebi*	Improved thermal stability and activity (Increase in T_m up to 11.2 °C)	Directed evolution (random mutagenesis, reassembly and activity screening)	Industrial production of l-tyrosine and its derivatives	[51]
Phytase	*Penicilium sp.*	Increased thermal stability	Random mutation and selection	Feed additives	[52]
L-Asparaginase	*Erwinia carotovora*	Increase in half-life from 2.7 to 159.7 h	*In vitro* directed evolution	Therapeutic agent	[53]
Endoglucanase CelA	*Clostridium thermocellum*	10-fold increase in half-life of inactivation at 86 °C	Saturation mutagenesis	Bioconversion of cellulosic biomass	[54]
β-glucosidase BglC	*Thermobifida fusca*	Increase in half-life from 12 to 1244 min	Family shuffling, site saturation, and site-directed mutagenesis	Bioconversion of cellulosic biomass	[55]
Phospholipase D	*Streptomyces*	Improved thermal stability and activity	Semi-rational, site-specific saturation mutagenesis	Phosphatidylinositol synthesis	[56]
β-glucosidase	*Trichoderma reesei*	Enhanced k_{cat}/K_m and k_{cat} values by 5.3- and 6.9-fold	Site-directed mutagenesis	Hydrolysis of cellobiose and cellodextrins	[57]
Lipases		144-fold enhanced thermostability	Error prone PCR	Synthesis and hydrolysis of long chain fatty acids	[58]
Laccase	*Pycnoporus cinnabarinus*	8000-fold increase in k_{cat}/K_m	Directed evolution and semi-rational engineering	Lignocellulose biorefineries, organic synthesis, and bioelectrocatalysis	[59]
Feruloyl esterase A	*Aspergillus niger*	Increase in half-life from 15 to >4000 min	Random and site-directed mutagenesis	Degradation of lignocellulose	[60]

Activity

Improving the activity of an industrial enzyme is often a primary goal. This is partly because naturally available enzymes are usually not optimally suited for many processes in industrial applications. Many industrial enzymes, such cellulases, amylases, lipases, and even proteases, act on insoluble substrates. Therefore, the rate of substrate turnover may be limited by diffusion, and controlled by enzyme mobility at the surface or by on/off enzyme desorption rates [61]. These, in turn, are often related to the surface properties of the enzyme and the conditions at the interface between the enzyme and substrate [62]. A comparative study of the experimental results from several site-directed variants with structural modeling of fungal lipase from *Rhizopuss oryzae* has provided much insight into the molecular mechanism of catalysis [63]. Substitutions at Glu87 and Trp89 in the lid region have been suggested to alter the activity of the lipase from *Humicola lanuginosa* (lipolase) [64].

Cellulases and xylanase have become major focus in recent years due to their ability to provide the soft feel of stone-washed jeans in textile processing, fabric care benefits (such as color crispness) when used in laundry detergents [65], and reduction of the quantity of chemicals required for bleaching in the pulp and paper industry, thereby minimizing environmental impact [66]. Tyr169 in the *Trichoderma reesei* cellobiohydrolase II catalytic domain plays an important role in distorting the glucose ring into a more reactive conformation [67]. A detailed discussion of different families of xylanases and their structure and activity is provided in a review article [68] in *Current Opinion in Biotechnology*.

Thermal Stability

Enhanced thermostability is one of the most common properties desired as output from a protein engineering study and is often an important economic factor. The stability of an enzyme is affected by many factors, such as temperature, pH, solvent, and the presence of surfactants. Among all possible deactivating factors, temperature is the best studied. At elevated temperatures, many enzymes tend to become (partly) unfolded and/or inactivated, meaning that they are no longer able to perform the desired tasks. There are two types of protein stability, thermodynamic and long-term, that are crucial from an applied perspective. Numerous protein engineering strategies have been reported in the last 10 years. Site-directed mutagenesis (SDM) and directed evolution have been exploited to engineer catalysts with improved thermostability (Figure 2). However, a combination of both strategies is becoming popular among researchers. Cherry *et al.* (1999) reported a combination of rational engineering and directed evolution techniques to improve the resistance of a fungal peroxidase towards hydrogen peroxide and high temperature, at high pH, with the aim of making the enzyme better suited for laundry applications [69]. They obtained an enzyme variant with drastically improved thermal stability (200-fold greater than that of the wild-type enzyme) under conditions that mimic those in a washing machine. In another early study, Martin *et al.* (2001) described the stabilization of a cold-shock protein from *B. subtilis* (Bs-Csp) [70]. The size of the library was limited by the transformation efficiency of *Escherichia coli*, but still amounted to a respectable 107 variants. After six rounds of phage selection from 107 variants, five variants displayed increased stability. The most stable variant, displaying a remarkable improvement in T_m (a 22 °C increase), differed at six randomized positions from its mesophilic parent. Very few charged residues were found in the selection, which might be related to the use of guanidine hydrochloride (GdmHCl) in the selection scheme. After applying several selection regimes (temperature, amount of protease), mutants displaying up to a 28 °C increase in T_m were obtained. The best mutant from the thermal screen differed at all six positions from the wild-type, the thermostable counterpart, Bc-Csp, and from the variant obtained by GdmHCl selection. Palackal *et al.* (2004) reported one of the highest stabilizations ever obtained by enzyme engineering (increase in T_m of over 30 °C) [71]. Their starting point was a xylanase that was discovered by screening

50,000 plaques from a complex environmental DNA library derived from a sample of fresh bovine manure. The use of site-saturation mutagenesis and the screening of approximately 70,000 clones led to the identification of nine interesting mutations, which, when combined, increased the Tm by 34.2 °C.

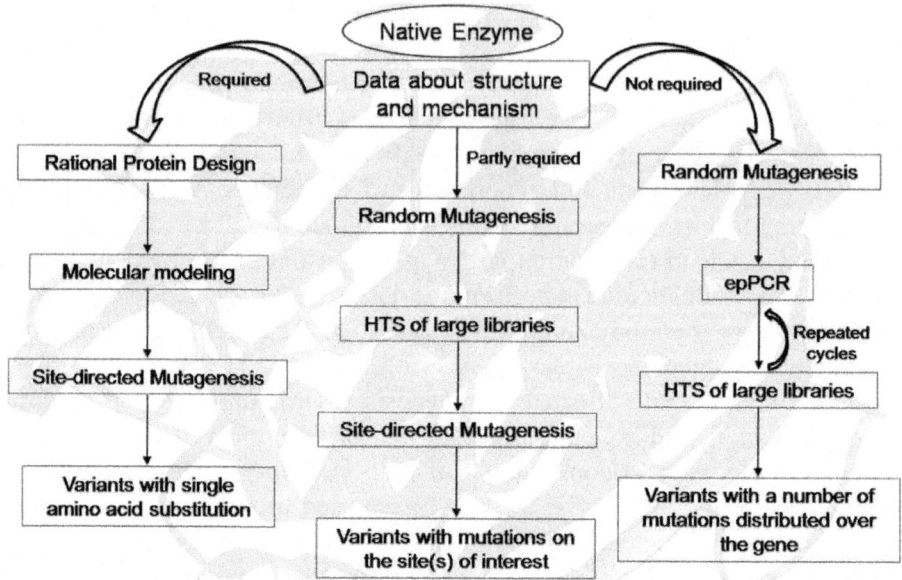

Figure 2: Schematic representation of protein engineering strategies. Engineering method should be selected on the basis of the structural and mechanistic information and the feasibility of a high-throughput screening (HTS) system for screening or selection.

Directed evolution is a powerful engineering method, and it is often used to design enzymes with increased thermostability [72]. By screening for initial activity and residual activity at an elevated temperature, both the thermostability and activity of mesophilic subtilisin E [73], psychrophilic subtilisin S41 [74] and mesophilic *p*-nitrobenzyl esterase [75] were significantly increased using directed evolution strategies. Enzymes that have been improved by directed evolution have already been commercialized [72]. A major advantage of this engineering method over SDM is that no knowledge about enzyme structure is necessary. Since there is still much to be learned about thermostabilization mechanisms, SDM approaches often yield disappointing results. *Bacillus subtilis* subtilisin E was converted into an equivalent of its thermophilic homolog thermitase through the successive application of one round of error-prone PCR, one step of DNA shuffling (to combine the properties of the best variants), and four additional rounds of error-prone (ep) PCR. The evolved enzyme was 15 times more active than subtilisin E at 37 °C, it showed a 16 °C increase in T_{opt}, and its T_m at 65 °C was more than 200 times that of subtilisin E [73]. In another experiment, the thermostability of *B. subtilis p*-nitrobenzyl esterase was enhanced through five cycles of epPCR followed by one step of DNA shuffling [76]. The evolved esterase showed a 14 °C increase in T_{opt} and a 10 °C increase in T_m, and it was more active than the wild-type enzyme at any temperature.

Using a directed evolution approach, Oshima and coworkers (2001) obtained four variants of a thermophilic 3-isopropylmalate dehydrogenase with enhanced specific activities at low temperatures [77]. Two of these variants exhibited wild-type thermostability, while the other two exhibited decreased thermostability, with an inverse correlation between activity and stability. A similar result was obtained in case of thermostable *S. cerevisiae* 3-isopropylmalate dehydrogenase variants evolved using genetic selection in an extreme thermophile [78]. On the other hand, mutations targeted at areas whose unfolding is limiting in the protein denaturation process can provide extensive stabilization. Good illustrations can be found in the stability studies of *Bacillus stearothermophilus* thermolysin-like protease. Stabilizing mutations were all located on the surface, around one flexible loop located in the β-pleated *N*-terminal domain [79–81]. The association of eight mutations in the same area resulted in a 340-fold kinetic stabilization of *B. stearothermophilus* thermolysin-like protease at 100 °C and did not affect catalytic activity at 37 °C [82].

Successes in substituting left-handed helical residues with Gly or Asn, in introducing prolines in surface turns or loops, in introducing non-local surface ion pairs, and in creating disulfide bridges that dock loops to the protein surface are well documented [82–86]. Two types of mutations, Gly to Xaa and Xaa to Pro, can be introduced. In the first case, the newly introduced β-carbon should not interfere with neighboring atoms. In the second case, the substitution site should have specific dihedral angles (φ and ψ) in the regions −50 to −80 and 120 to 180 or −50 to −70 and −10 to −50, and the residue preceding the potential proline should also have a specific conformation. In addition, the proline ring should not interfere with neighboring atoms, and the substitution should not eliminate stabilizing non-covalent interactions. The most promising strategies for thermostabilization using SDM should focus on the surface areas, particularly loops and turns, and on creating additional non-local ion pairs. The introduction of disulfide bonds, chemical crosslinks, and salt bridges has been widely used to increase stability, although not all disulfide bonds increase stability [87]. Loops can be made more rigid by decreasing their intrinsic entropy of unfolding. There are now several examples of proteins that have been stabilized by the introduction of numerous mutations with cumulative small stabilizing effects (Table 1) [76,88–91]. However, a clear conclusion to be drawn from others is that very large stability differences in some cases are due to only one or a few point mutations [39,45,46,92–94]. Another method is to anchor the loops to the protein surface, either by non-covalent interactions or using a disulfide bridge. Introducing a disulfide bridge in a semiflexible area of the protein should help compensate for any conformational strain created by the disulfide bridge [80].

Solvent Stability

The use of organic solvents as reaction media for biocatalytic reactions has proven to be an extremely useful approach to expand the range and efficiency of the practical applications of biocatalysis [95]. Unfortunately, the majority of naturally available biocatalysts are usually not optimally suited for catalysis in non-aqueous solvents in industrial processes. Polar solvents of practical interest, such as acetone or dimethylformamide (DMF), interact with the enzyme and associated water molecules and drastically reduce catalytic activity [96]. The ability to use enzymes in non-aqueous solvents has increasingly drawn the attention of researchers worldwide to the problems and potential of non-aqueous biocatalysis in chemical transformations that are useful for many industries. There are many potential advantages of enzyme catalysis in non-aqueous solvents, including (1) enhanced solubility of substrates such as lipids and phospholipids; (2) novel chemistry in synthetic applications; (3) altered substrate specificity; (4) easy product recovery; and (5) reduced microbial contamination [95,97]. Despite the many limitations of non-aqueous environments, particularly the poor stability of enzymes in polar organic solvents, research in this area has made tremendous progress in recent years, focusing in particular on the elucidation of enzyme structure and improvement of stability and catalysis in organic solvents, for synthetic applications.

Enzymes may be redesigned to enhance catalysis and stability in non-aqueous solvents by engineering their amino acid sequences, thereby altering their functional properties to adapt the new solvent environment. Many strategies to engineer proteins for non-aqueous environments have been developed [98–102]. Among them, directed evolution and rational design approaches are widely utilized. Directed evolution approaches involving SDM are more efficient when detailed structural information and the molecular basis for the property of interest are poorly understood. Impressive work has been done using an evolutionary approach consisting of multiple steps of random mutagenesis and screening to improve the activity and stability of subtilisin E in high concentrations of organic solvent [99]. Random mutagenesis by

PCR techniques combined with screening resulted in enhanced activity in the presence of dimethylformamide (DMF). The triple mutant (D60N + Q103R + N218S) is 38 times more active than wild-type subtilisin E in 85% DMF. Single amino acid substitutions increase the activity and stability of mutant enzyme in mixtures of organic solvents and water, and the effects of these mutations are additive. The N218S substitution is reported to cause improved activity and stability of subtilisin BPN, as well as improved activity and stability of subtilisin E in the presence of DMF. The double mutant Q103R + N218S is 10 times more active than the wild-type enzyme in 20% (v/v) DMF and twice as stable in 40% DMF. Similar examples of enzyme stabilization also involve subtilisin E in polar organic solvents (DMF) by rational engineering. The substitution of Asp248 with three amino acids of increasing hydrophobicity, Asn, Ala, and Leu, resulted in stabilized variants with respect to wild type in 80% DMF. This stabilization was only observed at high concentrations of organic solvent, and not at low organic acid concentrations (40% DMF). In contrast, the mutant N218S stabilizes subtilisin E at both low (40%) and high (80%) concentrations of DMF. The double mutant (D248N + N218S) protein is 3.4 times more stable than the wild type in 80% DMF [103]. This study provides additional evidence that substitution of surface-charged residues is generally additive and useful for stabilizing enzymes in organic solvent. Using the directed approach, random mutagenesis, recombination, and screening, Song and Rhee (2001) obtained three variants of phospholipase A with enhanced stability and activity in organic solvents [104]. Using a similar strategy, Arnold and coworkers (2001) reported a variant of horseradish peroxidase with enhanced stability in the presence of H_2O_2, sodium dodecyl sulfate (SDS), and salts [105].

Substrate Specificity

Engineering novel enzyme specificity using a directed evolution approach is of increasing importance to the chemical and pharmaceutical industries. Several recent reports describe significant advances made toward this goal. Using epPCR, followed by saturation mutagenesis and screening, Reetz and coworkers [106,107] considerably modified the enantioselectivity of a *Pseudomonas aeruginosa* lipase towards 2-methyldecanoate from $E = 1.04 \times$ (2% enantiomeric excess [ee]) to $E = 25 \times$ (90%–93% ee) (where E is the enantioselective factor). None of the five amino acid substitutions in the best variant was located near the substrate-binding pocket [108]. Using a similar approach, Arnold and coworkers (2000) successfully inverted the enantioselectivity of a hydantoinase from D selectivity (40% ee) to moderate L preference (20% ee) [7]. An impressive example of switching enzyme substrate specificities can be seen in the DNA shuffling of two highly homologous triazine hydrolases [109]. The two enzymes, AtzA and TriA, are distinguished by nine amino acids, and hydrolyze s-triazines by dechlorination and deamination, respectively, with little overlap in substrate preference. Permutations of the nine amino acid differences by DNA shuffling resulted in a set of variants that hydrolyzed five of eight triazines that were not substrates for either starting enzyme. The *E. coli*d-2-keto-3-deoxy-6-phosphogluconate (KDPG) aldolase catalyzes a highly specific reversible aldol reaction on d-configurated KDPG substrate. Using directed evolution approach, Wong and coworkers (2000) obtained a variant capable of accepting both d- and l-glyceraldehyde as substrates in a non-phosphorylated form [110]. Notably, all the six substitutions found in the resulting variant were far away from the active site. A double mutant (Lys133Gln/Thr161Lys) of the same enzyme with a considerably altered substrate profile was reported using epPCR and SDM [111]. The directed evolution of oxygenases exhibited similar results [112–115].

A Way Forward: Hybrid Approaches

Over the past decade, advances in DNA technologies and bioinformatics have substantially accelerated the redesign of proteins with novel or desired characteristics. Proteins can be rationally engineered if information about the catalytic mechanism and structure of a protein is known. However, practical experience shows that protein dynamics is complex and it is understood that substitutions distant from the active site can alter the characteristics of a protein [116,117]. Directed evolution would be a more suitable approach for the proteins whose structures and mechanisms are known. However, this method is driven by multiple rounds of selection and screening and each step improves the enzyme. In cases where high-throughput screens are unavailable, a semi-rational method involving site saturation mutagenesis may be applied to identify target residues through computational methods. This way, the library size can be reduced to a manageable size to increase the chance of discovering desired variants.

Recently, molecular modeling has been employed to predict protein structures and various algorithms are being used to predict secondary and tertiary structures based on amino acid sequence [118–120]. This advancement immensely supports rational design of proteins with unknown structures [118,120]. Furthermore, simulation and molecular docking of small molecules to proteins is another field that has advanced tremendously over the years [121,122]. Molecular docking can facilitate understanding of ligand–protein interactions and hence rational designing of proteins for desired properties [123,124].

Assimilation of unnatural amino acids (uAAs) into proteins has opened new possibilities for creating proteins with novel functions and improved properties [125]. Substitution of natural amino acids with a uAA at multiple specific sites in a protein and insertion of uAAs into proteins to expand the genetic code are two different approaches. With such progress, assimilation of uAAs into proteins may soon be a routine practice in the field of protein engineering, and may become a powerful technique for designing novel enzymes to meet the demands of synthetic biology [126].

IMMOBILIZATION TO UPGRADE INDUSTRIAL ENZYMES

Since the first industrial application of immobilized amino acylase in 1967 for the resolution of amino acids, enzyme immobilization technology has attracted increasing attention and considerable progress has been made in recent decades. Enzymes are exploited as catalysts in many industrial, biomedical, and analytical processes. There has been considerable interest in the development of enzyme immobilization techniques because immobilized enzymes have enhanced stability compared to soluble enzymes, and can easily be separated from the reaction. Approaches used for the design of immobilized enzymes have become increasingly more rational and are employed to generate improved catalysts for industrial applications.

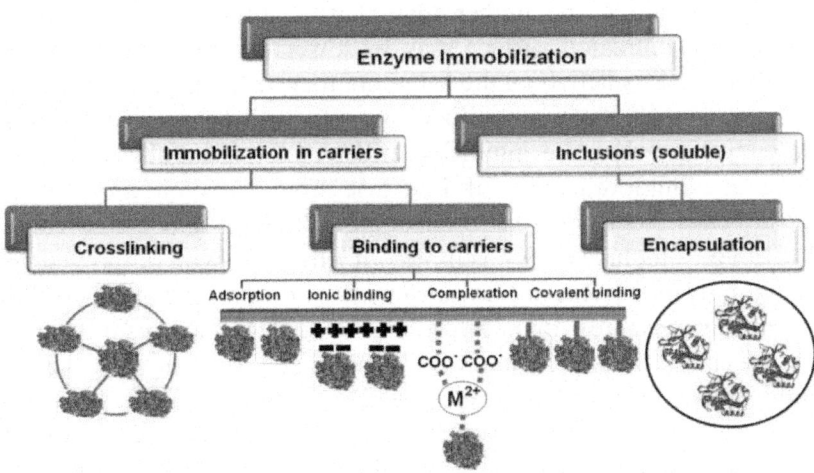

Figure 3: Immobilization of enzyme via different routes.

There are a variety of methods used to immobilize enzymes, the three of the most common being adsorption, entrapment, and crosslinking or covalently binding to a support (Figure 3). Recently, the major focus of enzyme immobilization is the development of robust enzymes that are not only active but also stable and selective in organic solvents. The ideal immobilization procedure for a given enzyme is one that permits a high turnover rate of the enzyme while retaining high catalytic activity over time. Proteins are immobilized either by physical adsorption to the surface of the nanoparticle or by covalent bonding to previously functionalized nanoparticles.

Activity

In the early 1960s Goldstein *et al.* observed the enhancement of enzyme activity upon immobilization because of the microenvironment effect [127]. In general, enhancement of enzyme activity upon immobilization depends on the microenvironment, partition effect, diffusion effect, conformational change, molecular orientation, conformational flexibility, conformation induction, and binding mode. It has been observed that many immobilized enzymes exhibit higher activity than the corresponding native enzyme. For example, immobilized lipase was 50-fold more active than the native enzyme [128]. In literature, there are few reports of enzymes that exhibit decreased K_m and increased V_{max} upon the formation of crosslinked enzyme aggregates (CLEA) [73,129–131] compared the covalent immobilization and physical adsorption of a cellulase enzyme mixture on plasma immersion ion implantation (PIII)-treated and untreated polystyrene, respectively. Activity on the PIII-treated surface was found to be higher than that on the untreated surface. The activity on the untreated surface may be lower because enzyme aggregates are generally not as active as the same number of unaggregated enzymes. It has been described that immobilized enzymes can be more active than the native enzymes, when the inhibiting effect of the substrate is reduced. For example, the immobilization of invertase from *Candida utilis* on porous cellulose beads led to decreased substrate inhibition and increased activity [132]. Table 2 includes the examples of enzymes with improved activity via immobilization. Recent advancements in immobilization techniques, particularly in oriented immobilization, have generally resulted in higher activity relative to their randomly immobilized counterparts, because of favorable accessibility or avoidance of active site modification [133]. These methods focus on the improvement of immobilized enzyme properties without compromising activity. In case of trivial immobilization techniques, activity retention was often marginal, requiring laborious screening of immobilization conditions such as enzyme loading, pH, carrier, and binding chemistry [134]. Higher enzyme activity can occasionally be achieved, especially for allosteric enzymes such as lipase. Lipases exist in two different conformations

[135,136]: the closed (inactive) form with lids covering the active site, and the open form, where the active site is fully exposed to the reaction medium [137–139], and enhancement of enzyme activity relative to that of the native enzyme can be observed. Conformation-controlled activity was found to be strongly dependent on the nature of the support used immobilization. For example, lipase PS (*Pseudomonas cepacia*) immobilized on toyonite, celite, glass, and amberlite, exhibited the highest activity on toyonite (37.2 μmol·min^{-1}·mg^{-1}) and the lowest on amberlite (0.4 μmol·min^{-1}·mg^{-1}) [140].

Table 2: Examples of immobilized enzymes with enhanced activity

Enzyme	Applications	Kinetic parameters	Reference
α-Chymotrypsin	Proteolysis (cleave Peptide amide bonds)	Immobilized enzyme: $K_m = 31.7$ μM, $k_{cat} = 20.0$ s^{-1}; soluble enzyme: $K_m = 47.8$ μM, $k_{cat} = 17.8$ s^{-1}	[141]
β-glucosidase	Lignocellulose hydrolysis	Immobilized enzyme: $K_m = 10.8$ mM, $V_{max} = 2430$ μmol·min^{-1}·mg^{-1}, soluble enzyme: $K_m = 1.1$ mM, $V_{max} = 296$ μmol·min^{-1}·mg^{-1}	[142]
Glucose oxidase	Estimation of glucose level up to 300 mg·mL^{-1}	Immobilized enzyme: $K_m = 3.74$ mM, soluble enzyme = 5.85 mM	[143]
Diastase	Starch hydrolysis	Immobilized enzyme: $K_m = 8414$ mM, $V_{max} = 4.92$ μmol min^{-1} mg^{-1}; soluble enzyme: $K_m = 10,176$ mM, $V_{max} = 2.71$ μmol min^{-1} mg^{-1}	[144]
β-galactosidase	GOS synthesis	Immobilized enzyme: $k_1 = 1.41$ h^{-1}; soluble enzyme: $k_1 = 1.16$ h^{-1}	[145]
Keratinase	Synthesis of keratin	Immobilized enzyme: specific activity = 129.0 U·mg^{-1}; soluble enzyme: specific activity = 37 U·mg^{-1}	[146]
Horseradish peroxidase		Immobilized enzyme: $K_m = 0.8$ mM, $V_{max} = 0.72$ μmol min^{-1} mg^{-1}; soluble enzyme: $K_m = 0.43$ mM, $V_{max} = 0.35$ μmol min^{-1} mg^{-1}	[147]
Glucose oxidase	Estimation of glucose level	Immobilized enzyme: $K_m = 2.7$ mM, $V_{max} = 28.6$ U·μg^{-1}; soluble enzyme: $K_m = 9$ mM, $V_{max} = 6.2$ μmol·min^{-1} mg^{-1}	[148]

β-1,4-glucosidase (*Agaricus arvensis*)	Lignocellulose hydrolysis	Immobilized enzyme: K_m = 3.8 mM, V_{max} = 3,347 μmol min^{-1} mg^{-1}; soluble enzyme: K_m = 2.5 mM, V_{max} = 3,028 μmol min^{-1} mg^{-1}	[149]
L-arabinose isomerase (*B. licheniformis*)		Immobilized enzyme: K_m = 352 mM, V_{max} = 326 μmol min^{-1} mg^{-1}; soluble enzyme: K_m = 369 mM, V_{max} = 232 μmol min^{-1} mg^{-1}	[150]
Diastase α-amylase	Hydrolyzing soluble starch	Immobilized enzyme: K_m = 10.3 mg/mL; V_{max} = 4.36 μmol min^{-1} mg^{-1} mg^{-1}; soluble enzyme: K_m = 8.85 mg mL^{-1}, V_{max} = 2.81 μmol·min^{-1}·mg^{-1}	[151]
Cellobiase	Bioethanol production	Immobilized enzyme: K_m = 0.30 mM, V_{max} = 6.77 μM min^{-1}; soluble enzyme: K_m = 2.48 mM, V_{max} = 2.38 μM min^{-1}	[152]
Laccase	Bioremediation of environmental pollutants	Immobilized enzyme: K_m (10^{-2} mM) = 10.7, V_{max} (10^{-2} mM min^{-1}) = 14.0; soluble enzyme: K_m (10^{-2} mM) = 5.69, V_{max} (10^{-2} mM min^{-1}) = 7.7	[153]
Keratinase	Synthesis of keratin	Immobilized enzyme: specific activity = 129 U mg^{-1}; soluble enzyme: specific activity = 37 U mg^{-1}	[146]
Raw starch digesting amylases	Starch hydrolysis	Immobilized enzyme: K_m (10^{-1}) = 3.8 mg mL^{-1}, V_{max} = 27.3 U·mg^{-1}; soluble enzyme: K_m (10^{-1}) = 3.5 mg mL^{-1}, V_{max} = 23.8 U·mg^{-1}	[154]
Aldolase		Immobilized enzyme: K_m = 0.10 mM; k_{cat}/K_m = 584 min^{-1}·mM^{-1}, soluble enzyme K_m = 0.12 mM; k_{cat}/K_m = 540 min^{-1}·mM^{-1}	[155]
α-galactosidase (*Aspergillus terreus*$_{GR}$)	Animal feed	Immobilized enzyme: K_m = 1.40 mM, V_{max} = 20.16 U mL^{-1}; soluble enzyme: K_m = 4.2 mM, V_{max} = 16.33 U·mL^{-1}	[156]
Laccase	Textile wastewater treatment	Immobilized enzyme: K_m = 0.0717 mM, V_{max} = 0.247 mM·min^{-1}; soluble enzyme: K_m = 0.0044 mM, V_{max} = 0.024 mM·min^{-1}	[157]
Papain	Food, pharmaceutical, leather, cosmetic, and textile industries	Immobilized enzyme: K_m = 0.308 × 10^5 g·mL^{-1}; V_{max} = 5.4 g mL^{-1} s^{-1}; soluble enzyme: K_m = 0.236 × 10^5 g·mL^{-1}; V_{max} = 4.08 g·mL^{-1}·s^{-1}	[158]

An enzyme immobilized in an oriented manner using a covalent procedure preserves its functionality and leads to the assemblage of a homogeneous layer. It was demonstrated that the lipase from *Mucor risopus* immobilized in organic solvent was more active in organic solvent whereas the same immobilized in an aqueous medium had almost no activity in organic solvents. It might be that the position of binding of the enzymes to the carrier in organic solvents is different from when immobilization is performed in an aqueous medium [159]. Recently, analysis of cytochrome c reductase assay revealed that oriented enzyme samples resulted in about a three-fold higher activity in solution than randomly bound samples because of favorable accessibility [160]. Orderly oriented enzyme molecules generally have higher activity or stability relative to their randomly immobilized counterpart [133].

When the enzyme is covalently immobilized in porous materials using the conventional protocol, diffusion limitations of the enzyme inside the carriers and slow binding reactions between the enzyme and activated carrier groups slow down the covalent immobilization of the enzyme [155]. Activity retention by carrier-bound immobilized enzymes is usually approximately 50% [161]. At high enzyme loading, especially, diffusion limitation might occur as a result of the unequal distribution of the enzyme within a porous carrier, leading to a reduction of pore volume available to the substrate and product diffusion. Bezbradica *et al.* (2009) reported the specific activity of microwave-assisted immobilization of lipase from*Candida rugosa* on Eupergit C250L was higher because of higher diffusion rate [162]. Microwave irradiation technology has been reported by several researchers to overcome the diffusion limitation. For instance, penicillin acylase from *Bacillus megaterium* was covalently immobilized on mesocellular silica foams [155].

It is a common belief that conformational flexibility provides a protein structure with the essential mechanism for carrying enzymatic reactions. The effect of conformational flexibility on the enzyme activity was initially justified by the observation that enzyme immobilized on a carrier via a suitable spacer often resulted in better retention of activity than the enzyme immobilized without a spacer [163–166]. Bigdeli *et al.* (2008) reported the retention of conformational flexibility of glutamate dehydrogenase, used as a model allosteric protein, upon its immobilization on self-assembled monolayers-modified gold [167].

Activity of immobilized enzyme is often influenced by binding mode. The effect of binding mode may be reflected by the number of bonds formed between the carrier and the enzyme molecules, the position of the bonds and the nature of the bonds. Immobilization of β-galactosidase from *E. coli* and *Kluyveromyces lactis* on thiolsulphinate-agarose and glutaraldehyde-agarose clearly demonstrates the relation between the number of bonds formed between the enzyme and the carrier and enzyme activity [168]. These two enzymes are richer in the lysine residues resulting in more bonds formation with glutaraldehyde-agarose and less retention of activity. Interestingly, α-amylase immobilized on thionyl chloride ($SOCl_2$) and carbodiimide (CDI) activated poly (Me methacrylate-acrylic acid) microspheres retained 67.5% and 80.4% of activity, respectively [169].

Conformational changes of the enzyme induced by immobilization usually decrease the affinity to the substrate (increase of K_m). Furthermore, a partial inactivation of all, or the complete inactivation of a part of the enzyme molecules may occur (decrease of V_{max}). Hanefeld et al. (2008) reported that the difference in kinetics between lipases immobilized on different supports was ascribable to conformational changes induced upon enzyme-polymer interaction [170]. A similar observation had been reported for immobilized invertase on white and black lahar (volcanic mudflow) by the silane–glutarldehyde method [171]. Immobilization brought about an increase in the K_m but a decrease in the V_{max} and these changes were correlated to immobilization induced conformational changes in the enzyme [172].

Thermal Stability

Enormous efforts have made by several research groups to enhance the stability of enzymes with the help of immobilization techniques [170], for use in industrial processes, in which the cost-contribution of the immobilized enzyme is often the indicator of process viability [173]. Random immobilization may not always applicable to improve enzyme rigidity; in some cases, the enzyme stability may even be reduced upon immobilization [174]. During random immobilization, the support may establish undesirable interactions with the enzyme resulting in the destabilization of the enzyme structure. Many useful strategies have been developed for enzyme stabilization by immobilization, such as crosslinking, multipoint attachment, and covalent and non-covalent immobilization. The stability of the immobilized enzyme depends on various factors such as the interaction with the support, binding position, the number of the bonds, conformational freedom, microenvironment, structure of the support, properties of the spacer (charged or neutral, hydrophilic or hydrophobic, size), and immobilization conditions. However, the enhanced stability resulting from immobilization is often ascribed to the intrinsic features of individual immobilization processes. The stability of immobilized enzyme is often determined by an individual stabilization factor or the cumulative effect of several factors. The same immobilization method leads to different stabilities depending on the selected support and immobilization conditions [134]. For example, lipase from Candida rugosa was found to be more stable when entrapped in alginate gel than when covalently bound on Eupergit C or encapsulated in a sol–gel matrix [175]. Another striking example is that immobilized glucoamylase entrapped in polyacrylamide gels was found to be more stable than that covalently bound to SP-Sephadex C-50 [176]. Enzyme stabilization by immobilization is currently no longer an exception, because of our increasing understanding of immobilization processes. Thermophilic and hyperthermophilic enzymes are further stabilized by immobilization [177–180], suggesting the additive nature of enzyme stabilization by immobilization.

Physical adsorption and covalent binding both reduce or avoid enzyme leaching, but binding to a planar surface can lead to decreased stability or even protein denaturation. Crosslinking of enzymes usually increases their stability at the expense of decreased activity. Microencapsulation into micelles or micellar polymers offers the highest potential to significantly increase enzyme lifetime and stop enzyme leaching, although mass transfer problems may occur. The binding of an enzyme to an external organic or inorganic support introduces additional covalent or non-covalent interactions. This additional interaction decreases the structural flexibility of the enzyme and provides rigidity to the immobilized enzyme, which in turn decreases its liability for denaturation. Figure 4 illustrates the different strategies used to immobilize enzymes at support surfaces to enhance stability. Multiple-point attachment (or crosslinking) for stabilization of carrier-bound immobilized enzyme was introduced in the beginning of the 1970s and further explored to design stable carrier-bound immobilized enzymes. The extent of stabilization of immobilized enzyme by multipoint attachment depends on the number of bonds between the enzyme and the support. The characteristics of the support, reactive groups, and immobilization conditions need to be carefully selected to involve the maximum number of enzyme groups in the immobilization. In fact, it has been possible to correlate the enzyme stabilization reached with the number of enzyme-support linkages [181].

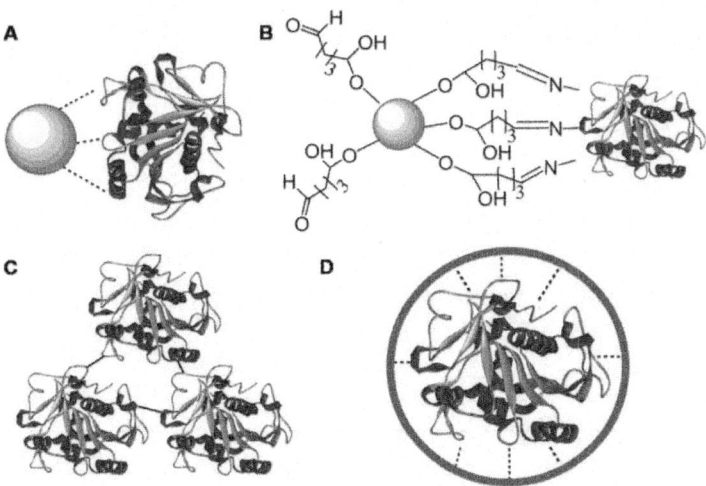

Figure 4: Enzyme stabilization by immobilization introduces additional covalent and non-covalent forces to an external matrix. (**A**) Non-covalent physical adsorption of an

enzyme on the nanoparticle; (**B**) covalent binding of an enzyme to the nanoparticle (multipoint attachment); (**C**) covalent crosslinking of enzymes; and (**D**) microencapsulation of an enzyme by a micelle.

Covalent binding of an enzyme to a carrier has the advantage that the enzyme is tightly fixed. This is due to the fact that the formation of multiple covalent bonds between the enzyme and the carrier reduces conformational flexibility and thermal vibrations, thus preventing protein unfolding and denaturation [149,170,182–184]. The amino groups of the enzyme can initiate nucleophilic attack on, for instance, an epoxide or an aldehyde on the support, forming covalent bonds. As in the case of lipase immobilization, shorter spacers confer higher thermal stability, because they restrict enzyme mobility and prevent unfolding. For example, immobilization of β-1,4-glucosidase (BGL) from *Agaricus arvensis* on silicon oxide nanoparticles by covalent binding confers a 288-fold enhancement in half-life over free BGL at 65 °C [149]. α-Amylase was covalently immobilized onto phthaloyl chloride-containing amino group-functionalized glass beads. The immobilized α-amylase exhibited better thermostability than free amylase [185,186]. The covalent immobilization of cellulase and invertase has been shown to improve stability with respect to pH, temperature, and storage, compared to free enzyme in solution [131,187].

Table 3: Examples of enzyme stabilization by immobilization

Enzyme	Recovered activity (%)	Stabilization factor [a]	Reference
Lipase (*C. rugosa*)	50	150 [a]	[192]
Penicillin G acylase (*E. coli*)	70	8000 [a]	[193]
Chymotrypsin	70	60,000 [a]	[194]
Penicillin G acylase (*K. citrophila*)	70	7000 [a]	[195]
Esterase (*B. stearothermophilus*)	70	1000 [a]	[178]
Thermolysin (*B. thermoproteolyticus*)	100	100 [a]	[191]
Cholesterol oxidase	nd	2.5 (50 °C)	[196]
Alcalase	54	500	[197]
Urokinase	80	10	[198]
α-Amylase (*B. licheniformis*)	nd	2 (70 °C)	[199]
Invertase	nd	2 (70 °C)	[200]
Dextransucrase (*L. mesenteriodes*)	nd	40 (30 °C)	[201]
Formate dehydrogenase (*Pseudomonas* sp. 101)	50	>5000 [a]	[188]
Alcohol dehydrogenase (H. Liver)	90	>3000	[202]
Cyclodextrin glycosyltransferase (*B. circulans*)	70	>100	[203]
Formate dehydrogenase (*C. boidini*)	15	150 [a]	[204]
Laccase (*Rhus vernicifera*)	80	6.4 (65 °C)	[205]

Xylitol dehydrogenase (*Rhizobium etli*)	92	2.2 (60 °C)	[206]
Laccase (*Trametes versicolor*)	69	2.5 (45 °C)	[207]
β-1,4-glucosidase (*Agaricus arvensis*)	158	288 (65 °C)	[149]
Cellulase (*Trichoderma viride*)	nd	2 (55 °C)	[208]
β-Galactosidase	nd	17 (55 °C)	[209]
Lipase G (*Penicillium camembertii*)	nd	1.7 (40 °C)	[210]
Phytases (*Aspergillus niger*)	66	7 (60 °C)	[211]
Phytases (*Escherichia coli*)	74	9.7 (60 °C)	[211]
L-arabinose isomerase (*Bacillus licheniformis*)	145	137.5 (50 °C)	[150]
Protease (*Aspergillus oryzea*)	85	3.5 (70 °C)	[212]
Papain	40	4.2 (70 °C)	[213]
Cellobiase	284	1.2 (60 °C)	[152]
Invertase	NR	3.5 (55 °C)	[214]
α-Amylase (*Bacillus amyloliquifaciens* TSWK1-1)	91	3.75 (60 °C)	[215]
α-Galactosidase (*Aspergillus terreus*$_{GR}$)	74	3.5 (65 °C)	[156]

Table 3 shows some of the results obtained after immobilizing many different proteins [178,181,188,189]. The stabilization factors are in many cases extremely high (1000- to 10,000-fold) with activity recoveries usually over 60%. Moreover, it should be noted that any other technique employed to obtain a stable enzyme (protein engineering, screening, *etc.*) should be compatible with the stabilization of the enzyme by multipoint covalent attachment, because immobilization will be almost a necessary step for the preparation of an industrial biocatalyst. Thus, enzymes from extremophiles have also been stabilized by multipoint covalent attachment [178,189–191].

Solvent Stability

Enzymes tend to form aggregates in organic solvents and hence tend to be poorly accessible for the substrate. The effects of organic solvents on enzymes are usually detrimental and their adverse influences can be mitigated by using immobilized enzyme preparations. Immobilization of enzymes has improved their activity in organic solvents 100-fold [216,217]. Mesoporous materials have attracted much attention due to their porous structure, which may permit full dispersal of enzyme molecules without the possibility of interacting with any external interface. Moreover, the immobilized enzyme molecules will not be in contact with any external hydrophobic interface [218,219] and cannot inactivate the enzymes immobilized on a porous solid [188]. In the presence of an organic solvent, the immobilized enzymes may be in contact with molecules that are soluble in the aqueous phase, but not with the molecules in the organic phase. Thus, any technique that immobilizes the enzyme inside a porous solid provides operational stability to the enzyme, even without affecting the native structure of the enzyme. However, this stabilization is not universally associated with immobilization. The activity of immobilized enzymes is also influenced by the properties of the support, e.g., the aquaphilicity [220]. For example, laccase immobilized on silica gel was 20-fold and 72-fold more active in diethyl ether than in ethyl acetate and methylene chloride, respectively [221]. Furthermore, overall, the Nylon 66 membrane support resulted in the best laccase activity in these three solvents. This shows that the effect of the support matrix on enzyme activity can be more pronounced than solvent-dependent differences. Wang *et al.* reported the activities of different enzymes immobilized in the molecular hydrogels and their superactivity in toluene relative to unconfined enzymes in water. α-glucosidase from baker's yeast in organic cosolvents was stabilized by an order of magnitude by immobilization onto macroporous poly(glycidyl methacrylate-co-ethylene glycol dimethacrylate) [222]. Lipase immobilized on Amberlite XAD-7 maintained 100% of its synthetic activity even after 30 hours of incubation in n-hexane, heptanes, or isooctane. In contrast, free lipase showed too low or no hydrolytic activity after incubation in organic solvents [223].

Selectivity

The selectivity of enzymes is nowadays becoming a powerful asset of enzyme-mediated asymmetric synthesis, because of the increasing need of the pharmaceutical industry for optically pure intermediates [163]. The enzyme shows high specificity towards the natural substrate, but this value may be significantly reduced when the enzyme is intended to be used against compounds that are unlike their natural substrate. The selectivity of enzymes includes substrate selectivity, stereoselectivity, and regioselectivity [224]. Modulation of enzyme substrate selectivity has been achieved by selecting the immobilization technique [174]. *Candida rugosa* lipase (CRL) is an important industrial lipase; due to its wide substrate specificity it is successfully utilized in a variety of hydrolysis and esterification reactions. The synthesis of several pharmaceuticals is made possible due to its high stereoselectivity and regioselectivity [186,225]. Genetic engineering has been extensively used to alter enzyme selectivity [226]. However, there are also many interesting examples in which enzyme selectivity has been altered by a variety of immobilization techniques such as covalent bonding, entrapment, and simple adsorption. On many occasions, it has been found that a non-selective enzyme such as chloroperoxidase is transformed into a stereoselective enzyme after immobilization an *S*-selective CRL has also been converted to an *R*-selective CRL by covalent immobilization [227].

Selectivity that can be influenced by the immobilization technique can be classified into two major groups: support-controlled (pore size-controlled and diffusion-controlled) and conformation-controlled (microenvironment-controlled and active site-controlled), on the basis of the effect of the source. For example, the product pattern of controlled pore glass (CPG)-immobilized subtilisin-catalyzed digestion of proteins can be affected by the pore size of the carrier used [228]. Urokinase covalently immobilized on glyoxal agarose exhibited changed selectivity towards glutathione S-transferase [198,229]. α-Amylase immobilized on silica [230] or covalently bound to CNBr-activated carboxymethyl cellulose [231] afforded products whose composition differed from that obtained with the native enzyme. This was largely attributed to the fact that the size of the pores where the enzyme molecules are located determines the accessibility of the substrates, depending on their size. Diffusion-controlled enantioselectivity was reported after a study of the enantioselectivity of the lipase CaL-B in transesterification in organic solvents [232]. A relevant example worth mentioning is that simple adsorption of CRL on Celite not only enhanced its stability in acetaldehyde but also enhanced its enantioselectivity up to three-fold [233].

Immobilization may produce some distortion in the enzyme structure, particularly in the active site region. The immobilization of a protein via different regions may therefore confer different rigidities and distort the enzyme in very different ways; it may be even possible to generate different microenvironments around the enzyme with very different physical properties [174]. The enantiopreference of lipases were successfully modulated using the above strategy [227,234–238]. It has been reported that the same lipase immobilized on different supports may exhibit very different enantioselectivities (in some cases even an inversion) in the same experimental conditions. A detailed analysis of the enantiopreferences of lipases has been reported by Mateo *et al.* (2007) [174]. In some cases, a change in experimental conditions can cause opposing effects on different immobilized preparations [227,234,237]. However, these enantioselectivity modulations are examples of random immobilization, which produces alteration even if it is not desirable. The use of this strategy may sometimes result in properties that are less desirable than those of the free enzyme. Therefore, directed immobilization is a powerful tool to modulate enzyme properties.

Directed immobilization involves the selection of the point of attachment through specific interactions between functional groups on the support and the enzyme, based on detailed structural information of the enzyme and the structure of the support. The properties of the immobilized enzyme depend on the modification (the immobilization conditions), the nature of the modifier (nature of support), and the nature of the enzyme (source, purity, and strain). Lipases are the best example and have been most extensively studied for the modulation of enantioselectivity by immobilization. In another successful example reported by Cabrera *et al.* (2009), the stereoselectivity of CaL-B lipase for hydrolysis of racemic 2-*O*-butyryl-2-phenylacetic acid changed both quantitatively (% ee) and qualitatively (R or S enantiomer product) when bound hydrophobically to different supports, shifting from 99% ee (S) on Lewatit to 95% (R) on octyl agarose [239]. Yilmaz et al.(2011) reported that sporopollenin-based encapsulated lipase in particular had higher conversion and enantioselectivity compared to sol–gel free lipase [186]. These results reveal that sol–gel encapsulated lipase has high enantioselectivity (E) and conversion (x) compared with covalently immobilized lipase (Candida rugosa lipase). In this study, excellent enantioselectivity ($E > 400$) was noticed for most lipase preparations ($E = 166$ for the free enzyme) with an ee value of 98% for S-Naproxen. Recently (Sahin et al., 2009), CRL was encapsulated by polycondensation of tetraethoxysilane and octyltriethoxysilane in the presence of the compounds calix[n]arene, calix[n]–NH, and calix[n]–COOH ($n = 4, 6$, and 8) [240]. For encapsulated CRL-catalyzed hydrolysis of racemic Naproxen methyl ester in aqueous phase/isooctane biphasic medium, the temperature, pH of the aqueous phase, and calix[n]arene-based additives were found to have important effects on the conversion and enantioselectivity. Wang et al. (2008) found that the esterase of Kelbsiella oxytoca provided a much higher enantiomeric ratio in the hydrolysis of (R,S)-ethyl mandelate when the enzyme was immobilized on Eupergit C 250 L [241].

With regard to the improvement of enzyme selectivity, there are many exciting examples, as noted above, of immobilized enzymes for which selectivity, e.g., reaction selectivity, substrate selectivity, stereoselectivity, or chemical selectivity, can be affected by the immobilization procedure [242,243], perhaps combined with reaction medium engineering [237]; however, the improvement of enzyme selectivity by immobilization is fundamentally still a new endeavor, lacking guidelines that can be used to guide practical experiments. Nevertheless, with the increasing understanding of the relationship between enzyme selectivity and the structural changes resulting from genetic engineering or other chemical modifications, increasing interest in the improvement of enzyme selectivity by immobilization can be expected in the near future.

Substrate Tolerance

Inhibition of enzymes during the reaction by substrates, reaction products, or components of the bulk medium is major issue in the application of biocatalysts in industrial processes. Depending on the type of inhibition, specific inhibitors tend to interact with the enzyme structure at specific positions to produce the inhibition. Thus, immobilization may reduce inhibition based on the type of inhibition mechanism in different situations. First, when the inhibitor acts allosterically, interacting with the protein in a place different from the active center, the allosteric site of enzyme may be blocked during the course of the immobilization procedure, significantly reducing the inhibition caused by this mechanism. Second, when the inhibitor interacts with the active site of the protein, immobilization may slightly distort the enzyme active site, and in some cases a higher distortion of the enzyme may be achieved in the inhibitor binding site than in the substrate binding site. This is more likely if the substrate is larger than the inhibitor and interacts with a larger number of groups in the enzyme.

A lactase from *Kluyveromyces lactis* shows competitive inhibition by galactose (χ = 45 mM) and non-competitive by glucose (χ = 750 mM) [244]; upon immobilization of this lactase, the inhibition constant for galactose was improved from 45 mM to 40 M. Similar enhancement was obtained in the case of lactase from *Thermus sp.* by galactose (χ = 3.1 mM) and non-competitive inhibition by glucose (χ was 50 mM) [245]. The immobilization of lactase on different supports permitted the screening of the preparation, where the χ by glucose was increased by a two-fold factor, while the competitive χ by galactose was increased by a four-fold factor [245]. Another example of reduced substrate inhibition and increased activity upon immobilization on porous cellulose beads has been reported for invertase from *Candida utilis*[132].

Multi-Step Reactions

Co-immobilized multi-enzymatic systems are increasingly driven by economic and environmental constraints that also permit multi-enzyme reactions through artificial enzymatic cascade processes through compartmentalization of the individual catalysts [217]. For example, Brazeau *et al.* (2008) have developed a multi-enzyme pathway, immobilized on Eupergit C, for the synthesis of the monatin [246]. Co-immobilization of three cellulases on Au-doped magnetic silica nanoparticles for the degradation of cellulose has been reported by Cho *et al.*[247]. The simultaneous co-immobilized three cystein-tagged cellulases, including endo-glucanase (EGIVCBDII), exo-glucanase (CBHII), and β-glucosidase (BglB) co-immobilized on AuNP (gold nanoparticles) and Au-MSNP (gold-doped magnetic silica nanoparticles) was successfully demonstrated for the production of cellobiose and glucose. Co-immobilization of coupled enzyme systems can enhance activity and stability [248] such as where nitrobenzene nitroreductase and glucose-6-phosphate dehydrogenase were co-encapsulated in silica particles, wherein the G6PD allowed regeneration of NADPH. St. Clair *et al.* (2000) also demonstrated that the CLEC could also retain cofactors for redox reactions [249]. As such, co-immobilization provides benefits that span numerous biotechnological applications, from biosensing of molecules to cofactor recycling and to combination of multiple biocatalysts for the synthesis of valuable products. Selective compartmentalization of enzymes during immobilization could provide advantages. An interesting example of multi-enzyme compartmentalization was reported by Dongen *et al.*[250]. The co-polymer of isocyanopeptides and styrene was used to form porous polymersomes immobilized with horseradish peroxidize, CaL-B and glucose oxidase anchored to the membrane surface, bilayer membrane, and in the polymersome lumen, respectively. The three enzymes were able to perform a three sequential steps reaction using glucose acetate as the initial substrate (which was subsequently deacetylated by the lipase and oxidized by the glucose oxidase), yielding peroxide that was subsequently used by the peroxidize to oxidize ABTS [2,20-azinobis (3-ethylbenzothiazoline-6-sulfonic acid)]. Similar uses of compartmentalization and multiple enzymes (laccase and lipase) have been achieved using Spherezymes [251]. α-Amylase, cellulase, protease, and lipase have been immobilized individually on various supports. However, Pundir *et al.* (2012) reported the covalent coimmobilization of commercial α-amylase, cellulase, protease, and lipase onto the inner side of a plastic beaker and bristles of a plastic brush, their properties, and use in removal of stain from cloth [252]. Similarly, commercial lipase, glycerol kinase (GK), glycerol-3-phosphate oxidase (GPO) and peroxidase were co-immobilized covalently on to arylamine glass beads [253]. Starch-converting enzymes, α-amylase and glucoamylase, were immobilized on surface-modified carriers using a co-immobilized, as well as a single system. The co-enzymes immobilized on hydrophilic silica gel and DEAE-cellulose entrapped in alginate beads exhibited 92.3 and 88.9% of the remaining activity even after 10 times of reuse [254]. Co-immobilized enzymes have been used extensively in expensive commercial and biological approaches in recent years to develop more economic and environmentally friendly processes.

Advances in Enzyme Immobilization

Recent advancements in nano and hybrid technology have made various materials more affordable hosts for enzyme immobilization. As a result, various nanostructured materials based on combined organic and inorganic species have received attention as enzyme immobilizing supports due to their intrinsic large surface areas and physicochemical properties such as pore size, hydrophilic/hydrophobic balance, aquaphilicity and surface chemistry [255]. Large surface area and controlled pore size often result in improved enzyme loading, which increases enzyme activity per unit mass or volume, compared to that of conventional materials [256]. Nanostructured materials with ability to control size and shape enables better interaction with the enzyme, increases immobilization efficiency, and enhances the long-term storage and recycling stability of the enzyme [257]. In addition, these organic, inorganic or hybrid materials may provide specific features such as enhanced strength, elasticity, plasticity, and chemical bonding in an appropriate microenvironment. Various organic or inorganic nanomaterials such as multiwall carbon nanotubes (MWCNTs), magnetic nanoparticles, silica nanoparticles, and quantum dots have been exploited to generate hybrid composite nanofibers with additional physical properties and mechanical stability [258,259].

Magnetite mesoporous silica hybrid support was fabricated by depositing magnetite and MCM-41 nanoparticles onto polystyrene beads using the layer-by-layer (LBL) method. The incorporation of magnetite gives an additional magnetic property to the hollow mesoporous silica shells to improve the enzyme immobilization. Magnetic/silica nanoparticles with a core of magnetic clusters were applied to stabilization of His-tagged *Bacillus stearothermopilus* L1 lipase [260]. Xia's group reported on encapsulation of enzymes in nanofibers by direct coelectrospinning [261].

New Technology for Enzyme Immobilization

In the last decade, more attention has been paid to the regulation of the microenvironment for the enzyme on a support surface in order to obtain significant stabilization of the immobilized enzyme and development of new support. Unfortunately, there have been few studies in enzyme immobilization that have focused on mimicking the environment of a cell to improve the properties of the enzyme *in vitro*. Moreover, many attempts have been focused on the multipoint covalent attachment of proteins on supports, which has been reported to clearly increase the thermal stability of immobilized enzymes. However, the stabilizing effect increases with the number of covalent bonds between enzymes and the support until some critical value (a limit) is achieved, and a further increase in the number of bonds does not result in further stabilization [262]. Excess activated groups present on the support surface would increase the possibility of enzyme deactivation due to their subsequent slow reaction with the enzyme [168].

Microwave Irradiation

Recently, controlled microwave irradiation has been shown to dramatically accelerate chemical reactions and reduce reaction times required for immobilization [263]. It is thought that microwave irradiation provides an additional driving force for mass transport and accelerates the mass transfer [264]. Microwave irradiation has been used to simplify and improve the reaction conditions for many classic organic reactions. Reactions performed under microwave irradiation proceed faster, are cleaner, present much better yields and are more reproducible than those performed under conventional conditions. Microwave assisted heterogeneous oxidation and reduction reactions have received attention in recent years. Varma *et al.* have demonstrated that the reaction rate of oxidation and reduction reaction under microwave irradiation could be greatly enhanced under microwave irradiation than that under conventional heating [265,266].

Wang *et al.* have immobilized papain and penicillin acylase in mesocellular siliceous foams (MCFs) by using the microwave irradiation technology [267,268]. 80 and 140 seconds were enough for immobilization of papain and penicillin acylase, respectively. The maximum loading of papain reached 984.1 mg/g, 1.26 times of that obtained from the conventional method which was non-microwave-assisted. The activities of microwave-assisted immobilized papain and penicillin acylase were 1.86 and 1.39 times of those with conventional method. Immobilization of lipases and horseradish peroxidase under microwave-assisted method were reported [269].

Photoimmobilization Technology

The protein molecules (e.g., horseradish peroxidase or glucose oxidase), when exposed to ultraviolet (UV) light at 365 nm along with a photoreactive polymer, were immobilized through covalent bonding by the highly reactive nitrene of the polymer in about 10 to 20 min [270]. The photoreactive polymer under UV light generates highly reactive nitrene, which has a property of inserting into C–H bond and is capable of binding with the biomolecule thus serving important for the immobilization of biomolecules irrespective of their functional groups. However, Kumar and Nahar (2007) reported that horseradish peroxidase and glucose oxidase showed efficient immobilization when placed on the photoreactive cellulose membranes and exposed to sunlight [271]. No detectable increase in immobilization was observed beyond 21,625 lux, which was considered as the sunlight intensity required for optimum immobilization. Thus, sunlight could be a very good alternative compared to 365 nm UV light for photoimmobilization. Therefore, photoimmobilization is a green technique and suitable for large-scale, as well as small-scale, immobilization.

Enzymatic Immobilization of Enzyme

The use of green chemistry rather than harsh chemicals is one of the main goals in enzyme industries to avoid the partial denaturation of enzyme protein. Enzyme-assisted immobilization of the enzyme is an emerging and novel technology to fabricate solid protein formulations [272,273]. Tanaka *et al.* (2007) reported the immobilization of enhanced green fluorescent protein (EGFP) and glutathione S-transferase (GST) onto the casein-coated polystyrene surface as model proteins. EGFP and GST were tagged with a neutral Gln-donor substrate peptide for MTG (Leu-Leu-Gln-Gly, LLQG-tag) at their *C*-terminus. This strategy has been also applied to immobilize Luciferase (Luc) and GST ybbR-fusion proteins and thioredoxin fusion proteins [273].

Controlled Immobilization of Enzyme onto Porous Materials

Protein distribution (heterogeneous or homogeneous) across a porous material depends on whether the immobilization rate is controlled by either protein/ support attachment or mass transference. Bolivar *et al.* (2011) described different strategies to control the distribution of fluorescent proteins by altering their immobilization rates under low protein loads conditions focusing on the chemical nature of protein-support attachment [274]. The control of protein immobilization by modulation of apparent immobilization rate was extended to the immobilization of two fluorescent proteins onto the same carrier, creating four different distribution patterns. These tailor-made distributions of two proteins inside one single porous carrier are pioneering in immobilization technology. One of these groups is highly reactive under alkaline conditions, thereby promoting intense covalent protein-support attachments, whereas the other group reversibly immobilizes the protein under mild pH conditions (pH 6–8). Another interesting approach to control the distribution of protein is the immobilization of multi-enzyme systems on the same solid surface that is activated by different reactive groups. Recently, Rocha-Martín *et al.* (2012) reported the co-immobilization of three bio-redox orthogonal cascades that involved a main and a recycling dehydrogenase onto a heterogeneously activated agarose-type support [275]. The agarose-type support was heterogeneously activated with glyoxyl groups and metal chelates to enable the immobilization of proteins through two different techniques.

Recommendation for the Future of Immobilization Technology

At present, a vast number of methods for immobilization of enzymes are available. Unfortunately, there is no universal immobilization method, nor a single preferred method of enzyme immobilization. Immobilization method or the choice of support differs from enzyme to enzyme, from application to application and from carrier to carrier. Cao (2005), in his book*Carrier Bound Immobilized Enzymes*, suggested that the major problem in enzyme immobilization is not only the selection of the right carrier for the enzyme immobilization, but it is how to design the performance of the immobilized enzyme [168]. In the future, information derived from protein sequences, 3D-structures, reaction mechanism, and working environment should be further combined with the properties of supports and immobilization methods in order to produce the immobilized enzyme with enhanced properties for a biochemical application. As methods for the immobilization of enzymes continue to improve and become commercially widespread, the cascade enzymatic reaction and *in vitro* synthetic biology, multi-enzyme immobilization will be one of next goals [276–278]. In addition, with the growing attention paid to the use of *in silico* technology in various fields of sciences, including protein engineering, the *in silico* model can be designed to validate the probability of success and the efficiency of the immobilization process before starting the immobilization.

Integration of Different Techniques

There are many reports showing that the enzyme properties may be directly improved by: chemical or genetic modification of enzymes, screening of the most suitable enzyme, and immobilization of enzymes. An excellent example of the coupling of engineering and immobilization is demonstrated by the variants of formate dehydrogenase whereby *Candida boidinii* preserved 4.4-fold higher activity after entrapment in polyacrylamide gels than the wild-type enzyme [159]. Godoy*et al.* (2011) reported the site-directed immobilization/rigidification of genetically modified enzymes through multipoint covalent attachment on bifunctional disulfide-glyoxyl supports [279]. Genetically engineered lipase 2 from *Geobacillus thermocatenulatus* and penicillin G acylase from *E. coli* were immobilized and uniquely rigidified by a multipoint covalent attachment onto tailor-made bifunctional disulfide-glyoxyl supports. Therefore, the use of coupled strategy (protein engineering and immobilization) seems to improve the properties of the final immobilized biocatalyst, and, moreover, modulates multiple properties of enzymes for industrial application.

CONCLUSIONS

In this review, we have shown some examples of where the modulation of biochemical and physical properties of engineered and/or immobilized enzymes has been permitted to solve problems such as stability, selectivity, and substrate or solvent tolerance of enzymes. It is clear from this report that protein engineering or immobilization individually cannot make an ideal catalyst for industrial processes. Until now, there has been no universal method to modulate a particular property of an enzyme. Therefore, the combination of protein engineering and immobilization of enzymes seems to be a powerful tool to greatly improve a number of industrial processes, and more effort may be expected in the coming years in this regard. When properly designed, the immobilization of enzymes in combination with protein engineering techniques can be a very successful strategy to improve enzymes of industrial significance.

ACKNOWLEDGMENTS

This research was supported by the Converging Research Center Program through the National Research Foundation of Korea, funded by the Ministry of Education, Science and Technology (Grant 2011-50210). This study was also supported by Brain Pool 2012 of Konkuk University, Republic of Korea.

REFERENCES

1.   Adamczak, M.; Sajja, K.H. Strategies for improving enzymes for efficient biocatalysis. *Food Technol. Biotechnol* **2004**, *42*, 251–264.

2.   Reetz, M.T.; Carballeira, J.D.; Vogel, A. Iterative saturation mutagenesis on the basis of B factors as a strategy for increasing protein thermostability. *Angew. Chem. Int. Ed* **2006**, *45*, 7745–7751.

3.   Bommarius, A.S.; Riebel-Bommarius, B.R. *Biocatalysis: Fundamentals and Applications*; Wiley-VCH: Weinheim, Germany, 2004; p. 611.

4.   Katchalski-Katzir, E.; Kraemer, D.M. Eupergit C, a carrier for immobilization of enzymes of industrial potential. *J. Mol. Catal. B* **2000**, *10*, 157–176.

5.   Straathof, A.J.J.; Panke, S.; Schmid, A. The production of fine chemicals by biotransformations. *Curr. Opin. Biotechnol* **2002**, *13*, 548–556.

6.   Polizzi, K.M.; Bommarius, A.S.; Broering, J.M.; Chaparro-Riggers, J.F. Stability of biocatalysts. *Curr. Opin. Chem. Biol* **2007**, *11*, 220–225.

7.   May, O.; Nguyen, P.T.; Arnold, F.H. Inverting enantioselectivity by directed evolution of hydantoinase for improved production of l-methionine. *Nat. Biotechnol* **2000**, *18*, 317–320.

8.   Lee, S.H.; Kim, Y.W.; Lee, S.; Auh, J.H.; Yoo, S.S.; Kim, T.J.; Kim, J.W.; Kim, S.T.; Rho, H.J.; Choi, J.H.; *et al.* Modulation of cyclizing activity and thermostability of cyclodextrin glucanotransferase and its application as an antistaling enzyme. *J. Agric. Food Chem* **2002**, *50*, 1411–1415.

9.   Zhang, N.; Suen, W.C.; Windsor, W.; Xiao, L.; Madison, V.; Zaks, A. Improving tolerance of *Candida antarctica* lipase B towards irreversible thermal inactivation through directed evolution. *Protein Eng* **2003**, *16*, 599–605.

10.   Williams, G.J.; Domann, S.; Nelson, A.; Berry, A. Modifying the stereochemistry of an enzyme-catalyzed reaction by directed evolution. *Proc. Natl. Acad. Sci. USA* **2003**, *100*, 3143–3148.

11.   Sriprapundh, D.; Vieille, C.; Zeikus, J.G. Directed evolution of *Thermotoga neapolitana* xylose isomerase: High activity on glucose at low temperature and low pH. *Protein Eng* **2003**, *16*, 683–690.

204   Immobilized Enzyme Principles

12. Van der Veen, B.A.; Potocki-Veronese, G.; Albenne, C.; Joucla, G.; Monsan, P.; Remaud-Simeon, M. Combinatorial engineering to enhance amylosucrase performance: Construction, selection, and screening of variant libraries for increased activity. *FEBS Lett* **2004**, *560*, 91–97.

13. Wilkinson, D.; Akumanyi, N.; Hurtado-Guerrero, R.; Dawkes, H.; Knowles, P.F.; Phillips, S.E.; McPherson, M.J. Structural and kinetic studies of a series of mutants of galactose oxidase identified by directed evolution. *Protein Eng. Des. Sel* **2004**, *17*, 141–148.

14. Hao, J.; Berry, A. A thermostable variant of fructose bisphosphate aldolase constructed by directed evolution also shows increased stability in organic solvents. *Protein Eng. Des. Sel* **2004**, *17*, 689–697.

15. Wen, T.N.; Chen, J.L.; Lee, S.H.; Yang, N.S.; Shyur, L.F. A truncated *Fibrobacter succinogenes* 1,3–1,4-β-d-glucanase with improved enzymatic activity and thermotolerance. *Biochemistry* **2005**, *44*, 9197–9205.

16. Fujii, R.; Nakagawa, Y.; Hiratake, J.; Sogabe, A.; Sakata, K. Directed evolution of *Pseudomonas aeruginosa* lipase for improved amide-hydrolyzing activity. *Protein Eng. Des. Sel* **2005**, *18*, 93–101.

17. Tobe, S.; Shimogaki, H.; Ohdera, M.; Asai, Y.; Oba, K.; Iwama, M.; Irie, M. Expression of *Bacillus* protease (Protease BYA) from *Bacillus* sp. Y in *Bacillus subtilis* and enhancement of its specific activity by site-directed mutagenesis-improvement in productivity of detergent enzyme. *Biol. Pharm. Bull* **2006**, *29*, 26–33.

18. Suemori, A.; Iwakura, M. A systematic and comprehensive combinatorial approach to simultaneously improve the activity, reaction specificity, and thermal stability of *p*-hydroxybenzoate hydroxylase. *J. Biol. Chem* **2007**, *282*, 19969–19978.

19. Fenel, F.; Zitting, A.J.; Kantelinen, A. Increased alkali stability in *Trichoderma reesei* endo-1,4-β-xylanase II by site directed mutagenesis. *J. Biotechnol* **2006**, *121*, 102–107.

20. Jones, A.; Lamsa, M.; Frandsen, T.P.; Spendler, T.; Harris, P.; Sloma, A.; Xu, F.; Nielsen, J.B.; Cherry, J.R. Directed evolution of a maltogenic α-amylase from *Bacillus* sp. TS-25. *J. Biotechnol* **2008**, *134*, 325–333.

21. Dumon, C.; Varvak, A.; Wall, M.A.; Flint, J.E.; Lewis, R.J.; Lakey, J.H.; Morland, C.; Luginbuhl, P.; Healey, S.; Todaro, T.; *et al.* Engineering hyperthermostability into a GH11 xylanase is mediated by subtle changes to protein structure. *J. Biol. Chem* **2008**, *283*, 22557–22564.

22. Hirokawa, K.; Ichiyanagi, A.; Kajiyama, N. Enhancement of thermostability of fungal deglycating enzymes by directed evolution. *Appl.*

Microbiol. Biotechnol **2008**, *78*, 775–781.

23. Belien, T.; Joye, I.J.; Delcour, J.A.; Courtin, C.M. Computational design-based molecular engineering of the glycosyl hydrolase family 11 *B. subtilis* XynA endoxylanase improves its acid stability. *Protein Eng. Des. Sel* **2009**, *22*, 587–596.

24. Zhong, C.Q.; Song, S.; Fang, N.; Liang, X.; Zhu, H.; Tang, X.F.; Tang, B. Improvement of low-temperature caseinolytic activity of a thermophilic subtilase by directed evolution and site-directed mutagenesis. *Biotechnol. Bioeng* **2009**, *104*, 862–870.

25. Gupta, N.; Farinas, E.T. Directed evolution of CotA laccase for increased substrate specificity using *Bacillus subtilis*spores. *Protein Eng. Des. Sel* **2010**, *23*, 679–682.

26. Spadiut, O.; Nguyen, T.T.; Haltrich, D. Thermostable variants of pyranose 2-oxidase showing altered substrate selectivity for glucose and galactose. *J. Agric. Food Chem* **2010**, *58*, 3465–3471.

27. Zhang, Z.G.; Yi, Z.L.; Pei, X.Q.; Wu, Z.L. Improving the thermostability of *Geobacillus stearothermophilus* xylanase XT6 by directed evolution and site-directed mutagenesis. *Bioresour. Technol* **2010**, *101*, 9272–9278.

28. Bustos-Jaimes, I.; Mora-Lugo, R.; Calcagno, M.L.; Farres, A. Kinetic studies of Gly28:Ser mutant form of *Bacillus pumilus* lipase: Changes in k_{cat} and thermal dependence. *Biochim. Biophys. Acta* **2010**, *1804*, 2222–2227.

29. Sun, J.; Wang, H.; Lv, W.; Ma, C.; Lou, Z.; Dai, Y. Construction and characterization of a fusion β-1,3–1,4-glucanase to improve hydrolytic activity and thermostability. *Biotechnol. Lett* **2011**, *33*, 2193–2199.

30. Lee, S.; Lee, D.G.; Jang, M.K.; Jeon, M.J.; Jang, H.J.; Lee, S.H. Improvement in the catalytic activity of β-agarase AgaA from *Zobellia galactanivorans* by site-directed mutagenesis. *J. Microbiol. Biotechnol* **2011**, *21*, 1116–1122.

31. Theriot, C.M.; Semcer, R.L.; Shah, S.S.; Grunden, A.M. Improving the catalytic activity of hyperthermophilic *Pyrococcus horikoshii* prolidase for detoxification of organophosphorus nerve agents over a broad range of temperatures. *Archaea***2011**, *2011*.

32. Shih, T.W.; Pan, T.M. Substitution of Asp189 residue alters the activity and thermostability of *Geobacillus* sp. NTU 03 lipase. *Biotechnol. Lett* **2011**, *33*, 1841–1846.

33. Hokanson, C.A.; Cappuccilli, G.; Odineca, T.; Bozic, M.; Behnke, C.A.; Mendez, M.; Coleman, W.J.; Crea, R. Engineering highly

thermostable xylanase variants using an enhanced combinatorial library method. *Protein Eng. Des. Sel* **2011**, *24*, 597–605.

34. Ben Mabrouk, S.; Aghajari, N.; Ben Ali, M.; Ben Messaoud, E.; Juy, M.; Haser, R.; Bejar, S. Enhancement of the thermostability of the maltogenic amylase MAUS149 by Gly312Ala and Lys436Arg substitutions. *Bioresour. Technol* **2011**, *102*, 1740–1746.

35. Sun, Y.; Yang, H.; Wang, W. Improvement of the thermostability and enzymatic activity of cholesterol oxidase by site-directed mutagenesis. *Biotechnol. Lett* **2011**, *33*, 2049–2055.

36. Le, Q.A.; Joo, J.C.; Yoo, Y.J.; Kim, Y.H. Development of thermostable *Candida antarctica* lipase B through novel in silico design of disulfide bridge. *Biotechnol. Bioeng* **2012**, *109*, 867–876.

37. Mollania, N.; Khajeh, K.; Ranjbar, B.; Hosseinkhani, S. Enhancement of a bacterial laccase thermostability through directed mutagenesis of a surface loop. *Enzyme Microb. Technol* **2011**, *49*, 446–452.

38. Choi, J.G.; Ju, Y.H.; Yeom, S.J.; Oh, D.K. Improvement in the thermostability of d-psicose 3-epimerase from*Agrobacterium tumefaciens* by random and site-directed mutagenesis. *Appl. Environ. Microbiol* **2011**, *77*, 7316–7320.

39. Huang, J.W.; Cheng, Y.S.; Ko, T.P.; Lin, C.Y.; Lai, H.L.; Chen, C.C.; Ma, Y.; Zheng, Y.; Huang, C.H.; Zou, P.; *et al.* Rational design to improve thermostability and specific activity of the truncated *Fibrobacter succinogenes* 1,3–1,4-β-d-glucanase. *Appl. Microbiol. Biotechnol* **2012**, *94*, 111–121.

40. Liu, Y.H.; Hu, B.; Xu, Y.J.; Bo, J.X.; Fan, S.; Wang, J.L.; Lu, F.P. Improvement of the acid stability of *Bacillus licheniformis*α amylase by error-prone PCR. *J. Appl. Microbiol* **2012**, *113*, 541–549.

41. Yang, H.; Liu, L.; Wang, M.; Li, J.; Wang, N.S.; Du, G.; Chen, J. Structure-based engineering of methionine residues in the catalytic cores of alkaline amylase from *Alkalimonas amylolytica* for improved oxidative stability. *Appl. Environ. Microbiol* **2012**, *78*, 7519–7526.

42. Srikrishnan, S.; Randall, A.; Baldi, P.; Da Silva, N.A. Rationally selected single-site mutants of the *Thermoascus aurantiacus* endoglucanase increase hydrolytic activity on cellulosic substrates. *Biotechnol. Bioeng* **2012**, *109*, 1595–1599.

43. Nishioka, T.; Yasutake, Y.; Nishiya, Y.; Tamura, T. Structure-guided mutagenesis for the improvement of substrate specificity of *Bacillus megaterium* glucose 1-dehydrogenase IV. *FEBS J* **2012**, *279*, 3264–3275.

44. Qi, X.; Guo, Q.; Wei, Y.; Xu, H.; Huang, R. Enhancement of pH stability and activity of glycerol dehydratase from*Klebsiella pneumoniae* by rational design. *Biotechnol. Lett* **2012**, *34*, 339–346.

45. Goh, P.H.; Illias, R.M.; Goh, K.M. Rational mutagenesis of cyclodextrin glucanotransferase at the calcium binding regions for enhancement of thermostability. *Int. J. Mol. Sci* **2012**, *13*, 5307–5323.

46. Ye, X.; Zhang, C.; Zhang, Y.H. Engineering a large protein by combined rational and random approaches: Stabilizing the *Clostridium thermocellum* cellobiose phosphorylase. *Mol. Biosyst* **2012**, *8*, 1815–1823.

47. Kumar, A.; Dutt, S.; Bagler, G.; Ahuja, P.S.; Kumar, S. Engineering a thermo-stable superoxide dismutase functional at sub-zero to >50 °C, which also tolerates autoclaving. *Sci. Rep.* **2012**, *2*, 387:1–387:8.

48. Anbar, M.; Gul, O.; Lamed, R.; Sezerman, U.O.; Bayer, E.A. Improved thermostability of *Clostridium thermocellum*endoglucanase Cel8A by using consensus-guided mutagenesis. *Appl. Environ. Microbiol* **2012**, *78*, 3458–3464.

49. Wang, Y.; Yuan, H.; Wang, J.; Yu, Z. Truncation of the cellulose binding domain improved thermal stability of endo-β-1,4-glucanase from *Bacillus subtilis* JA18. *Bioresour. Technol* **2009**, *100*, 345–349.

50. Spadiut, O.; Radakovits, K.; Pisanelli, I.; Salaheddin, C.; Yamabhai, M.; Tan, T.C.; Divne, C.; Haltrich, D. A thermostable triple mutant of pyranose 2-oxidase from *Trametes multicolor* with improved properties for biotechnological applications. *Biotechnol. J* **2009**, *4*, 525–534.

51. Rha, E.; Kim, S.; Choi, S.L.; Hong, S.P.; Sung, M.H.; Song, J.J.; Lee, S.G. Simultaneous improvement of catalytic activity and thermal stability of tyrosine phenol-lyase by directed evolution. *FEBS J* **2009**, *276*, 6187–6194.

52. Zhao, Q.; Liu, H.; Zhang, Y. Engineering of protease-resistant phytase from *Penicillium* sp.: High thermal stability, low optimal temperature and pH. *J. Biosci. Bioeng* **2010**, *110*, 638–645.

53. Kotzia, G.A.; Labrou, N.E. Engineering thermal stability of l-asparaginase by *in vitro* directed evolution. *FEBS J* **2009**,*276*, 1750–1761.

54. Yi, Z.L.; Pei, X.Q.; Wu, Z.L. Introduction of glycine and proline residues onto protein surface increases the thermostability of endoglucanase CelA from *Clostridium thermocellum*. *Bioresour. Technol* **2011**, *102*, 3636–3638.

55. Pei, X.Q.; Yi, Z.L.; Tang, C.G.; Wu, Z.L. Three amino acid changes

contribute markedly to the thermostability β-glucosidase BglC from *Thermobifida fusca. Bioresour. Technol* **2011**, *102*, 3337–3342.

56. Damnjanovic, J.; Takahashi, R.; Suzuki, A.; Nakano, H.; Iwasaki, Y. Improving thermostability of phosphatidylinositol-synthesizing *Streptomyces* phospholipase D. *Protein Eng. Des. Sel* **2012**, *25*, 415–424.

57. Lee, H.L.; Chang, C.K.; Jeng, W.Y.; Wang, A.H.; Liang, P.H. Mutations in the substrate entrance region of β-glucosidase from *Trichoderma reesei* improve enzyme activity and thermostability. *Protein Eng. Des. Sel* **2012**, *25*, 733–740.

58. Sharma, P.K.; Kumar, R.; Mohammad, O.; Singh, R.; Kaur, J. Engineering of a metagenome derived lipase toward thermal tolerance: Effect of asparagine to lysine mutation on the protein surface. *Gene* **2012**, *491*, 264–271.

59. Camarero, S.; Pardo, I.; Canas, A.I.; Molina, P.; Record, E.; Martinez, A.T.; Martinez, M.J.; Alcalde, M. Engineering platforms for directed evolution of Laccase from *Pycnoporus cinnabarinus. Appl. Environ. Microbiol* **2012**, *78*, 1370–1384.

60. Zhang, S.B.; Pei, X.Q.; Wu, Z.L. Multiple amino acid substitutions significantly improve the thermostability of feruloyl esterase A from *Aspergillus niger. Bioresour. Technol* **2012**, *117*, 140–147.

61. Brode, P.F., III; Erwin, C.R.; Rauch, D.S.; Barnett, B.L.; Armpriester, J.M.; Wang, E.S.; Rubingh, D.N. Subtilisin BPN' variants: Increased hydrolytic activity on surface-bound substrates via decreased surface activity. *Biochemistry* **1996**,*35*, 3162–3169.

62. Rubingh, D.N. The influence of surfactants on enzyme activity. *Curr. Opin. Coll. Int. Sci* **1996**, *1*, 598–603.

63. Beer, H.D.; Wohlfahrt, G.; McCarthy, J.E.; Schomburg, D.; Schmid, R.D. Analysis of the catalytic mechanism of a fungal lipase using computer-aided design and structural mutants. *Protein Eng* **1996**, *9*, 507–517.

64. Martinelle, M.; Holmquist, M.; Clausen, I.G.; Patkar, S.; Svendsen, A.; Hult, K. The role of Glu87 and Trp89 in the lid of*Humicola lanuginosa* lipase. *Protein Eng* **1996**, *9*, 519–524.

65. Pedersen, S.; Lange, N.K.; Nissen, A.M. Novel industrial enzyme applications. *Ann. N. Y. Acad. Sci* **1995**, *750*, 376–390.

66. Jeffries, T.W. *Enzyme Technology for Pulp Bleaching and Deinking*; International Business Communications: Southbourough, MA, USA, 1996; pp. 81–97.

67. Koivula, A.; Reinikainen, T.; Ruohonen, L.; Valkeajarvi, A.; Claeyssens, M.; Teleman, O.; Kleywegt, G.J.; Szardenings, M.; Rouvinen, J.; Jones, T.A.; *et al.* The active site of *Trichoderma reesei* cellobiohydrolase II: The role of tyrosine 169.*Protein Eng* **1996**, *9*, 691–699.

68. Jeffries, T.W. Biochemistry and genetics of microbial xylanases. *Curr. Opin. Biotechnol* **1996**, *7*, 337–342.

69. Cherry, J.R.; Lamsa, M.H.; Schneider, P.; Vind, J.; Svendsen, A.; Jones, A.; Pedersen, A.H. Directed evolution of a fungal peroxidase. *Nat. Biotechnol* **1999**, *17*, 379–384.

70. Martin, A.; Sieber, V.; Schmid, F.X. *In-vitro* selection of highly stabilized protein variants with optimized surface. *J. Mol. Biol* **2001**, *309*, 717–726.

71. Palackal, N.; Brennan, Y.; Callen, W.N.; Dupree, P.; Frey, G.; Goubet, F.; Hazlewood, G.P.; Healey, S.; Kang, Y.E.; Kretz, K.A.; *et al.* An evolutionary route to xylanase process fitness. *Protein Sci* **2004**, *13*, 494–503.

72. Schmidt-Dannert, C.; Arnold, F.H. Directed evolution of industrial enzymes. *Trends Biotechnol* **1999**, *17*, 135–136.

73. Zhao, H.; Arnold, F.H. Directed evolution converts subtilisin E into a functional equivalent of thermitase. *Protein Eng* **1999**, *12*, 47–53.

74. Miyazaki, K.; Wintrode, P.L.; Grayling, R.A.; Rubingh, D.N.; Arnold, F.H. Directed evolution study of temperature adaptation in a psychrophilic enzyme. *J. Mol. Biol* **2000**, *297*, 1015–1026.

75. Gershenson, A.; Schauerte, J.A.; Giver, L.; Arnold, F.H. Tryptophan phosphorescence study of enzyme flexibility and unfolding in laboratory-evolved thermostable esterases. *Biochemistry* **2000**, *39*, 4658–4665.

76. Giver, L.; Gershenson, A.; Freskgard, P.O.; Arnold, F.H. Directed evolution of a thermostable esterase. *Proc. Natl. Acad. Sci. USA* **1998**, *95*, 12809–12813.

77. Suzuki, T.; Yasugi, M.; Arisaka, F.; Yamagishi, A.; Oshima, T. Adaptation of a thermophilic enzyme, 3-isopropylmalate dehydrogenase, to low temperatures. *Protein Eng* **2001**, *14*, 85–91.

78. Tamakoshi, M.; Nakano, Y.; Kakizawa, S.; Yamagishi, A.; Oshima, T. Selection of stabilized 3-isopropylmalate dehydrogenase of *Saccharomyces cerevisiae* using the host-vector system of an extreme thermophile, *Thermus thermophilus*. *Extremophiles* **2001**, *5*, 17–22.

79. Hardy, F.; Vriend, G.; Veltman, O.R.; van der Vinne, B.; Venema, G.; Eijsink, V.G. Stabilization of *Bacillus stearothermophilus* neutral protease by introduction of prolines. *FEBS Lett* **1993**, *317*, 89–92.

80. Mansfeld, J.; Vriend, G.; Dijkstra, B.W.; Veltman, O.R.; van den Burg, B.; Venema, G.; Ulbrich-Hofmann, R.; Eijsink, V.G. Extreme stabilization of a thermolysin-like protease by an engineered disulfide bond. *J. Biol. Chem* **1997**, *272*, 11152–11156.

81. Veltman, O.R.; Vriend, G.; Middelhoven, P.J.; van den Burg, B.; Venema, G.; Eijsink, V.G. Analysis of structural determinants of the stability of thermolysin-like proteases by molecular modelling and site-directed mutagenesis.*Protein Eng* **1996**, *9*, 1181–1189.

82. Van den Burg, B.; Vriend, G.; Veltman, O.R.; Venema, G.; Eijsink, V.G. Engineering an enzyme to resist boiling. *Proc. Natl. Acad. Sci. USA* **1998**, *95*, 2056–2060.

83. Kawamura, S.; Kakuta, Y.; Tanaka, I.; Hikichi, K.; Kuhara, S.; Yamasaki, N.; Kimura, M. Glycine-15 in the bend between two α-helices can explain the thermostability of DNA binding protein HU from *Bacillus stearothermophilus.Biochemistry* **1996**, *35*, 1195–1200.

84. Kimura, S.; Kanaya, S.; Nakamura, H. Thermostabilization of *Escherichia coli* ribonuclease HI by replacing left-handed helical Lys95 with Gly or Asn. *J. Biol. Chem* **1992**, *267*, 22014–22017.

85. Matthews, B.W.; Nicholson, H.; Becktel, W.J. Enhanced protein thermostability from site-directed mutations that decrease the entropy of unfolding. *Proc. Natl. Acad. Sci. USA* **1987**, *84*, 6663–6667.

86. Sanz-Aparicio, J.; Hermoso, J.A.; Martínez-Ripoll, M.; González, B.; López-Camacho, C.; Polaina, J. Structural basis of increased resistance to thermal denaturation induced by single amino acid substitution in the sequence of β-glucosidase A from *Bacillus polymyxa. Proteins Struct. Funct. Bioinforma* **1998**, *33*, 567–576.

87. Tiwari, M.K.; Moon, H.J.; Jeya, M.; Lee, J.K. Cloning and characterization of a thermostable xylitol dehydrogenase from *Rhizobium etli* CFN42. *Appl. Microbiol. Biotechnol* **2010**, *87*, 571–581.

88. Wintrode, P.L.; Arnold, F.H. Temperature adaptation of enzymes: Lessons from laboratory evolution. *Adv. Protein Chem* **2000**, *55*, 161–225.

89. Zhang, X.J.; Baase, W.A.; Shoichet, B.K.; Wilson, K.P.; Matthews, B.W. Enhancement of protein stability by the combination of point mutations in T4 lysozyme is additive. *Protein Eng* **1995**, *8*, 1017–1022.

90. Lehmann, M.; Loch, C.; Middendorf, A.; Studer, D.; Lassen, S.F.; Pasamontes, L.; van Loon, A.P.; Wyss, M. The consensus concept for thermostability engineering of proteins: Further proof of concept. *Protein Eng* **2002**, *15*, 403–411.

91. D'Amico, S.; Marx, J.C.; Gerday, C.; Feller, G. Activity-stability relationships in extremophilic enzymes. *J. Biol. Chem* **2003**, *278*, 7891–7896.

92. Sandgren, M.; Gualfetti, P.J.; Shaw, A.; Gross, L.S.; Saldajeno, M.; Day, A.G.; Jones, T.A.; Mitchinson, C. Comparison of family 12 glycoside hydrolases and recruited substitutions important for thermal stability. *Protein Sci* **2003**, *12*, 848–860.

93. Williams, J.C.; Zeelen, J.P.; Neubauer, G.; Vriend, G.; Backmann, J.; Michels, P.A.; Lambeir, A.M.; Wierenga, R.K. Structural and mutagenesis studies of leishmania triosephosphate isomerase: A point mutation can convert a mesophilic enzyme into a superstable enzyme without losing catalytic power. *Protein Eng* **1999**, *12*, 243–250.

94. Hasegawa, J.; Shimahara, H.; Mizutani, M.; Uchiyama, S.; Arai, H.; Ishii, M.; Kobayashi, Y.; Ferguson, S.J.; Sambongi, Y.; Igarashi, Y. Stabilization of *Pseudomonas aeruginosa* cytochrome c_{551} by systematic amino acid substitutions based on the structure of thermophilic *Hydrogenobacter thermophilus* cytochrome c_{552}. *J. Biol. Chem* **1999**, *274*, 37533–37537.

95. Khmelnitsky, Y.L.; Rich, J.O. Biocatalysis in nonaqueous solvents. *Curr. Opin. Chem. Biol* **1999**, *3*, 47–53.

96. Arnold, F.H. Engineering enzymes for non-aqueous solvents. *Trends. Biotechnol* **1990**, *8*, 244–249.

97. Dordick, J.S. Enzymatic catalysis in monophasic organic solvents. *Enzyme Microb. Technol* **1989**, *11*, 194–211.

98. Wong, C.H.; Chen, S.T.; Hennen, W.J.; Bibbs, J.A.; Wang, Y.F.; Liu, J.L.C.; Pantoliano, M.W.; Whitlow, M.; Bryan, P.N. Enzymes in organic synthesis: Use of subtilisin and a highly stable mutant derived from multiple site-specific mutations. *J. Am. Chem. Soc* **1990**, *112*, 945–953.

99. Chen, K.Q.; Arnold, F.H. Enzyme engineering for nonaqueous solvents: Random mutagenesis to enhance activity of subtilisin E in polar organic media. *Biotechnology* **1991**, *9*, 1073–1077.

100. Economou, C.; Chen, K.; Arnold, F.H. Random mutagenesis to enhance the activity of subtilisin in organic solvents: Characterization of Q103R subtilisin E. *Biotechnol. Bioeng* **1992**, *39*, 658–662.

101. Pantoliano, M.W. Proteins designed for challenging environments and catalysis in organic solvents. *Curr. Opin. Struct. Biol* **1992**, *2*, 559–568.

102. Moore, J.C.; Arnold, F.H. Directed evolution of a *p*-nitrobenzyl esterase for aqueous-organic solvents. *Nat. Biotechnol* **1996**, *14*, 458–467.

103. Martinez, P.; van Dam, M.E.; Robinson, A.C.; Chen, K.; Arnold,

F.H. Stabilization of substilisin E in organic solvents by site-directed mutagenesis. *Biotechnol. Bioeng* **1992**, *39*, 141–147.

104. Song, J.K.; Rhee, J.S. Enhancement of stability and activity of phospholipase A_1 in organic solvents by directed evolution. *Biochim. Biophys. Acta* **2001**, *1547*, 370–378.

105. Morawski, B.; Quan, S.; Arnold, F.H. Functional expression and stabilization of horseradish peroxidase by directed evolution in *Saccharomyces cerevisiae*. *Biotechnol. Bioeng* **2001**, *76*, 99–107.

106. Liebeton, K.; Zonta, A.; Schimossek, K.; Nardini, M.; Lang, D.; Dijkstra, B.W.; Reetz, M.T.; Jaeger, K.E. Directed evolution of an enantioselective lipase. *Chem. Biol* **2000**, *7*, 709–718.

107. Jaeger, K.E.; Eggert, T.; Eipper, A.; Reetz, M.T. Directed evolution and the creation of enantioselective biocatalysts.*Appl. Microbiol. Biotechnol* **2001**, *55*, 519–530.

108. Nardini, M.; Lang, D.A.; Liebeton, K.; Jaeger, K.E.; Dijkstra, B.W. Crystal structure of *Pseudomonas aeruginosa* lipase in the open conformation. The prototype for family I.1 of bacterial lipases. *J. Biol. Chem* **2000**, *275*, 31219–31225.

109. Raillard, S.; Krebber, A.; Chen, Y.; Ness, J.E.; Bermudez, E.; Trinidad, R.; Fullem, R.; Davis, C.; Welch, M.; Seffernick, J.;*et al.* Novel enzyme activities and functional plasticity revealed by recombining highly homologous enzymes. *Chem. Biol* **2001**, *8*, 891–898.

110. Fong, S.; Machajewski, T.D.; Mak, C.C.; Wong, C. Directed evolution of d-2-keto-3-deoxy-6- phosphogluconate aldolase to new variants for the efficient synthesis of d- and l-sugars. *Chem. Biol* **2000**, *7*, 873–883.

111. Wymer, N.; Buchanan, L.V.; Henderson, D.; Mehta, N.; Botting, C.H.; Pocivavsek, L.; Fierke, C.A.; Toone, E.J.; Naismith, J.H. Directed evolution of a new catalytic site in 2-keto-3- deoxy-6-phosphogluconate aldolase from*Escherichia coli*. *Structure* **2001**, *9*, 1–9.

112. Suenaga, H.; Goto, M.; Furukawa, K. Emergence of multifunctional oxygenase activities by random priming recombination. *J. Biol. Chem* **2001**, *276*, 22500–22506.

113. Lee, J.K.; Simurdiak, M.; Zhao, H. Reconstitution and characterization of aminopyrrolnitrin oxygenase, a rieske *N*-oxygenase that catalyzes unusual arylamine oxidation. *J. Biol. Chem* **2005**, *280*, 36719–36727.

114. Lee, J.K.; Zhao, H. Mechanistic studies on the conversion of arylamines into arylnitro compounds by aminopyrrolnitrin oxygenase: Identification of intermediates and kinetic studies. *Angew. Chem. Int. Ed* **2006**, *45*,

622–625.

115. Lee, J.K.; Ang, E.L.; Zhao, H. Probing the substrate specificity of aminopyrrolnitrin Oxygenase (PrnD) by mutational analysis. *J. Bacteriol* **2006**, *188*, 6179–6183.

116. Jackson, C.J.; Foo, J.L.; Tokuriki, N.; Afriat, L.; Carr, P.D.; Kim, H.K.; Schenk, G.; Tawfik, D.S.; Ollis, D.L. Conformational sampling, catalysis, and evolution of the bacterial phosphotriesterase. *Proc. Natl. Acad. Sci. USA* **2009**,*106*, 21631–21636.

117. Foo, J.L.; Jackson, C.J.; Carr, P.D.; Kim, H.K.; Schenk, G.; Gahan, L.R.; Ollis, D.L. Mutation of outer-shell residues modulates metal ion co-ordination strength in a metalloenzyme. *Biochem. J* **2010**, *429*, 313–321.

118. Saunders, R.; Deane, C.M. Protein structure prediction begins well but ends badly. *Proteins* **2010**, *78*, 1282–1290.

119. Zhang, Y. Protein structure prediction: When is it useful? *Curr. Opin. Struct. Biol* **2009**, *19*, 145–155.

120. Zhang, Y. Progress and challenges in protein structure prediction. *Curr. Opin. Struct. Biol* **2008**, *18*, 342–348.

121. Taylor, R.D.; Jewsbury, P.J.; Essex, J.W. A review of protein-small molecule docking methods. *J. Comput. Aided Mol. Des* **2002**, *16*, 151–66.

122. Foo, J.L.; Ching, C.B.; Chang, M.W.; Leong, S.S. The imminent role of protein engineering in synthetic biology.*Biotechnol. Adv* **2012**, *30*, 541–549.

123. Jackson, C.J.; Weir, K.; Herlt, A.; Khurana, J.; Sutherland, T.D.; Horne, I.; Easton, C.; Russell, R.J.; Scott, C.; Oakeshott, J.G. Structure-based rational design of a phosphotriesterase.*Appl. Environ. Microbiol* **2009**, *75*, 5153–5156.

124. Jackson, C.J.; Foo, J.L.; Kim, H.K.; Carr, P.D.; Liu, J.W.; Salem, G.; Ollis, D.L. In crystallo capture of a Michaelis complex and product-binding modes of a bacterial phosphotriesterase. *J. Mol. Biol* **2008**, *375*, 1189–1196.

125. Ugwumba, I.N.; Ozawa, K.; Xu, Z.Q.; Ely, F.; Foo, J.L.; Herlt, A.J.; Coppin, C.; Brown, S.; Taylor, M.C.; Ollis, D.L.; *et al.* Improving a natural enzyme activity through incorporation of unnatural amino acids. *J. Am. Chem. Soc* **2011**, *133*, 326–333.

126. Voloshchuk, N.; Montclare, J.K. Incorporation of unnatural amino acids for synthetic biology. *Mol. Biosyst* **2010**, *6*, 65–80.

127. Goldstein, L.; Levin, Y.; Katchalski, E. A water-insoluble polyanionic derivatives of trypsin, effect of the polyelectrolyte carrier on the kinetic

behaviour of the bound trypsin. *Biochemistry* **1964**, *3*, 1914–1919.

128. Goto, M.; Hatanaka, C.; Goto, M. Immobilization of surfactant-lipase complexes and their high heat resistance in organic media. *Biochem. Eng. J* **2005**, *24*, 91–94.

129. Cabana, H.; Jones, J.P.; Agathos, S.N. Preparation and characterization of cross-linked laccase aggregates and their application to the elimination of endocrine disrupting chemicals. *J. Biotechnol* **2007**, *132*, 23–31.

130. Sangeetha, K.; Abraham, T.E. Preparation and characterization of cross-linked enzyme aggregates (CLEA) of subtilisin for controlled release applications. *Int. J. Biol. Macromol* **2008**, *43*, 314–319.

131. Hirsh, S.L.; Bilek, M.M.; Nosworthy, N.J.; Kondyurin, A.; dos Remedios, C.G.; McKenzie, D.R. A comparison of covalent immobilization and physical adsorption of a cellulase enzyme mixture. *Langmuir* **2010**, *26*, 14380–14388.

132. Dickensheets, P.A.; Chen, L.F.; Tsao, G.T. Characteristics of yeast invertase immobilized on porous cellulose beads.*Biotechnol. Bioeng* **1977**, *19*, 365–375.

133. Turkova, J. Oriented immobilization of biologically active proteins as a tool for revealing protein interactions and function. *J. Chromatogr. B* **1999**, *722*, 11–31.

134. Taylor, R.F. A comparison of various commercially-available liquid chromatographic supports for immobilization of enzymes and immunoglobulins. *Analytica. Chimica. Acta* **1985**, *172*, 241–248.

135. Aloulou, A.; Rodriguez, J.A.; Fernandez, S.; Oosterhout, D.V.; Puccinelli, D.; Carrière, F. Exploring the specific features of interfacial enzymology based on lipase studies. *Biochim. Biophys. Acta* **2006**, *1761*, 995–1013.

136. Palomo, J.M. Lipases enantioselectivity alteration by immobilization techniques. *Curr. Bioact. Compd* **2008**, *4*, 126–138.

137. Brady, L.; Brzozowski, A.M.; Derewenda, Z.S.; Dodson, E.; Dodson, G.; Tolley, S.; Turkenburg, J.P.; Christiansen, L.; Huge-Jensen, B.; Norskov, L.; *et al.* A serine protease triad forms the catalytic centre of a triacylglycerol lipase. *Nature* **1990**, *343*, 767–770.

138. Brzozowski, A.M.; Derewenda, U.; Derewenda, Z.S.; Dodson, G.G.; Lawson, D.M.; Turkenburg, J.P.; Bjorkling, F.; Huge-Jensen, B.; Patkar, S.A.; Thim, L. A model for interfacial activation in lipases from the structure of a fungal lipase-inhibitor complex. *Nature* **1991**, *351*, 491–494.

139. Derewenda, U.; Brzozowski, A.M.; Lawson, D.M.; Derewenda, Z.S.

Catalysis at the interface: The anatomy of a conformational change in a triglyceride lipase. *Biochemistry* **1992**, *31*, 1532–1541.

140. Kanori, M.; Hori, T.; Yamashita, Y.; Hirose, Y.; Naoshima, Y. A new inorganic ceramic support, Toyonite and the reactivity and enantioselectivity of the immobilised lipase. *J. Mol. Catal* **2000**, *9*, 269–274.

141. Jia, H.; Zhu, G.; Wang, P. Catalytic behaviors of enzymes attached to nanoparticles: The effect of particle mobility.*Biotechnol. Bioeng* **2003**, *84*, 406–414.

142. Tu, M.; Zhang, X.; Kurabi, A.; Gilkes, N.; Mabee, W.; Saddler, J. Immobilization of β-glucosidase on Eupergit C for lignocellulose hydrolysis. *Biotechnol. Lett* **2006**, *28*, 151–156.

143. Pandey, P.; Singh, S.P.; Arya, S.K.; Gupta, V.; Datta, M.; Singh, S.; Malhotra, B.D. Application of thiolated gold nanoparticles for the enhancement of glucose oxidase activity. *Langmuir* **2007**, *23*, 3333–3337.

144. Prakasham, R.S.; Devi, G.S.; Laxmi, K.R.; Rao, C.S. Novel synthesis of ferric impregnated silica nanoparticles and their evaluation as a matrix for enzyme immobilization. *J. Phys. Chem. C* **2007**, *111*, 3842–3847.

145. Neri, D.F.M.; Balcão, V.M.; Costa, R.S.; Rocha, I.C.A.P.; Ferreira, E.M.F.C.; Torres, D.P.M.; Rodrigues, L.R.M.; Carvalho, L.B.; Teixeira, J.A. Galacto-oligosaccharides production during lactose hydrolysis by free *Aspergillus oryzae*β-galactosidase and immobilized on magnetic polysiloxane-polyvinyl alcohol. *Food Chem* **2009**, *115*, 92–99.

146. Konwarh, R.; Karak, N.; Rai, S.K.; Mukherjee, A.K. Polymer-assisted iron oxide magnetic nanoparticle immobilized keratinase. *Nanotechnology* **2009**, *20*, 225107–225117.

147. Qiu, H.; Lu, L.; Huang, X.; Zhang, Z.; Qu, Y. Immobilization of horseradish peroxidase on nanoporous copper and its potential applications. *Bioresour. Technol* **2010**, *101*, 9415–9420.

148. Hashemifard, N.; Mohsenifar, A.; Ranjbar, B.; Allameh, A.; Lotfi, A.S.; Etemadikia, B. Fabrication and kinetic studies of a novel silver nanoparticles-glucose oxidase bioconjugate. *Anal. Chim. Acta* **2010**, *675*, 181–184.

149. Singh, R.K.; Zhang, Y.W.; Nguyen, N.P.; Jeya, M.; Lee, J.K. Covalent immobilization of β-1,4-glucosidase from *Agaricus arvensis* onto functionalized silicon oxide nanoparticles. *Appl. Microbiol. Biotechnol* **2010**, *89*, 337–344.

150. Zhang, Y.W.; Jeya, M.; Lee, J.K. Enhanced activity and stability of

l-arabinose isomerase by immobilization on aminopropyl glass. *Appl. Microbiol. Biotechnol* **2011**, *89*, 1435–1442.

151. Singh, V.; Ahmed, S. Silver nanoparticle (AgNPs) doped gum acacia-gelatin-silica nanohybrid: An effective support for diastase immobilization. *Int. J. Biol. Macromol* **2011**, *50*, 353–361.

152. Wang, F.; Su, R.X.; Qi, W.; Zhang, M.J.; He, Z.M. Preparation and activity of bubbling-immobilized cellobiase within chitosan-alginate composite. *Prep. Biochem. Biotechnol* **2010**, *40*, 57–64.

153. Ansari, S.A.; Husain, Q. Potential applications of enzymes immobilized on/in nano materials: A review. *Biotechnol. Adv* **2012**, *30*, 512–523.

154. Nwagu, T.N.; Okolo, B.N.; Aoyagi, H. Stabilization of a raw starch digesting amylase from *Aspergillus carbonarius* via immobilization on activated and non-activated agarose gel. *World J. Microbiol. Biotechnol* **2012**, *28*, 335–345.

155. Wang, A.; Wang, M.; Wang, Q.; Chen, F.; Zhang, F.; Li, H.; Zeng, Z.; Xie, T. Stable and efficient immobilization technique of aldolase under consecutive microwave irradiation at low temperature. *Bioresour. Technol* **2011**, *102*, 469–474.

156. Shankar, S.K.; Praveen Kumar, S.K.; Mulimani, V.H. Calcium alginate entrapped preparation of α-galactosidase: Its stability and application in hydrolysis of soymilk galactooligosaccharides. *J. Ind. Microbiol. Biotechnol* **2011**, *38*, 1399–1405.

157. Cristovao, R.O.; Silverio, S.C.; Tavares, A.P.; Brigida, A.I.; Loureiro, J.M.; Boaventura, R.A.; Macedo, E.A.; Coelho, M.A. Green coconut fiber: A novel carrier for the immobilization of commercial laccase by covalent attachment for textile dyes decolourization. *World J. Microbiol. Biotechnol* **2012**, *28*, 2827–2838.

158. Sahoo, B.; Sahu, S.K.; Bhattacharya, D.; Dhara, D.; Pramanik, P. A novel approach for efficient immobilization and stabilization of papain on magnetic gold nanocomposites. *Colloids Surf. B* **2012**, *101*, 280–289.

159. Stark, M.B.; Holmberg, K. Covalent immobilization of lipase in organic solvents. *Biotechnol. Bioeng* **1989**, *34*, 942–950.

160. Marcuello, C.; de Miguel, R.; Gomez-Moreno, C.; Martinez-Julvez, M.; Lostao, A. An efficient method for enzyme immobilization evidenced by atomic force microscopy. *Protein Eng. Des. Sel* **2012**, *25*, 715–723.

161. Janssen, M.H.A.; van Langen, L.M.; Pereira, S.R.M.; van Rantwijk, F.; Sheldon, R.A. Evaluation of the performance of immobilized penicillin G acylase using active-site titration. *Biotechnol. Bioeng* **2002**, *78*, 425–432.

162. Bezbradica, D.; Mijin, D.; Mihailović, M.; Knežević-Jugović, Z. Microwave-assisted immobilization of lipase from*Candida rugosa* on Eupergit® supports. *J. Chem. Technol. Biotechnol* **2009**, *84*, 1642–1648.

163. Itoyama, K.; Tokura, S.; Hayashi, T. Lipoprotein lipase immobilization onto porous chitosan beads. *Biotechnol. Prog* **1994**, *10*, 225–229.

164. Hayashi, T.; Ikada, Y. Protease immobilization onto polyacrolein microspheres. *Biotechnol. Bioeng* **1990**, *35*, 518–524.

165. Itoyama, K.; Tanibe, H.; Hayashi, T.; Ikada, Y. Spacer effects on enzymatic activity of papain immobilized onto porous chitosan beads. *Biomaterials* **1994**, *15*, 107–112.

166. Abian, O.; Wilson, L.; Mateo, C.; Fernández-Lorente, G.; Palomo, J.M.; Fernández-Lafuente, R.; Guisán, J.M.; Re, D.; Tam, A.; Daminatti, M. Preparation of artificial hyper-hydrophilic micro-environments (polymeric salts) surrounding enzyme molecules: New enzyme derivatives to be used in any reaction medium. *J. Mol. Catal. B* **2002**, (19–20), 295–303.

167. Bigdeli, S.; Talasaz, A.H.; Ståhl, P.; Persson, H.H.J.; Ronaghi, M.; Davis, R.W.; Nemat-Gorgani, M. Conformational flexibility of a model protein upon immobilization on self-assembled monolayers. *Biotechnol. Bioeng* **2008**, *100*, 19–27.

168. Cao, L.; Schmid, R.D. *Carrier-Bound Immobilized Enzymes: Principles, Application and Design*; Wiley-VCH Verlag: Weinheim, Germany, 2005.

169. Aksoy, S.; Tumturk, H.; Hasirci, N. Stability of α-amylase immobilized on poly (methyl methacrylate-acrylic acid) microspheres. *J. Biotechnol* **1998**, *60*, 37–46.

170. Hanefeld, U.; Gardossi, L.; Magner, E. Understanding enzyme immobilisation. *Chem. Soc. Rev* **2009**, *38*, 453–468.

171. Sabularse, V.C.; Tud, M.T.; Lacsamana, M.S.; Solivas, J.L. Black and white lahar as inorganic support for the immobilization of yeast invertase. *ASEAN J. Sci. Technol. Dev* **2005**, *22*, 331–344.

172. Kotwal, S.M.; Shankar, V. Immobilized invertase. *Biotechnol. Adv* **2009**, *27*, 311–322.

173. Clark, D.S. Can immobilization be exploited to modify enzyme activity? *Trends Biotechnol* **1994**, *12*, 439–443.

174. Mateo, C.; Palomo, J.M.; Fernandez-Lorente, G.; Guisan, J.M.; Fernandez-Lafuente, R. Improvement of enzyme activity, stability and selectivity via immobilization techniques. *Enzyme Microb. Technol* **2007**, *40*, 1451–1463.

175. Matsumoto, M.; Ohashi, K. Effect of immobilization on thermostability of lipase from *Candida rugosa*. *Biochem. Eng. J* **2003**, *14*, 75–77.

176. Moriyama, S.; Noda, A.; Nakanishi, K.; Matsuno, R.; Kamikubo, T. Thermal stability of immobilized glucoamylase entrapped in polyacrylamide gels and bound to SP-Sephadex C-50. *Agric. Biol. Chem* **1980**, *44*, 2024–2054.

177. Fruhwirth, G.O.; Paar, A.; Gudelj, M.; Cavaco-Paulo, A.; Robra, K.H.; Gübitz, G.M. An immobilised catalase peroxidase from the alkalothermophilic *Bacillus* SF for the treatment of textile-bleaching effluents. *Appl. Microbiol. Biotechnol* **2002**,*60*, 313–319.

178. Fernandez-Lafuente, R.; Cowan, D.A.; Wood, A.N.P. Hyperstabilization of a thermophilic esterase by multipoint covalent attachment. *Enzyme Microb. Technol* **1995**, *17*, 366–372.

179. Lee, D.C.; Lee, S.G.; Kim, H.S. Production of d-*p*-hydroxyphenylglycine from d,l-5-(4-hydroxyphenyl)hydantoin using immobilized thermostable d-hydantoinase from *Bacillus stearothermophilus* SD-1. *Enzyme Microb. Technol* **1996**, *18*, 35–40.

180. Kawakami, K.; Abe, T.; Tomoaki, Y. Silicone-immobilized biocatalysts effective for bioconversions in nonaqueous media. *Enzyme Microb. Technol* **1992**, *14*, 371–375.

181. Pedroche, J.; del Mar Yust, M.; Mateo, C.; Fernández-Lafuente, R.; Girán-Calle, J.; Alaiz, M.; Vioque, J.; Guisán, J.M.; Millán, F. Effect of the support and experimental conditions in the intensity of the multipoint covalent attachment of proteins on glyoxyl-agarose supports: Correlation between enzyme-support linkages and thermal stability. *Enzyme Microb. Technol* **2007**, *40*, 1160–1166.

182. Mateo, C.; Abian, O.; Fernández-Lafuente, R.; Guisan, J.M. Increase in conformational stability of enzymes immobilized on epoxy-activated supports by favoring additional multipoint covalent attachment. *Enzyme Microb. Technol* **2000**, *26*, 509–515.

183. Wang, P.; Hu, X.; Cook, S.; Hwang, H.M. Influence of silica-derived nano-supporters on cellobiase after immobilization. *Appl. Biochem. Biotechnol* **2009**, *158*, 88–96.

184. Wong, L.S.; Khan, F.; Micklefield, J. Selective covalent protein immobilization: Strategies and applications. *Chem. Rev* **2009**, *109*, 4025–4053.

185. Kahraman, M.V.; Bayramođlu, G.; Kayaman-Apohan, N.; Güngör, A. α-Amylase immobilization on functionalized glass beads by covalent attachment. *Food Chem* **2007**, *104*, 1385–1392.

186. Yilmaz, E.; Can, K.; Sezgin, M.; Yilmaz, M. Immobilization of *Candida rugosa* lipase on glass beads for enantioselective hydrolysis of racemic naproxen methyl ester. *Bioresour. Technol* **2011**, *102*, 499–506.

187. Garcia, A., III; Oh, S.; Engler, C.R. Cellulase immobilization on Fe_3O_4 and characterization. *Biotechnol. Bioeng.* **1989**, *33*, 321–326.

188. Bolivar, J.M.; Wilson, L.; Ferrarotti, S.A.; Fernandez-Lafuente, R.; Guisan, J.M.; Mateo, C. Stabilization of a formate dehydrogenase by covalent immobilization on highly activated glyoxyl-agarose supports. *Biomacromolecules* **2006**, *7*, 669–673.

189. Tardioli, P.W.; Zanin, G.M.; de Moraes, F.F. Characterization of *Thermoanaerobacter cyclomaltodextrin* glucanotransferase immobilized on glyoxyl-agarose. *Enzyme Microb. Technol* **2006**, *39*, 1270–1278.

190. Pessela, B.C.; Mateo, C.; Fuentes, M.; Vian, A.; Garcia, J.L.; Carrascosa, A.V.; Guisan, J.M.; Fernandez-Lafuente, R. Stabilization of a multimeric β-galactosidase from *Thermus* sp. strain T2 by immobilization on novel heterofunctional epoxy supports plus aldehyde-dextran cross-linking. *Biotechnol. Prog* **2004**, *20*, 388–392.

191. Guisán, J.M.; Polo, E.; Aguado, J.; Romero, M.D.; Álvaro, G.; Guerra, M.J. Immobilization-stabilization of thermolysin onto activated agarose gels. *Biocatal. Biotransform* **1997**, *15*, 159–173.

192. Otero, C.; Ballesteros, A.; Guisán, J.M. Immobilization/stabilization of lipase from *Candida rugosa*. *Appl. Biochem. Biotechnol* **1988**, *19*, 163–175.

193. Alvaro, G.; Fernandez-Lafuente, R.; Blanco, R.M.; Guisan, J.M. Immobilization-stabilization of penicillin G acylase from *Escherichia coli*. *Appl. Biochem. Biotechnol* **1990**, *26*, 181–195.

194. Guisan, J.M.; Bastida, A.; Cuesta, C.; Fernandez-Lufuente, R.; Rosell, C.M. Immobilization-stabilization of α-chymotrypsin by covalent attachment to aldehyde-agarose gels. *Biotechnol. Bioeng* **1991**, *38*, 1144–1152.

195. Guisan, J.M.; Alvaro, G.; Fernandez-Lafuente, R.; Rosell, C.M.; Garcia, J.L.; Tagliani, A. Stabilization of heterodimeric enzyme by multipoint covalent immobilization: Penicillin G acylase from *Kluyvera citrophila*. *Biotechnol. Bioeng* **1993**,*42*, 455–464.

196. Akgöl, S.; Bayramoðlu, G.; Kacar, Y.; Denizli, A.; Arýca, M.Y. Poly(hydroxyethyl methacrylateco- glycidyl methacrylate) reactive membrane utilised for cholesterol oxidase immobilisation. *Polym. Int* **2002**, *51*, 1316–1322.

197. Tardioli, P.W.; Pedroche, J.; Giordano, R.L.; Fernandez-Lafuente, R.; Guisan, J.M. Hydrolysis of proteins by immobilized-stabilized alcalase-glyoxyl agarose. *Biotechnol. Prog* **2003**, *19*, 352–360.

198. Suh, C.W.; Choi, G.S.; Lee, E.K. Enzymic cleavage of fusion protein using immobilized urokinase covalently conjugated to glyoxyl-agarose. *Biotechnol. Appl. Biochem* **2003**, *37*, 149–155.

199. Bayramoglu, G.; Yilmaz, M.; Arica, M.Y. Immobilization of a thermostable α-amylase onto reactive membranes: Kinetics characterization and application to continuous starch hydrolysis. *Food Chem* **2004**, *84*, 591–599.

200. Danisman, T.; Tan, S.; Kacar, Y.; Ergene, A. Covalent immobilization of invertase on microporous pHEMA-GMA membrane. *Food Chem* **2004**, *85*, 461–466.

201. De Segura, A.G.; Alcalde, M.; Yates, M.; Rojas-Cervantes, M.L.; López-Cortés, N.; Ballesteros, A.; Plou, F.J. Immobilization of dextransucrase from *Leuconostoc mesenteroides* NRRL B-512F on Eupergit C Supports. *Biotechnol. Prog* **2004**, *20*, 1414–1420.

202. Bolivar, J.M.; Wilson, L.; Ferrarotti, S.A.; Guisan, J.M.; Fernandez-Lafuente, R.; Mateo, C. Improvement of the stability of alcohol dehydrogenase by covalent immobilization on glyoxyl-agarose. *J. Biotechnol* **2006**, *125*, 85–94.

203. Ferrarotti, S.A.; Bolivar, J.M.; Mateo, C.; Wilson, L.; Guisan, J.M.; Fernandez-Lafuente, R. Immobilization and stabilization of a cyclodextrin glycosyltransferase by covalent attachment on highly activated glyoxyl-agarose supports. *Biotechnol. Prog* **2006**, *22*, 1140–1145.

204. Bolivar, J.M.; Wilson, L.; Ferrarotti, S.A.; Fernandez-Lafuente, R.; Guisan, J.M.; Mateo, C. Evaluation of different immobilization strategies to prepare an industrial biocatalyst of formate dehydrogenase from *Candida boidinii*. *Enzyme Microb. Technol* **2007**, *40*, 540–546.

205. Arica, M.Y.; Altintas, B.; Bayramoglu, G. Immobilization of laccase onto spacer-arm attached non-porous poly(GMA/EGDMA) beads: Application for textile dye degradation. *Bioresour. Technol* **2009**, *100*, 665–669.

206. Zhang, Y.W.; Tiwari, M.K.; Jeya, M.; Lee, J.K. Covalent immobilization of recombinant *Rhizobium etli* CFN42 xylitol dehydrogenase onto modified silica nanoparticles. *Appl. Microbiol. Biotechnol* **2010**, *90*, 499–507.

207. Corman, M.E.; Ozturk, N.; Bereli, N.; Akgol, S.; Denizli, A. Preparation of nanoparticles which contains histidine for immobilization of *Trametes*

versicolor laccase. *J. Mol. Catal. B* **2010**, *63*, 102–107.

208. Zhou, J. Immobilization of cellulase on a reversibly soluble-insoluble support: Properties and application. *J. Agric. Food Chem* **2010**, *58*, 6741–6746.

209. Fu, J.; Reinhold, J.; Woodbury, N.W. Peptide-modified surfaces for enzyme immobilization. *PLoS One* **2011**, *6*, e18692.

210. Mendes, A.A.; Freitas, L.; de Carvalho, A.K.; de Oliveira, P.C.; de Castro, H.F. Immobilization of a commercial lipase from *Penicillium camembertii* (lipase G) by different strategies. *Enzyme Res.* **2011**, *2011*, 967239:1–967239:8.

211. Menezes-Blackburn, D.; Jorquera, M.; Gianfreda, L.; Rao, M.; Greiner, R.; Garrido, E.; de la Luz Mora, M. Activity stabilization of *Aspergillus niger* and *Escherichia coli* phytases immobilized on allophanic synthetic compounds and montmorillonite nanoclays. *Bioresour. Technol* **2011**, *102*, 9360–9367.

212. Madadlou, A.; Iacopino, D.; Sheehan, D.; Emam-Djomeh, Z.; Mousavi, M.E. Enhanced thermal and ultrasonic stability of a fungal protease encapsulated within biomimetically generated silicate nanospheres. *Biochim. Biophys. Acta* **2010**,*1800*, 459–465.

213. Xue, Y.; Nie, H.; Zhu, L.; Li, S.; Zhang, H. Immobilization of modified papain with anhydride groups on activated cotton fabric. *Appl. Biochem. Biotechnol* **2010**, *160*, 109–121.

214. Tananchai, P.; Chisti, Y. Stabilization of invertase by molecular engineering. *Biotechnol. Prog* **2010**, *26*, 111–117.

215. Kikani, B.A.; Pandey, S.; Singh, S.P. Immobilization of the α-amylase of *Bacillus amyloliquifaciens* TSWK1–1 for the improved biocatalytic properties and solvent tolerance. *Bioprocess Biosyst. Eng.* **2012**.

216. Khalaf, N.; Govardhan, C.P.; Lalonde, J.J.; Persichetti, R.A.; Wang, Y.F.; Margolin, A.L. Cross-linked enzyme crystals as highly active catalysts in organic solvents. *J. Am. Chem. Soc* **1996**, *118*, 5494–5495.

217. Sheldon, R.A. Enzyme immobilization: The quest for optimum performance. *Adv. Synth. Catal* **2007**, *349*, 1289–1307.

218. Bommarius, A.S.; Karau, A. Deactivation of formate dehydrogenase (FDH) in solution and at gas-liquid interfaces.*Biotechnol. Prog* **2005**, *21*, 1663–1672.

219. Caussette, M.; Gaunand, A.; Planche, H.; Monsan, P.; Lindet, B. Inactivation of enzymes by inert gas bubbling: A kinetic study. *Ann. N. Y. Acad. Sci* **1998**, *864*, 228–233.

220. Orsat, B.; Drtina, G.J.; Williams, M.G.; Klibanov, A.M. Effect of support material and enzyme pretreatment on enantioselectivity of immobilized subtilisin in organic solvents. *Biotechnol. Bioeng* **1994**, *44*, 1265–1269.

221. Ruiz, A.I.; Malavé, A.J.; Felby, C.; Griebenow, K. Improved activity and stability of an immobilized recombinant laccase in organic solvents. *Biotechnol. Lett* **2000**, *22*, 229–233.

222. Prodanovic, R.M.; Milosavic, N.B.; Jovanovic, S.M.; Velickovic, T.C.; Vujcic, Z.M.; Jankov, R.M. Stabilization of α-glucosidase in organic solvents by immobilization on macroporous poly(GMA-co-EGDMA) with different surface characteristics. *J. Serb. Chem. Soc* **2006**, *71*, 339–347.

223. Lo, Y.S.; Ibrahim, C.O. Some characteristics of amberlite XAD-7-adsorbed lipase from *Pseudomonas* sp. AK. *Malays. J. Microbiol* **2005**, *1*, 53–56.

224. Rozzell, J.D. Commercial scale biocatalysis: Myths and realities. *Bioorg. Med. Chem* **1999**, *7*, 2253–2261.

225. Takac, S.; Bakkal, M. Impressive effect of immobilization conditions on the catalytic activity and enantioselectivity of*Candida rugosa* lipase toward *S*-Naproxen production. *Process Biochem* **2007**, *42*, 1021–1027.

226. Reetz, M.T.; Zonta, A.; Schimossek, K.; Jaeger, K.E.; Liebeton, K. Creation of enantioselective biocatalysts for organic chemistry by *in vitro* evolution. *Angew. Chem. Int. Ed* **1997**, *36*, 2830–2832.

227. Palomo, J.M.; Fernandez-Lorente, G.; Mateo, C.; Ortiz, C.; Fernandez-Lafuente, R.; Guisan, J.M. Modulation of the enantioselectivity of lipases via controlled immobilization and medium engineering: Hydrolytic resolution of mandelic acid esters. *Enzyme Microb. Technol* **2002**, *31*, 775–783.

228. Nishio, T.; Hayashi, R. Digestion of protein substrates by subtilisin: Immobilization changes the pattern of products.*Arch. Biochem. Biophys* **1984**, *229*, 304–311.

229. Hernaiz, M.J.; Crout, D.H.G. A highly selective synthesis of N-acetyllactosamine catalyzed by immobilised β-galactosidase from *Bacillus circulans*. *J. Mol. Catal. B* **2000**, *10*, 403–408.

230. Boundy, J.A.; Smiley, K.L.; Swanson, C.L.; Hofreiter, B.T. Exoenzymic activity of α-amylase immobilized on a phenol-formaldehyde resin. *Carbohydr. Res* **1976**, *48*, 239–244.

231. Linko, Y.Y.; Saarinen, P.; Linko, M. Starch conversion by soluble and immobilized α-amylase. *Biotechnol. Bioeng* **1975**, *17*, 153–165.

232. Rotticci, D.; Norin, T.; Hult, K. Mass transport limitations reduce the effective stereospecificity in enzyme-catalyzed kinetic resolution. *Org. Lett* **2000**, *2*, 1373–1376.

233. Ogino, S. Formation of the fructose-rich polymer by water-insoluble dextransucrease and presence of glycogen valuelowering factor. *Agric. Biol. Chem* **1970**, *34*, 1268–1271.

234. Palomo, J.M.; Segura, R.L.; Mateo, C.; Terreni, M.; Guisan, J.M.; Fernández-Lafuente, R. Synthesis of enantiomerically pure glycidol via a fully enantioselective lipase-catalyzed resolution. *Tetrahedron Asymmetry* **2005**, *16*, 869–874.

235. Palomo, J.M.; Segura, R.L.; Fernandez-Lorente, G.; Guisan, J.M.; Fernandez-Lafuente, R. Enzymatic resolution of (±)-glycidyl butyrate in aqueous media. Strong modulation of the properties of the lipase from *Rhizopus oryzae* via immobilization techniques. *Tetrahedron Asymmetry* **2004**, *15*, 1157–1161.

236. Palomo, J.M.; Fernández-Lorente, G.; Rúa, M.L.; Guisún, J.M.; Fernández-Lafuente, R. Evaluation of the lipase from*Bacillus thermocatenulatus* as an enantioselective biocatalyst. *Tetrahedron Asymmetry* **2003**, *14*, 3679–3687.

237. Palomo, J.M.; Muáoz, G.; Fernádez-Lorente, G.; Mateo, C.; Fuentes, M.; Guisan, J.M.; Fernández-Lafuente, R. Modulation of *Mucor miehei* lipase properties via directed immobilization on different hetero-functional epoxy resins: Hydrolytic resolution of (*R,S*)-2-butyroyl-2-phenylacetic acid. *J. Mol. Catal. B* **2003**, *21*, 201–210.

238. Palomo, J.M.; Segura, R.L.; Fernandez-Lorente, G.; Fernandez-Lafuente, R.; Guisán, J.M. Glutaraldehyde modification of lipases adsorbed on aminated supports: A simple way to improve their behaviour as enantioselective biocatalyst.*Enzyme Microb. Technol* **2007**, *40*, 704–707.

239. Cabrera, Z.; Fernandez-Lorente, G.; Fernandez-Lafuente, R.; Palomo, J.M.; Guisan, J.M. Novozym 435 displays very different selectivity compared to lipase from *Candida antarctica* B adsorbed on other hydrophobic supports. *J. Mol. Catal. B* **2009**, *57*, 171–176.

240. Sahin, O.; Erdemir, S.; Uyanik, A.; Yilmaz, M. Enantioselective hydrolysis of (*R/S*)-Naproxen methyl ester with sol-gel encapsulated lipase in presence of calixnarene derivatives. *Appl. Catal. A* **2009**, *369*, 36–41.

241. Wang, P.Y.; Tsai, S.W.; Chen, T.L. Improvements of enzyme activity and enantioselectivity via combined substrate engineering and covalent immobilization. *Biotechnol. Bioeng* **2008**, *101*, 460–469.

242. Partridge, J.; Halling, P.J.; Moore, B.D. Practical route to high activity enzyme preparations for synthesis in organic media. *Chem. Commun.* **1998**, 841–842.

243. Ursini, A.; Maragni, P.; Bismara, C.; Tamburini, B. Enzymatic method of preparation of opticallly active trans-2-amtno cyclohexanol derivatives. *Synth. Commun* **1999**, *29*, 1369–1377.

244. Mateo, C.; Monti, R.; Pessela, B.C.C.; Fuentes, M.; Torres, R.; Guisán, J.M.; Fernández-Lafuente, R. Immobilization of lactase from *Kluyveromyces lactis* greatly reduces the inhibition promoted by glucose. Full hydrolysis of lactose in milk. *Biotechnol. Prog* **2004**, *20*, 1259–1262.

245. Pessela, B.C.C.; Mateo, C.; Fuentes, M.; Vian, A.; Garcia, J.L.; Carrascosa, A.V.; Guisan, J.M.; Fernandez-Lafuente, R. The immobilization of a thermophilic β-galactosidase on Sepabeads supports decreases product inhibition: Complete hydrolysis of lactose in dairy products. *Enzyme Microb. Technol* **2003**, *33*, 199–205.

246. Brazeau, B.J.; de Souza, M.L.; Gort, S.J.; Hicks, P.M.; Kollmann, S.R.; Laplaza, J.M.; McFarlan, S.C.; Sanchez-Riera, F.A.; Solheid, C. Polypeptides and Biosynthetic Pathways for the Production of Stereoisomers of Monatin and Their Precursors. U.S. Patent 20080020434, 24 January 2008.

247. Cho, E.J.; Jung, S.; Kim, H.J.; Lee, Y.G.; Nam, K.C.; Lee, H.J.; Bae, H.J. Co-immobilization of three cellulases on Au-doped magnetic silica nanoparticles for the degradation of cellulose. *Chem. Commun* **2012**, *48*, 886–888.

248. Betancor, L.; Berne, C.; Luckarift, H.R.; Spain, J.C. Coimmobilization of a redox enzyme and a cofactor regeneration system. *Chem. Commun.* **2006**, 3640–3642.

249. St. Clair, N.; Wang, Y.F.; Margolin, A.L. Cofactor-bound cross-linked enzyme crystals (CLEC) of alcohol dehydrogenase. *Angew. Chem. Int. Ed* **2000**, *39*, 380–383.

250. Van Dongen, S.F.M.; Nallani, M.; Cornelissen, J.J.L.M.; Nolte, R.J.M.; van Hest, J.C.M. A three-enzyme cascade reaction through positional assembly of enzymes in a polymersome nanoreactor. *Chem. Eur. J* **2009**, *15*, 1107–1114.

251. Brady, D.; Jordaan, J.; Simpson, C.; Chetty, A.; Arumugam, C.; Moolman, F.S. Spherezymes: A novel structured self-immobilisation enzyme technology. *BMC Biotechnol* **2008**, *8*, 8.

252. Pundir, C.S.; Chauhan, N. Coimmobilization of detergent enzymes onto

a plastic bucket and brush for their application in cloth washing. *Ind. Eng. Chem. Res* **2012**, *51*, 3556–3563.

253. Minakshi; Pundir, C.S. Co-immobilization of lipase, glycerol kinase, glycerol-3-phosphate oxidase and peroxidase on to aryl amine glass beads affixed on plastic strip for determination of triglycerides in serum. *Indian J. Biochem. Biophys.* **2008**, *45*, 111–115.

254. Park, D.; Haam, S.; Jang, K.; Ahn, I.S.; Kim, W.S. Immobilization of starch-converting enzymes on surface-modified carriers using single and co-immobilized systems: Properties and application to starch hydrolysis. *Process Biochem* **2005**, *40*, 53–61.

255. Larsen, G.; Velarde-Ortiz, R.; Minchow, K.; Barrero, A.; Loscertales, I.G. A method for making inorganic and hybrid (organic/inorganic) fibers and vesicles with diameters in the submicrometer and micrometer range via sol-gel chemistry and electrically forced liquid jets. *J. Am. Chem. Soc* **2003**, *125*, 1154–1155.

256. Kim, J.; Grate, J.W.; Wang, P. Nanostructures for enzyme stabilization. *Chem. Eng. Sci* **2006**, *61*, 1017–1026.

257. Kim, J.; Grate, J.W.; Wang, P. Nanobiocatalysis and its potential applications. *Trends Biotechnol* **2008**, *26*, 639–646.

258. Hwang, E.T.; Tatavarty, R.; Lee, H.; Kim, J. Shape reformable polymeric nanofibers entrapped with QDs as a scaffold for enzyme stabilization. *J. Mater. Chem* **2011**, *21*, 5215–5218.

259. Lee, S.M.; Jin, L.H.; Kim, J.H.; Han, S.O.; Na, H.B.; Hyeon, T.; Koo, Y.M.; Kim, J.; Lee, J.H. β-glucosidase coating on polymer nanofibers for improved cellulosic ethanol production. *Bioprocess Biosyst. Eng* **2010**, *33*, 141–147.

260. Woo, E.; Ponvel, K.M.; Ahn, I.; Lee, C. Synthesis of magnetic/silica nanoparticles with a core of magnetic clusters and their application for the immobilization of His-tagged enzymes. *J. Mater. Chem* **2010**, *20*, 1511–1515.

261. Herricks, T.E.; Kim, S.; Kim, J.; Li, D. Direct fabrication of enzyme-carrying polymer nanofibers by electrospinning. *J. Mater. Chem* **2005**, *15*, 3241–3245.

262. Martinek, K.; Mozhaev, V.V. Practice importance of enzyme stability. 2. Increase of enzyme stability by immobilization and treatment with low-molecularweight reagents. *Pure Appl. Chem* **1991**, *63*, 1533–1537.

263. Dallinger, D.; Kappe, C.O. Microwave-assisted synthesis in water as solvent. *Chem. Rev* **2007**, *107*, 2563–2591.

264. Galinada, W.A.; Guiochon, G. Influence of microwave irradiation on the intraparticle diffusion of an insulin variant in reversed-phase liquid chromatography under linear conditions. *J. Chromatogr. A* **2007**, *1163*, 157–168.

265. Varma, R.S.; Saini, R.K. Microwave-assisted reduction of carbonyl compounds in solid state using sodium borohydride supported on alumina. *Tetrahedron Lett* **1997**, *38*, 4337–4338.

266. Varma, R.S.; Dahiya, R.; Saini, R.K. Lodobenzene diacetate on alumina: Rapid oxidation of alcohols to carbonyl compounds in solventless system using microwaves. *Tetrahedron Lett* **1997**, *38*, 7029–7032.

267. Wang, A.; Wang, H.; Zhu, S.; Zhou, C.; Du, Z.; Shen, S. An efficient immobilizing technique of penicillin acylase with combining mesocellular silica foams support and *p*-benzoquinone cross linker. *Bioprocess Biosyst. Eng* **2008**, *31*, 509–517.

268. Wang, A.; Liu, M.; Wang, H.; Zhou, C.; Du, Z.; Zhu, S.; Shen, S.; Ouyang, P. Improving enzyme immobilization in mesocellular siliceous foams by microwave irradiation. *J. Biosci. Bioeng* **2008**, *106*, 286–291.

269. Nahar, P.; Bora, U. Microwave-mediated rapid immobilization of enzymes onto an activated surface through covalent bonding. *Anal. Biochem* **2004**, *328*, 81–83.

270. Naqvi, A.; Nahar, P. Photochemical immobilization of proteins on microwave-synthesized photoreactive polymers.*Anal. Biochem* **2004**, *327*, 68–73.

271. Kumar, S.; Nahar, P. Sunlight-induced covalent immobilization of proteins. *Talanta* **2007**, *71*, 1438–1440.

272. Tanaka, Y.; Tsuruda, Y.; Nishi, M.; Kamiya, N.; Goto, M. Exploring enzymatic catalysis at a solid surface: A case study with transglutaminase-mediated protein immobilization. *Org. Biomol. Chem* **2007**, *5*, 1764–1770.

273. Wong, L.S.; Thirlway, J.; Micklefield, J. Direct site-selective covalent protein immobilization catalyzed by a phosphopantetheinyl transferase.*J. Am. Chem. Soc* **2008**, *130*, 12456–12464.

274. Bolivar, J.M.; Hidalgo, A.; Sanchez-Ruiloba, L.; Berenguer, J.; Guisan, J.M.; Lopez-Gallego, F. Modulation of the distribution of small proteins within porous matrixes by smart-control of the immobilization rate. *J. Biotechnol* **2011**,*155*, 412–420.

275. Rocha-Martín, J.; de las Rivas, B.; Muñoz, R.; Guisán, J.M.; López-Gallego, F. Rational co-immobilization of bi-enzyme cascades on

porous supports and their applications in bio-redox reactions with *in situ* recycling of soluble cofactors.*ChemCatChem* **2012**, *4*, 1279–1288.

276. Dalal, S.; Kapoor, M.; Gupta, M.N. Preparation and characterization of combi-CLEAs catalyzing multiple non-cascade reactions. *J. Mol. Catal. B* **2007**, *44*, 128–132.

277. Forster, A.C.; Church, G.M. Synthetic biology projects *in vitro*. *Genome Res* **2007**, *17*, 1–6.

278. Gore, J.; van Oudenaarden, A. Synthetic biology: The yin and yang of nature. *Nature* **2009**, *457*, 271–272.

279. Godoy, C.A.; de las Rivas, B.; Grazu, V.; Montes, T.; Guisan, J.M.; Lopez-Gallego, F. Glyoxyl-disulfide agarose: A tailor-made support for site-directed rigidification of proteins. *Biomacromolecules* **2011**, *12*, 1800–1809.

Chapter 10

PREPARATION OF A CU(II)-PVA/PA6 COMPOSITE NANOFIBROUS MEMBRANE FOR ENZYME IMMOBILIZATION

Quan Feng[1,2], Bin Tang[2], Qufu Wei[1], Dayin Hou[2], Songmei Bi[2] and Anfang Wei[1]

[1]Key Laboratory of Eco-Textiles (Ministry of Education), Jiangnan University, Wuxi 214122, China

[2]Textiles and Clothing Department, Anhui Polytechnic University, Wuhu 241000, China

ABSTRACT

PVA/PA6 composite nanofibers were formed by electrospinning. Cu(II)-PVA/PA6 metal chelated nanofibers, prepared by the reaction between PVA/PA6 composite nanofibers and Cu^{2+} solution, were used as the support for catalase immobilization. The result of the experiments showed that PVA/PA6 composite nanofibers had an excellent chelation capacity for Cu^{2+} ions, and the structures of nanofibers were stable during the reaction with Cu^{2+} solution. The adsorption of Cu(II) onto PVA/PA6 composite nanofibers was studied by the Langmuir isothermal adsorption model. The maximum amount of coordinated Cu(II) (q_m) was 3.731 mmol/g (dry fiber), and the binding constant (K_l) was 0.0593 L/mmol. Kinetic parameters were analyzed for both immobilized and free catalases. The value of V_{max} (3774 μmol/mg·min) for the immobilized catalases was smaller than that of the free catalases (4878 μmol/mg·min), while the K_m for the immobilized catalases was larger. The immobilized catalases showed better resistance to pH and temperature than that of free form, and the storage stabilities, reusability of immobilized catalases were significantly improved. The half-lives of free and immobilized catalases were 8 days and 24 days, respectively.

INTRODUCTION

Enzymes are well known green catalysts which are highly specific and efficient. However, the applications of enzymes suffer from various problems, e.g., non-reusability, high-cost and instability [1]. Enzyme immobilization has become an effective way to overcome these limitations to some extent. The enzyme immobilized on the surface of insoluble supports can be recycled much more easily than a soluble enzyme. On the other hand, the multiple-point attachment to the support can restrict the undesirable conformational changes of enzyme proteins in practical applications, and maintain the activity of immobilized enzymes as much as possible [2–4].

Immobilization of enzymes can be achieved through physical adsorption on solid supports, micro-encapsulation, chelation bonds, covalent bonds or matrix entrapment. Among these, chelation bond seems to have some advantages over the other methods such as simplicity of the process, more stability and less possibility of inactivation of the immobilized enzyme [5]. Most of the transition metal ions such as Cu^{2+}, Fe^{3+}, Ni^{2+}, Zn^{2+} can form stable complexes with electron-rich compounds and coordinate molecules containing O, N and S, for example, amine (NH_2), hydroxyl (OH) and thiol (SH) groups.

Catalases are enzymes that decompose hydrogen peroxide (H_2O_2) into water and oxygen, which are commonly used in various fields, including the food, textile, agriculture and detergent and so on. The use of catalases is very effective in the aspects of lower resource utilization and energy consumption [6]. Every catalase protein consists of four subunits, and each of them includes a ferriporphyrin as a prosthetic group. Immobilization of enzymes on a metal chelated matrix is based on chelation bonds between the chelated metal ions on the support and active sites such as indole groups of tryptophanes, imidazole groups of histidines and thiol groups of cysteines [7].

Application of nanofibrous membranes for enzyme immobilization is becoming popular because of their large surface area. Nanofibrous membranes can improve the binding capacity of immobilized enzymes and increase the mass transfer kinetics when biocatalytic reactions occur on the surface of the nanofibrous membrane [8–10]. Electrospinning is the most effective technique for preparing nanofibrous membranes, in which a polymer jet is ejected when the electrostatic force applied to the polymer liquid, which overcomes the surface tension of the polymer solution. The jet is elongated and accelerated in the electrostatic field, followed by solvent evaporation and deposition on a substrate at random [11,12]. Poly (vinyl alcohol) (PVA) is a cheap, nontoxic, hydrophilic, biocompatible synthetic polymer, and there are plenty of hydroxy (OH) groups in it. PVA has been widely used in cell and enzyme immobilization [13]. In our previous work, PVA nanofibers were formed by electrospinning, and metal chelated nanofibers were prepared by reaction with metal ions solution, but it was found that PVA nanofibers had undergone a swelling phenomenon in aqueous solution, and the nanofibrous structures were distorted to a certain extent. In addition, we found that PVA/PA6 composite nanofibers could be formed by electrospinning, and PVA/PA6 composite nanofibers showed excellent structural stability in aqueous solutions of metal ions [14]. In this work, PVA/PA6 composite nanofibrous membranes were formed by electrospinning. Then, Cu(II) ions were incorporated into the hydroxyl groups and carbonyl groups of the nanofibers via metal chelation, and the chelated reaction kinetics of PVA/PA6 composite nanofibers to Cu(II) were established. Likewise, catalases were immobilized on the surface of the Cu(II)-PVA/PA6 composited nanofibrous membrane by chelated groups, and the activity and stability of the immobilized catalases were studied by examining the decomposition of H_2O_2. At the same time, the kinetic parameters of free and immobilized catalases were studied.

RESULTS AND DISCUSSION

Adsorption Isotherm of Cu(II) on PVA/PA6 Composite Nanofibrous Membrane

In order to get the best values of the binding parameters from the experimental results, Equation 1 can be rearranged as below:

$$\frac{1}{q} = \frac{1}{q_m K_l C_e} + \frac{1}{q_m}$$

$$(1)$$

According to experimental results, plots of q vs. Ce and $1/q$ vs. $1/Ce$ are shown in Figures 1 and 2.

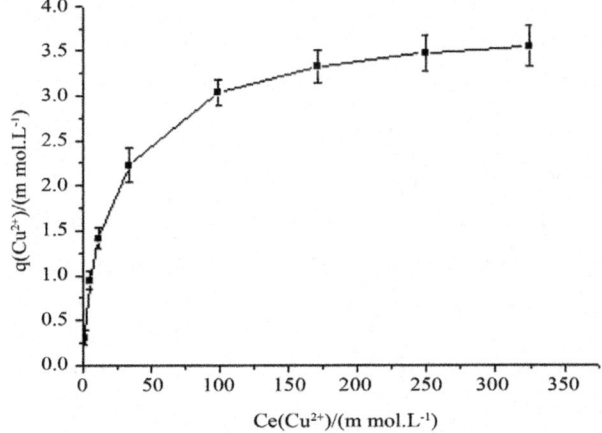

Figure 1: Adsorption isotherm of Cu^{2+} on the PVA/PA6 composite nanofibers. Bars represented standard deviations ($n = 3$).

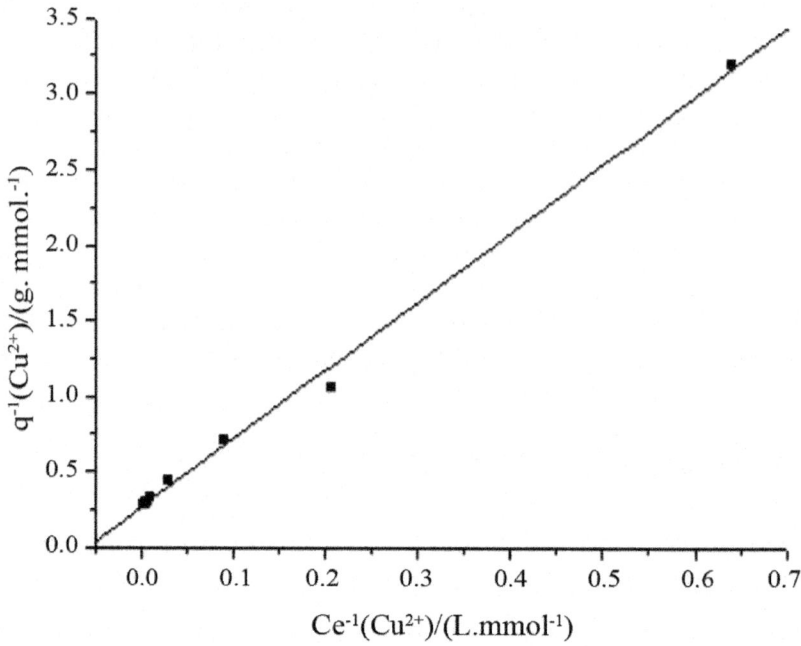

Figure 2: Relation of $1/q$ and $1/C_e$.

It is clearly observed that the adsorption of Cu(II) increased significantly with the rising concentration from 1.88 mmol/L to 327.79 mmol/L and gradually leveled off. The initial increase of Cu(II) adsorption might be due to many available chelating hydroxyl groups and carbonyl groups in PVA/PA6 composite nanofibers, then reached saturation gradually with the increase in the concentration of Cu(II).

The adsorption equilibrium data could be interpreted by the Langmuir absorption equation. The basic assumption of the Langmuir theory is once a metal ion occupies a reaction site, then no further adsorption occurs at the same location [15]. The reaction kinetics parameters could be calculated through straight line slope and ordinate intercept of plots. The reaction rate constant (K_l) was 0.0593 L/mmol, and the maximum coordinate capacity (q_m) of PVA/PA6 composite nanofibers to Cu(II) was 3.731 mmol/g (dry fiber). Comparing with our previous work, PVA/PA6 composite nanofibers showed obviously higher adsorption capability for Cu^{2+} ions than that of PVA nanofibers [2].

FTIR Spectra of Nanofibers

The FTIR spectra of the PVA/PA6 composite nanofibers and the Cu(II)-PVA/

PA6 metal chelated nanofibers are presented in Figure 3. The FTIR spectrum of the PVA/PA6 (Figure 3a) showed the following characteristic peaks: the bands of 3500 cm^{-1}–2800 cm^{-1} correspond to the N–H, O–H and C–H stretching vibrations. The band at 1173 cm^{-1} was assigned to the C–O stretching vibration in PVA, and the band at 1642 cm^{-1} indicated the C=O stretching vibration in PA6.

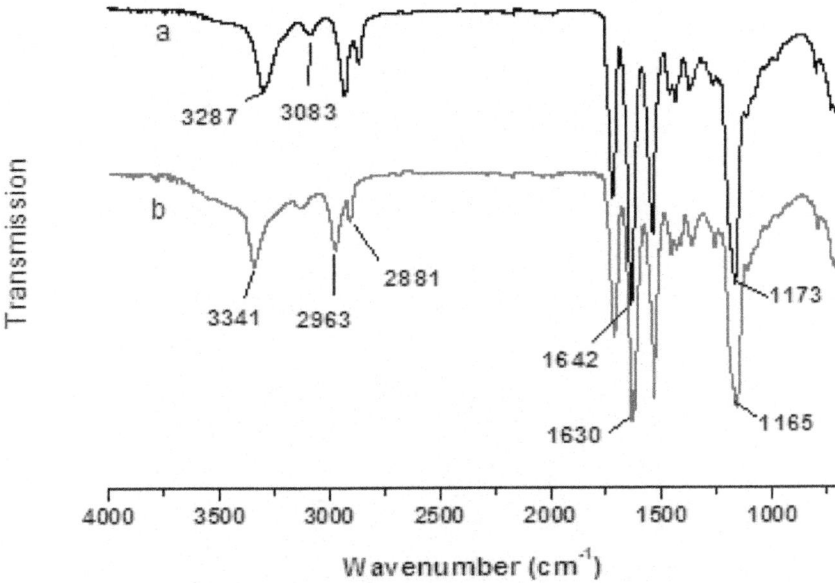

Figure 3: FTIR spectra of **(a)** PVA/PA6 composite nanofibers and **(b)** Cu(II)-PVA/PA6 metal chelated nanofibers.

The C-O band at 1173 cm^{-1} (Figure 3a) moved to 1165 cm^{-1} (Figure 3b), the C=O band at 1642 cm^{-1} (Figure 3a) moved to 1630 cm^{-1} (Figure 3b) due to the decrease of bond force constant coupled with decrease of the electron density, which indicated that the corresponding groups of PVA/PA6 coordinated with Cu(II). Moreover, the wide band of 3,287 cm^{-1}(Figure 3a) corresponding to the O–H stretching vibration moved to 3,341 cm^{-1} (Figure 3b) due to the increase of the relative mass of hydroxyl group, which was caused by the coordination of metal ions.

The Surface Morphologies of Nanofibrous Membrane

The SEM images of the original PVA/PA6 composite nanofibrous membranes are displayed in Figure 4a. The electrospun PVA/PA6 composite nanofibrous membranes formed a fibrous membrane with random orientations. The nanofibrous membranes looked very even and the average diameter of the electrospun nanofibers ranged between 90 nm and 110 nm.

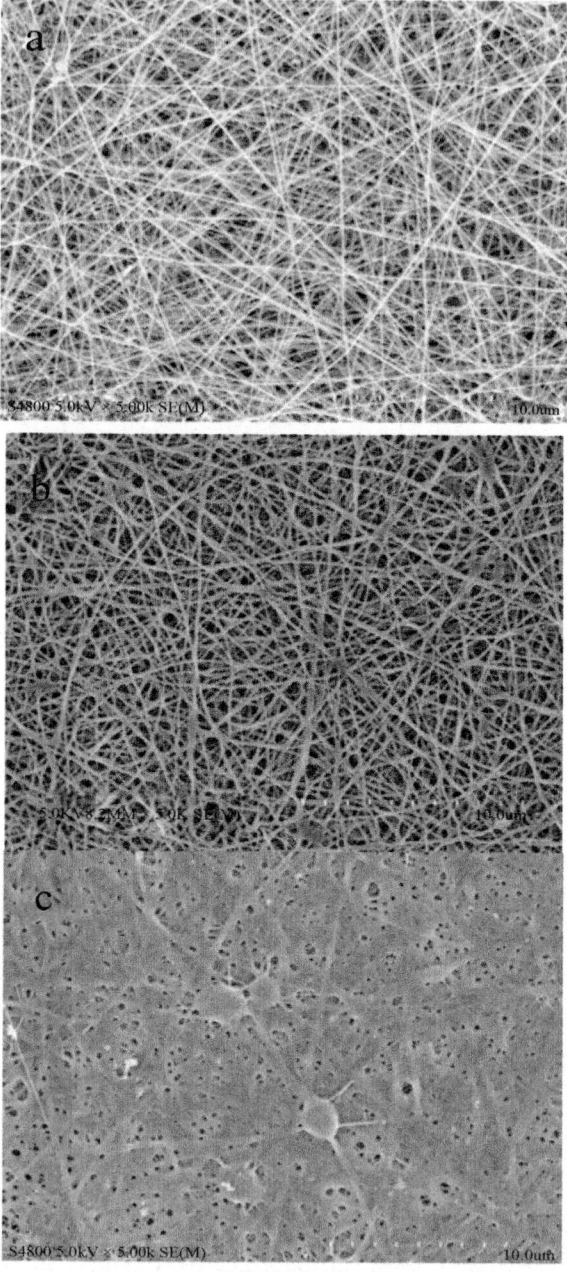

Figure 4: Micrographs of (**a**) original PVA/PA6 composite nanofibrous membrane; (**b**)

and (c) PVA/PA6 composite nanofibers and PVA nanofibers after reaction with aqueous Cu^{2+} ions solution for 24 h, respectively.

The diameter of composite nanofibrous membranes did not change substantially after the reaction with aqueous solution of Cu(II) ions for 24 h, and the fibrous structures was not obviously distorted, as revealed in Figure 4b. However, the constructions of PVA nanofibrous membranes showed obvious damages because of swelling after the reaction with aqueous solution of Cu(II) ions for 24 h. The stability of nanofibrous membranes was obviously improved when PA6 was added to the composite, and the nanofibrous structure of PVA/PA6 composite nanofibers was still maintained in the metal ions aqueous solution.

Kinetic Parameters of Immobilized and Free Catalases

Maximum reaction rate V_{max} and Michaelis-Menten constants K_m are shown in Table 1. K_m of the immobilized catalases was higher than that of free catalases, and V_{max} of the immobilized catalases was smaller than that of free catalases.

Table 1: The amount of bound enzyme and the kinetic parameters of the immobilized and free enzyme

	Amount of bound enzyme (mg/g fibers)	Specific activity (Units/mg)	K_m (mM)	V_{max} (μmol/mg·min)
Free catalase		3400	26.815	4878
Immobilized catalases	−64	2150	41.132	3774

These data indicated that affinity between immobilized catalases and substrate decreased comparing with free catalases. This might be affected by the structural changes of immobilized catalases, with less accessibility between substrate and active points of immobilized catalases caused by the space barriers presented by the supports [16,17].

Effect of pH and Temperature on the Enzyme Activity

The effect of pH on the activity of free and immobilized catalase is shown in Figure 5.

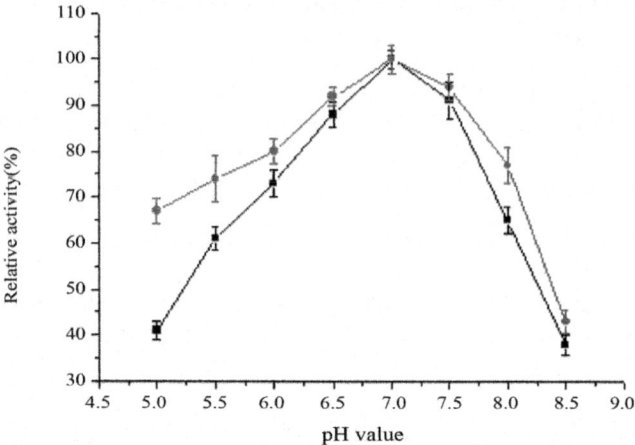

Figure 5: Effect of pH on the (●) immobilized and (■) free catalase. Bars represent standard deviations ($n = 3$).

According to the experimental results, no significant shift of the optimal pH was observed. The optimal pH value was about 7.0 for free and immobilized enzyme, but the residual relative activity of the immobilized catalase was higher than that of the free one in the pH range between 5.0 and 8.5. It was found that the immobilized enzyme showed less sensitivity to pH than that of the free enzyme, probably because of the production of oxygen, forming foams, and causing slight diffusion limitation on the surface of nanofibrous membrane [18]. The effect of temperature on the activity of free and immobilized enzyme is depicted in Figure 6.

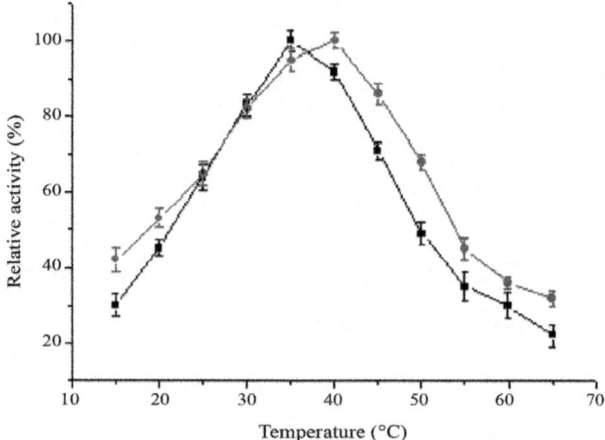

Figure 6: Effect of temperature on the (●) immobilized and (■) free catalases. Bars represent standard deviations ($n = 3$).

It is obvious that the initial relative activity was increased with the increase of temperature and decreased while the temperature was further increased for immobilized and free catalases. The optimum temperatures for the free and immobilized catalases were observed to lay at approximately 35 °C and 40 °C, respectively. At higher temperature range, the immobilized catalase exhibited higher stability than the free one, and it showed the anti-thermal properties of the immobilized enzyme. The multipoint interactions between enzyme and Cu(II)-PVA/PA6 support might reduce the degree of freedom of the spatial structure of enzyme, protecting it from deactivation at high temperatures.

Storage Stability and Reusability

The storage stability of immobilized and free enzyme is presented in Figure 7. The residual activity of immobilized and free catalase was 54% and 19% after 20 days, and the half-lives of free and immobilized catalase were 8 days and 24 days, respectively. The results indicated that the storage stability of immobilized catalases was better than that of free catalases, which might be attributed to the immobilization of enzyme to a matrix. The immobilization of enzyme could limit the freedom of conformational changes, resulting in increasing stability towards denaturalization [19].

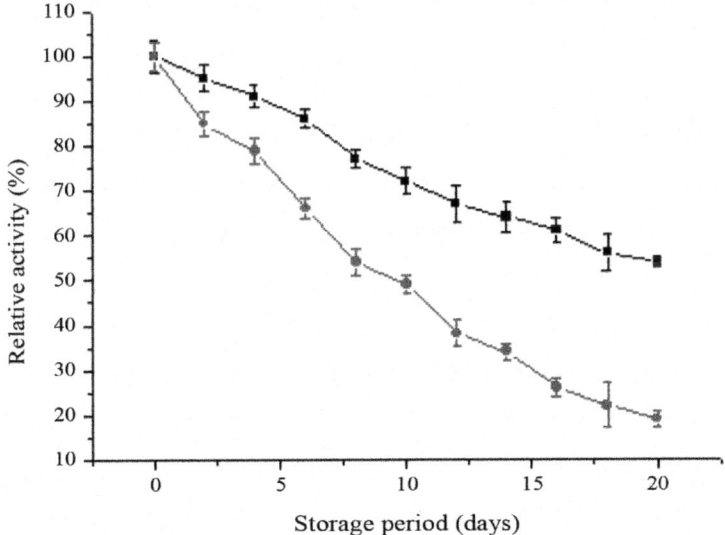

Figure 7: Storage stability of (■) immobilized catalases and (●) free catalases. Bars represent standard deviations ($n = 3$).

According to the effect of repeated use on activity of immobilized catalase (Figure 8), after five repeated uses, immobilized catalases retained about 74% of their initial activity, and the residual activity of the immobilized enzyme was about 51% of the initial activity after 10 repeated uses, which showed its potential value in actual applications.

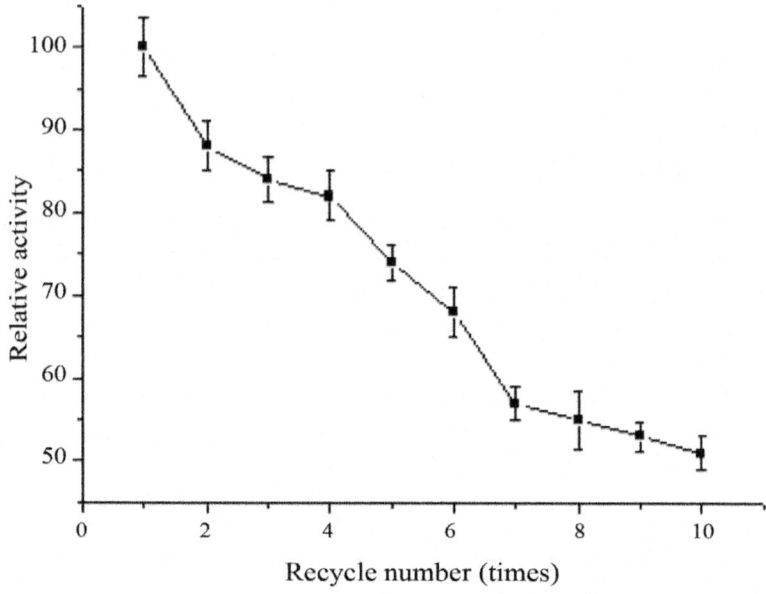

Figure 8: Reuse stability of the immobilized catalases. Bars represent standard deviations ($n = 3$).

MATERIALS AND METHODS

Materials

PA6 (weight-average molecular was 1800, characteristic viscosity was 2.8) and PVA (weight-average molecular was 84,000–89,000) obtained from Shanghai Kanghu Chemical (Shanghai, China). Bovine liver catalase (hydrogen peroxide oxidoreduction; EC. 1.11.1.6) were purchased from Sigma (Shanghai, China). Copper dichloride, hydrogen peroxide, formic acid and Coomassie Brilliant Blue (G250) for the Bradford protein assay were purchased from Shenyang Sinopharm Chemical Reagent (city, China). Hydrogen peroxide (30%) and the ingredients of phosphate buffer solution (PBS) such as NaCl, KCl, KH_2PO_4, K_2HPO_4 were analytical grade and used as received. Water used in all experiments was de-ionized.

Electrospinning

Electrospinning was carried out to fabricate PVA/PA6 composite nanofibers. PVA and PA6 were dissolved in formic acid (of 88% purity) at room temperature with stirring. The mass ratios of PVA:PA6 were 4%:12% in blended solutions. PVA/PA6 blended solutions were placed in a syringe (content of 20 mL) equipped with a 0.7 mm diameter spinner jet, and the solution flow rate was controlled by a microinfusion pump (WZ-50C2, Zhejiang, China). The high-voltage supplier (DW-P503-4AC, Tianjin, China) was used to connect the metal needles and the grounded collector to form the electrostatic fields. The blended polymer solutions were electrospun at the positive voltage of 15 kV with a collecting distance of 14 cm (the distance between the syringe needle tip and the collection roller covered with the aluminum foil). Nanofibers were collected by the roller for 24 h. On the other hand, PVA nanofibers were also prepared with the same parameters (weight ratio of PVA in solution was 8%) for comparison.

Chelation of Cu(II)

PVA/PA6 composite nanofibers (0.05 g) were put in flasks, each of them containing various concentrations of copper dichloride solution (50 mL). The flasks were stirred at 20 °C for 24 h. The nanofibrous membranes with adsorbed Cu(II) were washed with de-ionized water. The amount of adsorbed Cu(II) was calculated by using the concentrations of the Cu(II) in the initial solution and in the resulting solution. The Cu(II) concentration was determined by atomic absorption spectrophotometry (AAS). The Cu(II)-PVA/PA6 nanofibrous membranes were dried in a vacuum drying oven, at a temperature of 40 °C for 24 h. The reaction kinetics parameters were studied by the Langmuir isothermal adsorption model:

$$q = \frac{q_m K_l C_e}{1 + K_l C_e} \tag{2}$$

where q_m is the maximum amount of the coordinated Cu(II), K_l is the binding constant, C_e is the equilibrium concentration. Fourier transform infrared spectroscopy (FTIR) spectra of the PVA/PA6 composite nanofibers and Cu(II)-PVA/PA6 nanofibers were recorded using a FTIR spectrometer (NICOLET NEXUS 470). Surface morphologies of Cu(II)-PVA/PA6 and Cu(II)-PVA nanofibers were examined using FE-SEM (S-4800). All samples were coated with the platinum by sputtering before scanning electron microscope observation.

Enzyme Immobilization

Cu(II)-PVA/PA6 chelated nanofibrous membranes (coordinated Cu^{2+} ions = 3.03 mmol/g) were immersed in 50 mL of catalase solution (0.3 mg/mL) for 4 h at 20 °C in shakers while stirring continuously. Catalases were dissolved in 50 mM PBS (value of pH was 7.0). Then, the nanofibrous membranes were removed from the solution and rinsed with the same PBS until no soluble protein was detectable. The method of Bradford was applied to determine the concentration of enzyme, the amount of the catalase was determined by spectrophotometry according to the absorbance of Coomassie Brilliant Blue and catalases at 595 nm, and the amount of the bound enzyme was calculated as:

$$Q = \frac{(C_0 - C)V}{m}$$

(3)

where Q is the amount of catalases bound onto unit mass of nanofibrous membranes (mg/g), C_0 and C are the initial and equilibrium enzyme concentrations in the solution (mg/mL), V is the volume of the catalases solution, and m is the mass of the nanofibrous membranes.

Activity Assays

The activity of the free and immobilized catalases was determined spectrophotometrically by the measurement of decrease in the absorbance change of hydrogen peroxide at 240 nm. Immobilized catalases were mixed with hydrogen peroxide solution (100 mL, 50 mmol/L). The reaction was kept at 35 °C for 3 min, and the specific activity of enzyme was calculated by the following formula:

$$v = \frac{\Delta A \times V}{T \times K \times Ew}$$

(4)

where v is specific activity of enzyme and immobilized enzyme (μmol/mg·min), ΔA is the absorbance decrease of the solution at 240 nm, V is the volume of the hydrogen peroxide solution (mL), T is the time of reaction (min), K is the molar extinction coefficient of hydrogen peroxide at 240 nm (K = 0.033 L/mmol cm), and E_w is the amount of enzyme (mg).

The catalytic mechanism of catalases is presented as follows [20]:

$$CAT\,(Por\text{-}Fe^{III}) + H_2O_2 \rightarrow Cpd\,I(Por^+\text{-}Fe^{IV}{=}O) + H_2O$$

$$Cpd\,I(Por^+\text{-}Fe^{IV}{=}O) + H_2O_2 \rightarrow CAT\,(Por\text{-}Fe^{III}) + H_2O{+}O_2$$

Determination of Kinetic Parameters

The effect of hydrogen peroxide concentration on the activity was tested. V_{max} and K_m values of free and immobilized catalases were calculated by the Lineweaver-Burk plots:

$$\frac{1}{v} = \frac{K_m}{V_{max}} \times \frac{1}{[S]} + \frac{1}{V_{max}}$$

(5)

where K_m is the Michaelis constant, v and V_{max} represent the initial and maximal rate of the reaction, $[S]$ is the concentration of the hydrogen peroxide. Kinetics parameters for immobilized and free enzyme were investigated at 35 °C and the concentrations of hydrogen peroxide ranged from 20 to 200 mmol/L (pH 7.0).

Dependence of Temperature and Value of pH

The immobilized and free enzyme were mixed with hydrogen peroxide solution (100 mL, 50 mmol/L) respectively, at temperatures ranging from 15 °C to 65 °C. The optimum pH values of free and immobilized enzyme were investigated at 35 °C for 3 min. PBS (pH 5.0–8.5) was used for pH dependence study.

Storage Stability and Reusability

The stability of free and immobilized catalases at storage was measured by calculating their activity retention during 20 days at 4 °C in 50 mM PBS (pH 7.0), using 2 days intervals, then a sample was removed and determined for the enzyme activity as described above. The half-lives of free and immobilized catalases were calculated by the following formulas.

$$[E_t] = [E_0]\exp(-K_d t)$$

(6)

$$t_{1/2} = \frac{\ln 2}{K_d}$$

(7)

where K_d is the decay constant, E_0 is the initial activity of enzyme, E_t is the residual activity of enzyme, $t_{1/2}$ is the half-life of enzyme. In addition, the reusability of immobilized catalases was studied. Immobilized catalases were reused 10 times within 10 days (once a day). After each reaction run, the enzyme was washed with PBS to remove any residual substrate, and the retention of activity of the immobilized enzyme was conducted under the optimum conditions.

CONCLUSIONS

Electrospinning was carried out to fabricate PVA/PA6 composite nanofibrous membranes, and Cu(II)-PVA/PA6 nanofibrous membranes were prepared for catalase immobilization. The storage stability was significantly improved and the reusability of immobilized catalases was very high after immobilization onto the Cu(II)-PVA/PA6 nanofibrous membranes. These results indicated that the PVA/PA6 composite nanofibers showed excellent structural stability in aqueous solution, and the immobilized catalases had a high affinity with the support, which demonstrated that the catalases immobilized on Cu(II)-PVA/PA6 nanofibrous membranes may have potential in various applications.

ACKNOWLEDGMENTS

The authors wish to thank the National High-tech R & D Program of China (863 Program, NO.2012AA030313) and the Natural Science Foundation of Anhui Provincial Education Department (KJ2012A039, KJ2009-119) for their financial support in this research.

REFERENCES

1.   Hong, J.; Xu, D.M.; Gong, P.J.; Yu, J.H.; Ma, H.J.; Yao, S.D. Covalent-bonded immobilization of enzyme on hydrophilic polymercovering magneti cnanogels. *Microporou. Mesoporous Mater* **2008**, *109*, 470–477.

2.   Feng, Q.; Xia, X.; Wei, A.F.; Wang, X.Q.; Wei, Q.F. Preparation of Cu(II)-chelated poly(vinylalcohol) nanofibrous membranes for catalase immobilization. *J. Appl. Polymer. Sci* **2011**, *120*, 3291–3296.

3.   Ricca, E.; Calabrò, V.; Curcio, S.; Basso, A.; Gardossi, L.; Iorio, G. Fructose production by Inulinase covalently immobilized on sepabeads in batch and fluidized bed bioreactor. *Int. J. Mol. Sci* **2010**, *11*, 1180–1189.

4.   Park, H.; Ahn, J.; Lee, J.; Lee, H.; Kim, C.; Jung, J.-K.; Lee, H.; Lee, E.G. Expression, immobilization and enzymatic properties of glutamate decarboxylase fused to a cellulose-binding domain. *Int. J. Mol. Sci* **2012**, *13*, 358–368.

5.   Akgöl, S.; Öztürk, N.; Karagözler, A.A.; Aktas Uygun, D.; Uygun, M.; Denizli, A. A new metalchelated beads for reversible use in uricase adsorption. *J. Mol. Catal. B Enzym* **2008**, *51*, 36–41.

6.   Bayramoglu, G.; Arica, M.Y. Reversible immobilization of catalase on

fibrous polymer grafted and metal chelated chitosan membrane. *J. Mol. Catal. B Enzym* **2010**, *62*, 297–304.

7. Wu, Z.C.; Zhang, Y.; Tao, T.X.; Zhang, L.F.; Fong, H. Silver nanoparticles on amidoximebers for photo-catalytic degradation of organic dyes in waste water. *Appl. Surf. Sci* **2010**, *257*, 1092–1097.

8. Yew, P.L.; Lee, Y.H. A potentiometric formaldehyde biosensor based on immobilization of alcohol oxidase on acryloxysuccinimide-modified acrylic microspheres. *Sensors* **2010**, *10*, 9963–9981.

9. Liu, X.Q.; Guan, Y.P.; Liu, H.Z.; Ma, Z.Y.; Yang, Y.; Wu, X.B. Preparation and characterization of magnetic polymer nanospheres with high protein binding capacity. *J. Magn. Magn. Mater* **2005**, *293*, 111–118.

10. Wu, Z.C.; Zhang, B.; Yan, B. Regulation of enzyme activity through interactions with nanoparticles. *Int. J. Mol. Sci* **2009**, *10*, 4198–4209.

11. Wan, L.S.; Ke, B.B.; Xu, Z.K. Electrospun nanofibrous membranes filled with carbon nanotubes for redox enzyme immobilization. *Enzym. Microb. Tech* **2008**, *42*, 332–339.

12. Bayramoglu, G.; Arica, M.Y. Reversible immobilization of catalase on fibrous polymer grafted and metal chelated chitosan membrane. *J. Mol. Catal. B Enzym.* **2010**, *62*, 297–304.

13. Ayse, D.; Azmi, T. Improving the stability of cellulase by immobilizationon modified polyvinyl alcohol coated chitosan beads. *J. Mol. Catal. B Enzym* **2007**, *45*, 10–14.

14. Feng, Q.; Wang, X.Q.; Wei, Q.F.; Wang, X.; Hou, D. Preparation of Cd^{2+}-PVA/PA6 metal chelated nanofibers and analysis of reaction kinetics. *J. Textil. Res* **2012**, *33*, 5–8.

15. Saeed, K.; Haider, S.; Oh, T.J.; Park, S.Y. Preparation of amidoxime-modied polyacrylonitrile (PAN-oxime) nanofibers and their applications to metal ions adsorption. *J. Membr. Sci* **2008**, *322*, 400–405.

16. Tumturk, H.; Karaca, N.; Demirel, G. Preparation and application of poly(*N*, *N*dimethylacrylamide- co-acrylamide) and poly (*N*-isopropylacrylamide-co-acrylamide)/- carrageenan hydrogels for immobilization of lipase. *Int. J. Biol. Macromol* **2007**, *40*, 281–285.

17. Senay, A.C.; Ebru, S.; Dursun, S. Preparation of Cu(II) adsorbed chitosan beads for catalase immobilization. *Food Chem* **2009**, *114*, 962–969.

18. Sinan, A.; Nevra, O.; Adil, D. New generation polymeric nanospheres for catalase immobilization. *J. Appl. Polymer. Sci* **2009**, *114*, 962–970.

19. Nursel, P.; Bekir, S.; Olgun, G. Activity studies of glucose oxidase immobilized onto poly(*N*vinylimidazole) and metal ion-chelated poly(*N*-

vinylimidazole) hydrogels. *J. Mol. Catal. B: Enzym* **2003**, *21*, 273–282.

20. Wang, Z.G.; Xu, Z.K.; Wan, L.S.; Wu, J. Nanofibrous membranes containing carbon nanotubes: Electrospun for redox enzyme immobilization. *Macromol. Rapid Comm* **2006**, *27*, 516–521.

Chapter 11

ENZYME-IMMOBILIZED MICROFLUIDIC PROCESS REACTORS

Yuya Asanomi[1], Hiroshi Yamaguchi[2], Masaya Miyazaki[1,3,4,] and Hideaki Maeda[1,3]

[1]Measurement Solution Research Center, National Institute of Advanced Industrial Science and Technology (AIST), 807-1 Shuku, Tosu, Saga 841-0052, Japan

[2]Liberal Arts Education Center, Aso Campus, Tokai University, Minami-aso, Aso, Kumamoto 869 1404, Japan

[3]Department of Molecular and Material Sciences, Interdisciplinary Graduate School of Engineering Science, Kyushu University, 6-1 Kasuga-koen, Kasuga, Fukuoka 816-8580, Japan

[4]Department of Advanced Technology Fusion, Graduate School of Science and Engineering, Saga University, 1 Honjo, Saga 840-8502, Japan

ABSTRACT

Microreaction technology, which is an interdisciplinary science and engineering area, has been the focus of different fields of research in the past few years. Several microreactors have been developed. Enzymes are a type of catalyst, which are useful in the production of substance in an environmentally friendly way, and they also have high potential for analytical applications. However, not many enzymatic processes have been commercialized, because of problems in stability of the enzymes, cost, and efficiency of the reactions. Thus, there have been demands for innovation in process engineering, particularly for enzymatic reactions, and microreaction devices represent important tools for the development of enzyme processes. In this review, we summarize the recent advances of microchannel reaction technologies especially for enzyme immobilized microreactors. We discuss the manufacturing process of microreaction devices and the advantages of microreactors compared to conventional reaction devices. Fundamental techniques for enzyme immobilized microreactors and important applications of this multidisciplinary technology are also included in our topics.

INTRODUCTION

Microfluidic reaction devices, which can be prepared by microfabrication techniques, or by assembly and modification of microcapillaries, constitute reaction apparatus with small dimensions, large surface to volume ratios and well defined reaction times [1-3]. These systems take advantage of microfluidics or nanofluidics that enables use of micro- or nanoliter volumes of reactant solutions and offer the advantages of high efficiency and repeatability. Therefore, microchannel reaction systems are expected to be the new and promising technology in the fields of chemistry, chemical engineering and biotechnology [4-10]. They offer several advantages over traditional technologies in performing chemical reactions. The key advantages of microsystems include rapid heat exchange and rapid mass transfer that cannot be achieved by the conventional batch system. Unlike macro scale solutions, streams of solutions in a microfluidic system mainly form laminar flow which allows strict control of reaction conditions and time. In addition, microchannel reaction systems provide large surface and interface areas, which are advantageous for many chemical processes such as extractions and catalytic reactions. Several chemical reaction devices have been reported to demonstrate potential applications [4-10]. Moreover, many potential applications for miniaturized synthetic reactors require only small volumes of catalysts.

Enzymatic conversions have recently attracted considerable attention because of their environmentally-friendly nature. Several enzyme processes have been developed; however, improvement of the entire process is still required to obtain the benefits that can be derived from their use and for them to become a common or standard technology [11,12]. Reaction engineering might provide solutions to develop enzyme reaction processes at the commercial level [13], and microreaction engineering is one candidate for such technology. Therefore, several techniques have been developed, either in solution phase or by immobilizing enzymes, to realize enzyme microreaction processes [10,14]. In this review, we summarize recent advances in microchannel reaction technologies, focusing especially on enzyme-immobilized microreactors. We discuss the manufacturing process of microreaction devices and the advantages of microfluidic systems compared to conventional reaction devices. Fundamental techniques for enzyme-immobilized microreactors and important applications of this multidisciplinary technology are also presented

MICROREACTOR FUNDAMENTALS

For the beginner unfamiliar with microfluidic reaction techniques, we would like to start our review with a brief introduction to microreactors. Microfluidic reactions occur in a small space within a reaction apparatus. Continuous-flow systems are mainly employed, and in most cases mechanical pumping, commonly by syringe pumping, or electroosmotic flow, which is the motion of ions in a solvent environment through very narrow channels, where an applied potential across the channels causes ion migration, are used as the driving force of the reaction systems. Microreaction devices developed so far can be classified into two types: chip-type microreactors and microcapillary devices.

Chip-type microreactors which offer several advantages, including easy control of microfluidics, and integration of many processes into one reaction device. Chip-type microreactors have been mainly used for the development of bioanalytical devices. The manufacturing processes of such devices were adapted mainly from the microelectronics industry. Dry-or wet-etching processes have been used for creating channels on a silicone or glass plates. Polymer-based materials can be used for preparation of enzyme microreactors because most enzyme reactions are performed in aqueous solution, especially for bioanalytical use. Polydimethylsiloxane (PDMS), polymethylmethacrylate (PMMA), polycarbonate, and Teflon were used for preparation of microreaction devices. These plates could be processed by photolithography, soft lithography, injection molding, embossing, and micromachining with lasers or microdrilling. The LIGA (Lithographic Galvanoforming Abforming) process which consists of a combination of lithography, electrochemical technology and molding, can also be used for the production of microreactors.

In a microreactor, stable formation of laminar streams of different solutions is sometimes required, although some cases require better mixing by disrupting laminar streams. Methods for stabilizing multiple laminar flows and micromixing have been developed. Tokeshi et al. developed guide structures at the bottom of microchannels [15]. The structures were prepared by wet etching of a glass plate. Laminar streams of organic solvent and water in these microchannels were stabilized by these guide structures. Techniques for partial surface modification of microchannels were also developed. These techniques take advantage of the different surface properties. Organic solvents prefer hydrophobic regions, whereas aqueous solutions go to hydrophilic regions. Modification of glass by octadesylsilane was used to stabilize the flow of organic solvent and aqueous solutions [16]. In another study, a UV-sensitive self-assembled monolayer with fluorous chains was used for preparing partially-modified microchannel surfaces [17]. Our laboratory developed another method to stabilize solutions [18]. Microchannels were fabricated on both bottom and top plates. The microchannels of one of the two plates were coated with gold, then treated with alkanethiol to produce a hydrophobic surface. The resulting microreactor, which forms an upside-down laminar stream, was not only stabilized by interaction with surface, but also supported by gravity. Overall, such partial modification methods are useful to stabilize laminar streams under pressure below the critical value. Indeed, microfluidic phenomenon of laminar flow is one important aspect in the development of chip-type microreactors.

Micromixers which enhance mixing of two or more different solutions in microspace have also been constructed. Rapid mixing in microfluidics is difficult to achieve because under laminar flow mixing of fluids is principally limited to diffusion through the interface. Several micromixers were developed by adding devices or materials in the microchannel, such as electrokinetical mixing [19] and microbeads [20]. Various types of micromixers which only require structured microchannels were also developed. These include chaotic mixers with oriented ridges at the bottom of microchannels [21], repeated dividing and merging of fluids with a two-way separated serpentine flow path [22], zig-zag microchannels [23], and simple convergence to 32 layers with two solutions divided into 16 microchannels [24]. Detail of micromixing can be found in recent reviews [25]. Still, design and fabrication of highly efficient micromixers for effective functioning of microfluidic devices are desired research topics.

The other type of microreaction device consists of microcapillaries. This is the simplest method which does not require any control of microfluidics, rather, it uses a microchannel as the reaction space. The major advantage of this type of microreactors is in scaling up process which can be achieved by simply bundling more microcapillaries. Gas or liquid chromatography parts are chiefly used to prepare this type of microreactors. Capillary type microreactors are mainly used to develop manufacturing processes, especially catalytic reactions, to take advantage of the large surface area. The users should select the type of microreactor depending on the feature of reaction which they want to perform in a microreactor.

FUNDAMENTAL TECHNIQUES FOR ENZYME IMMOBILIZED MICROREACTOR

Enzyme-Immobilization within Microchannels

In the development of enzyme processes, the use of immobilized enzymes is preferable. Several methods have been used to immobilize enzymes on supports in conventional reaction apparatus, and these techniques have also been applied to immobilize enzyme within a microspace (Tables 1-3)

Table 1: Enzyme-immobilization within microchannel reactors by particle entrapment techniques

Media	Immobilization method	Enzyme	Advantage and disadvantage	Ref.
Glass	Cross-linking (3 aminopropylsilane/ glutaraldehyde)	Xantin oxidase Horseradish peroxidase	Ease in preparation Enable multistep reaction Limited number of enzymes are applicable due to denaturation Pressure gain	[26]
Polystyrene	Biotin-Avidin (Avidin-coated beads were used)	Horseradish peroxidase	Ease in preparation Enable multistep reaction Biotin-label is required Pressure gain	[20]
Agarose	Complex formation (Ni-NTA and His-tag)	Horseradish peroxidase	Ease in preparation Applicable for engineered enzymes Higher pressure by increasing flow rate and particles may be crushed	[27]
Polystyrene	Complex formation (Ni-NTA and His-tag)	Glucose oxidase	Ease in preparation Applicable for engineered enzymes Higher pressure by increasing flow rate and particles may be crushed	[28]
Magnetic bead	Cross-linking (3-aminopropylsilane/ glutaraldehyde)	Bacterial P450	Preparation is easy Enzyme can be immobilized on any place by placing a magnet Amount of enzyme particle is limited because of plugging	[29,30]
Polymer monolith	Entrapment(2-vinyl-4,4-dimethylazlactone, ethylenedimethacrylate, 2-hydroxyethyl methacrylate, acrylamide)	Benzaldehyde liase	Stabilization of enzyme structure and activity Requirement of skill in preparation Denaturation during entrapment process	[31]
Silica monolith	Entrapment within porous silica	p-Nitrobenzyl esterase	Stabilization of enzyme structure and activity Compatibility in organic solvent Requirement of skill in preparation Denaturation possible during entrapment process	[32,33, 35]

Aluminium oxide	Cross-linking (3-aminopropylsilane/ glutaraldehyde)	Glucose oxidase	Large surface area due to porous nature Applicable for heterogeneous reactions Complicated preparation Not applicable for large-scale processing	[34]
Porous polymer monolith	Multistep photografting	Trypsin LysC	Eliminate nonspecific adsorption of proteins and peptides	[36]
CIM-disk epoxy monolith	Entrapment within monolith	Glycosyltransferases	CIM® Epoxy Disk Monolithic Column is available for purchase	[37]
Caged mesoporous silica in Ca- alginate fiber	Entrapment within amine-modified mesoporous silica	Glucose oxidase	Reduced leakage and improved activity and stability of the immobilized enzyme	[38]
LTCC multilayer substrates	Cross-linking (Glyoxal-agarose gels)	β-galactosidase	Stable operation for 6 months	[39]

Table 2: Typical techniques for enzyme-immobilization on microchannel surfaces

Media	Immobilization method	Enzyme	Advantage and disadvantage	Ref.
SiO₂ surface	Physical adsorption of biotinylated poly-lysine /biotin-avidin	Alkaline phosphatase	Ease in preparation Requirement for avidin-conjugation Possible occurrence of detachment	[40]
PDMS (O₂ Plasma treated)	Physical adsorption of lipid bilayer/biotin-avidin	Alkaline phosphatase	Enable immobilization of enzyme on plastic surface Possible occurrence of detachment Expensive reagents Requirement for avidin-conjugation	[41]
PDMS	Physical adsorption of fibrinogen/Photochemical reaction of Fluorescein-biotin	Alkaline phosphatase	Enable partial modification of microchannel Special equipment is required	[42]
Silicon	Cross-linking (3-aminopropylsilane/ glutaraldehyde)	Trypsin	Simple operation Difficulty in channel preparation Poor reproducibility	[43]
Fused silica (Sol-gel modified)	Cross-linking (3-aminopropylsilane/ glutaraldehyde)	Cucumisin Lipase L-Lactic dehydrogenase	Simple operation Immobilize ~10 times more enzymes than single layer immobilization and therefore, performs with higher reaction efficiency Several chemistry is available (amide, disulfide, His-tag) Needs several steps for immobilization Reproducibility strongly affected by characteristics of silica surface	[44-47]
PMMA	Cross-linking (Si-O bond between modified surface and silica monolith)	Trypsin	Stabilize enzyme under denaturation condition Complicated preparation method	[48]
PDMS (O₂ Plasma treated)	Cross-linking (Si-O-Ti or Si-O-Al bond between titania or alumina monolith)	Trypsin	Stabilizes enzyme under denaturation condition Complicated preparation method	[49]
PET microchip	Entrapment within nanozeolite-assembled network	Trypsin	Large surface/volume network by layer-by-layer technique	[50]

Silicon rubber	Cross-linking (3-aminopropyltrieth-oxysilane and glutaraldehyde)	Thermophilic β- glycosidase	Reaction can be performed at 80°C Complicated preparation method Reaction is slow because not much enzyme can be immobilized	[51]
Fused silica	Cross-linking between physically-immobilized Silica particle (3-aminopropylsilane/ succinate)	Lipase	Much larger surface area (1.5 times greater than sol-gel modified surface) and higher efficiency Complicated preparation method Unstable withed physical force (bending etc.)	[52]
SiO₂ nanospring	Disulfide bond	β-galactosidase	High solvent-accessible surface area permeability and mechanical stability Repeatability of re-immobilization was poor	[53]
Photopatterning onto PEG-grafted surface	Cross-linking by photo-patterned vinylazlactone	Horseradish peroxidase Glucose oxidase	Reduced non-specific absorption Sequentially multistep reaction could be achieved Requires special equipment	[54]
PDMS	Entrapment within hydrogel formed on surface	Alkaline phosphatase Urease	Quite fast reaction (90% conversion at 10 min reaction) Immobilization of multiple enzyme Complicated preparation method Not applicable for higher flow rate	[55]

Table 3: Enzyme-immobilization techniques on a membrane

Media	Immobilization method	Enzyme	Advantage and disadvantage	Ref.
PDMS/Glass	Place PVDF membrane that adsorbs enzymes	Trypsin	Easy preparation Less efficiency Possibility of leakage at higher flow rate	[56]
Glass	Covalent cross-linking with Nylon membrane formed at liquid-liquid interface (glutaraldehyde)	Horseradish peroxidase	Integration of membrane permeation and enzyme reaction Preparation of multiple membrane Complicated preparation method Unstable membrane at higher flow rat	[57]
PTFE	Enzyme-embedded membrane formation using glutaraldehyde/ paraformaldehyde	α-Chimotrypsin Trypsin α-Aminoacylase Other various enzymes	Easy preparation Durable (>40days) Applicable in organic solvents Almost all enzymes can be immobilized by adding poly-Lys	[58-60]

In batchwise reactors, immobilization of enzymes on beads or monoliths has been used for separation and recycling of enzymes. This approach has also been applied to microreaction systems. Microreactors with enzymes immobilized on glass beads have been prepared by simply filling the reaction chamber with enzyme-immobilized particles. Such a device was used for the determination of xanthine using chemiluminescent detection [26]. Crooks and co-workers developed advanced analytical microreactors using enzyme-immobilized microbead-mixing [20], and efficiently performed multistep enzyme reactions using glucose oxidase- and horseradish peroxidase-immobilized polystyrene. In addition, immobilization of enzyme on Ni-NTA-agarose bead has also been reported [27]. This immobilized enzyme is less denaturated because binding of the enzyme is achieved using a His-tag. This method was applied to immobilize bacterial P450 [27]. A similar approach was applied to immobilize enzymes onto a Merrifield resin [28]. A tyrosine-based Ni-NTA linker was created on the surface of the resin to immobilize His-tagged enzymes. This matrix was loaded into a microstructured channel of a PASS flow™ system. Synthesis of (R)-benzoin, (R)-2-hydroxy-1-phenylpropan-1-one, and 6-O-acetyl-D-glucal were performed using this system. Magnetic beads were also used for enzyme immobilization within the microchannel. Glucose oxidase was immobilized within a Teflon tube by placing a magnet [29]. The enzyme-immobilized magnetic particles were stable and active for more than eight months. This approach was also applied for the preparation of a protease-immobilized microreactor for proteomic analysis [30]. A similar technique was used for preparation of enzymeimmobilized microfluidic reactors.

Monolithic microreactors can be prepared using several methods. A trypsin-immobilized microreactor was prepared by molding a porous polymer monolith, prepared from 2-vinyl-4,4- dimethylazlactone, ethylene dimethacrylate, and acrylamide or 2-hydroxyethyl methacrylate, with an enzyme, in microchannels [31]. This microreactor was used for mapping protein digested fragments. Preparation of a microreactor by filling a silica monolith made from tetraethoxysilane with an enzyme and entrapping it within a microchannel was also developed. Trypsin-encapsulated monolith was fabricated in situ on a PMMA microchip to produce an integrated bioreactor that can perform enzymatic digestion, electrophoretic separation and detection in one chip [32]. Another example is a protease-P-including monolith prepared from a mixture of tetramethoxysilane and methyltrimethoxysilane (1:4), used to fill in PEEK [poly(ether ether ketone)] microcapillary to produce a microreaction system [33]. Aluminum oxide powder can be used as a solid support. Horseradish peroxidase was immobilized on aluminium oxide with 3-aminopropylsilane, and then placed within the microdevice [34]. This method takes advantage of the porous nature of ceramic microstructures. Overall, preparation of immobilized enzymes with powdered materials or monoliths is significantly easier; however it is unfavorable in large scale processing because of increasing pressure.

Immobilization of Enzyme on Microchannel Surface

Methods for enzyme immobilization on the microchannel surface have also been developed because they can take advantage of the larger surface area of microreaction systems without pressure increases. Physical immobilization is an easy way to immobilize molecules. In microchannel systems, a biotin-avidin system has mainly been used to immobilize enzymes. The biotinylated polylysine was physically immobilized on a glass surface to immobilize streptavidin-conjugated alkaline phosphatase [40]. This microreactor was used for rapid determination of enzyme kinetics. Biotinylated lipid bilayer [41] and partial biotinylation by photo patterning on fibrinogen [42] were also used for immobilization. However, these methods are not suitable for long-term use because of their instability. Also, applications are limited to streptavidin-conjugated enzymes. The introduction of a functional group on the microchannel surface was used for covalent crosslinking. A trypsin-immobilized microreactor was prepared by modification with 3-aminopropylsilane and glutaraldehyde using the classical method [43]. Although this immobilization method is easy, fabrication of complex microstructures is required to achieve high performance. Our group developed a modified sol-gel technique to form nanostructures on a silica microchannel surface [44]. This method modifies the microchannel surface with polymerized copolymer of 3-aminopropylsilane/methylsilane. Using this method, increased surface area was obtained. At least 10 times more enzymes can be immobilized on these nanostructures by covalent cross-linking through amide-bond formation, disulfide or His-tag, by modifying succinate spacer, compared with single layer immobilization [45- 47]. A microreactor with immobilized cucumisin on the nanostructured surface could process substrate 15 times faster than the corresponding batchwise reaction [46]. Similar surface modification methods employing sol-gel techniques were also developed [48]. A PMMA surface was modified with a copolymer of butyl methacrylate/γ-(methylacryloxy)- propyltrimethoxysilicane and silica-sol-gel to immobilize enzymes. Using this method, a trypsinimmobilized microreactor was developed. In addition, a trypsin-encapsulated titania and alumina gel matrix was immobilized through SiOH group formed on a PDMS surface by plasma oxidation [49]. Using this device, digestion time was significantly shortened (ca. 2 s) and the application for highthroughput protein identification was realized. Ji et al. developed the layer-by-layer nanozeoliteassembled network to immobilize enzymes in the porous structure formed within zeolite (Figure 1a) [50]. Alternatively, silicone rubber material was used for the preparation of functional nanostructure on the microchannel surface (Figure 1b) [51].

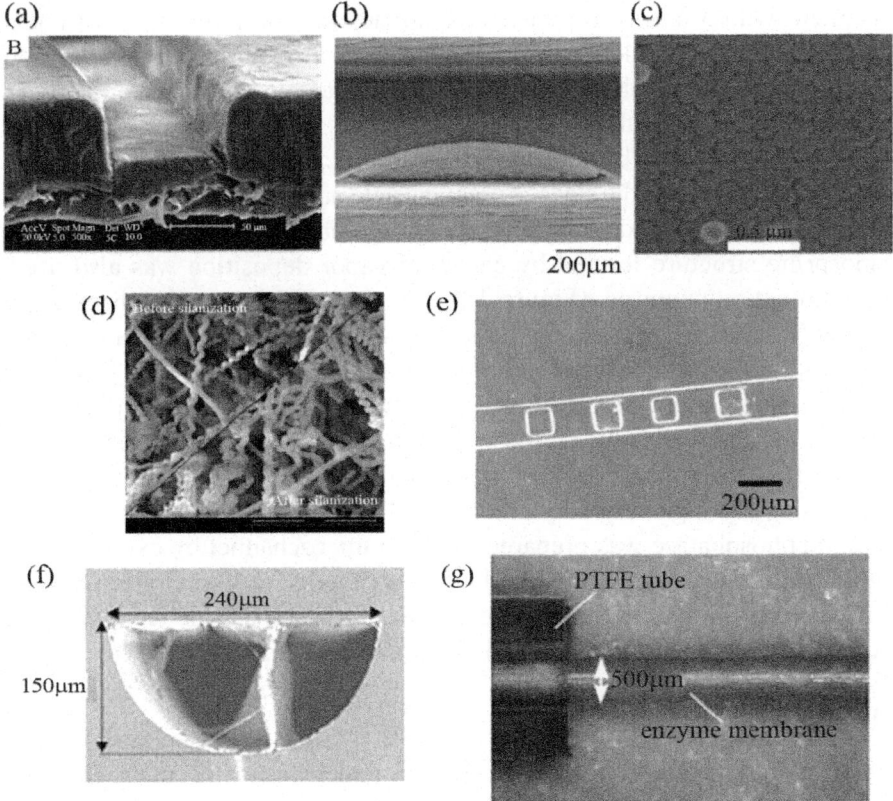

Figure 1: Images of surface modification and membrane formation techniques for

micro enzyme reactor. Modified surface obtained by functionalized microstructure fabricated from layer-by-layer nanozeolite-assembled network (a), silicone rubber (b), nanoparticle arrangement (c), SiO2 nanospring structures (d), and hydrogel formation (e). Membrane formed within the microchannel can also be used as support for enzyme immobilization. Nylon membrane formed at liquid-liquid interface (f), or membrane of cross-linking enzyme aggregate formed at microchannel surface (g) was used for immobilization. These images were reproduced with permission from references [50,51,53,55,57,58,61].

The structure was prepared by micromold fabrication using vinyl-group-containing PDMS and silicic acid, and enzyme immobilization by cross-linking with glutaraldehyde. Using this procedure, a microstructured enzyme reactor with immobilized thermophilic β-glycosidase capable of performing hydrolysis at 80°C was created. A particle-arrangement technique was also applied for enzyme immobilization. Silica nanoparticles were immobilized onto the surface using slow evaporation of particle suspension filled-in microchannel (Figure 1c) [61]. The obtained microchannel was subjected to treatment with 3-aminopropyltriethoxysilane, and immobilization of enzyme was achieved by covalent cross-linking through the amino groups. Although physical stability needs to be improved, a lipase-immobilized microreactor prepared by this method showed 1.5 times faster kinetics than that of microreactor obtained by sol-gel surface modification [52]. This result showed good correlation with the surface area; particle arrangement has approximately 1.5 times larger surface area and could immobilize more enzymes. A SiO_2 nanospring structure formed by chemical vapor deposition was also used as immobilization supports. (Figure 1d) [53]. Photochemistry has been applied to enable selective immobilization of enzymes on the microchannel surface [54]. In the procedure, vinyl azlactone was photografted onto a PEG-coated polymer surface as a reactive monomer and the enzymes were immobilized through their amino groups. This approach was applied for immobilization of horseradish peroxidase. Another approach for efficient enzyme immobilization is polymer coating. Poly(ethylene glycol)based-hydrogels which incorporate alkaline phosphatase was prepared within a microchannel by exposure to UV light (Figure 1e) [55]. This method was also applied to immobilize urease and different enzymes on microchannel surfaces. Overall, these techniques need expensive equipments and/or specialized fabrication skills.

Membrane-Formation

Enzymes can be immobilized on a membrane within the microchannel. A porous poly(vinylidene fluoride) membranes embedded within microchannel can be used for enzyme immobilization. Preparation of a miniaturized membrane reactor by absorption of enzymes onto the membrane has been reported [56]. Hisamoto et al. demonstrated that a nylon-membrane formation at the interface of two solutions formed in a microchannel (Figure 1f). Peroxidase was immobilized on this membrane which was then used as a chemicofunctional membrane [57]; however, immobilization of the membrane is technically difficult, and application of this method is limited because the nylon-membrane is unstable in organic solvents.

We have developed a technique that forms an enzyme-immobilizing membrane on the microchannel surface [58]. This is a modification of cross-linked enzyme aggregate (CLEA) formation, which is used in batchwise organic synthesis [62]. Simple loading of the enzyme solution and a mixture of glutaraldehyde and paraformaldehyde into the microchannel forms a CLEA membrane on the microchannel wall (Figure 1g). The resulting microreactor can be used for prolonged periods (>40 days), and shows excellent stability against organic solvents. Taking into account these advantages, this method is considered ideal for the development of an enzymatic reactor tailored for specific applications. However, this method requires amino groups for immobilization, and is difficult to apply to acidic enzymes with few amino groups on their surface. The application of the approach developed in our laboratory was expanded by adding poly-Lys to aid in membrane formation of acidic proteins [59]. By this method, almost all enzymes, including highly acidic proteins, can form crosslinked aggregates. We applied this technique for the preparation of enzyme microreactors, and demonstrated immobilization of several acidic enzymes by this method [59]. Our results indicate that almost all enzymes can be immobilized onto the microchannel surface by our method, and our approach is a robust way of enzyme-immobilized microreactor development.

ENZYME-IMMOBILIZED MICROFLUIDIC REACTOR PROCESSES

Hydrolysis and Esterification

Applications of enzyme-immobilized microreactors for processing for several important reactions in the synthetic organic chemistry field have been reported. As shown in Table 4, esterification and hydrolysis reactions are important processes in the industry that have also been performed in a microchannel system. Lipase-immobilized microreactors were prepared using a ceramic microreactor and glass microcapillaries [45], wherein hydrolysis of the ester was conducted. Both microreactors showed 1.5 times better yield than the batchwise reaction using the same volume/enzyme ratios. This could have resulted from an increase in contact due to the larger surface area of the microchannel systems.

Hydrolysis of triglyceride using a lipase-immobilized microreactor was also reported. The reaction yield was 10 times higher than that of the corresponding batchwise reaction [63]. Hydrolysis of vegetable oil to produce monoacyl glycerol was performed in an immobilized lipase microreactor [64]. Almost complete conversion was enabled by the microreaction system. Not only hydrolysis, but esterifications were also performed in microfluidic format. A microreaction using immobilized Novozym-435TM was also reported, where esterification of diglycerol with lauric acid was performed [65].

Esterases are also used as catalysts to produce esters and their hydrolysis products. Dräger et al. reported a regioselective hydrolysis reaction by p-nitrobenzyl esterase immobilized onto Ni-NTA agarose beads entrapped within microchannels [28]. Although an 80% yield was obtained with the microreactor, trace by-products were also detected. Proteases and aminoacylases are important tools for the preparation of chiral compounds. A monolithic microreactor tethered protease P was applied for bioconversion processes. Transesterification of (S)-(-)-glycidol and vinyl n-butyrate was performed using this microreaction device [33], but the conversion depended on the amount of immobilized enzymes.

Table 4: The use of enzyme-immobilized microreactors for hydrolysis and esterification

Immobilization technique	Enzyme	Reaction scheme	Results	Ref.
Surface modification of silica capillary by sol-gel technique/immobilized through amide bond formation using succinate linker	Lipase		1.5 time better yield was obtained compared with batchwise reaction	[45]
Entrapment within folded-sheet mesoporous silicas	Lipase		Reaction yield was 10 time higher than batchwise reaction	[63]
Covalently immobilized in silica micro structured fiber	Lipase		Almost complete conversion of a vegetable oil to monoacylglycerol	[64]
Entrapment of Novozym-435TM within microchannel	Lipase		Much less of the reactant was required compared with the batchwise test	[65]
Ni-NTA agarose bead immobilization	p-Nitrobenzyl esterase		80% yields were obtained along with traces of byproduct	[28]

Silica monolith entrapped within microchannels	Protease P		Conversion within microreactor was higher than that of the batchwise reaction at higher flow rates	[33]
Silica monolith entrapped within microchannels	Lipase		Optical resolution of products was achieved by connecting commercially available chiral column	[35]
Membrane formation with paraformaldehyde, glutaraldehyde, and poly-Lys	α-Amino-acylase		Optical resolution of D/L-amino acids were achieved by connecting to micro solvent extractor	[60]

Similarly, they separated the racemic product which was obtained by reaction in an entrapped lipase microreactor, by connecting a chiral column sequentially to the microreactor [35]. We developed a novel integrated microreaction system which combined an enzyme microreactor and a solvent extractor. The enzyme-immobilized microreactor was prepared by the membrane-formation technique using α-aminoacylase with poly-Lys [59]. This microreactor was connected with a microextractor which has a partially modified microchannel [18]. Using this microreaction system, optical resolution of D/L-phenylalanine analogs was performed. The D-phenylalanine analogs were obtained efficiently with high optical purity [60].

C-C Bond Formation, Condensation and Addition

Processing with C-C bond formation, condensation and addition reactions performed in a microchannel system are shown in Table 5.

Hydroxylation of macrolides in a microreactor was reported [27]. PikC Hydroxylase was immobilized on Ni-NTA agarose beads, and then filled into the microchannel. This microreactor was used for hydroxylation to produce methymycin and neomethymycin, and over 90% conversion was achieved at a flow rate of 70 nL/min. Such high efficiency might have resulted from the shorter residence time, which is preferable for enzymes with inherent stability. Similar immobilization technique was applied for the synthesis of (R)-benzoin using benzaldehyde lyase [28]. His-tagged protein was directly immobilized within the microstructured PASSflow reaction system through tyrosine-based Ni-NTA system.

Table 5: Processing with C-C bond formation, condensation and addition

Immobilization technique	Enzyme	Reaction scheme	Results	Ref.
Ni-NTA agarose bead immobilization	PikC hydroxylase (Bacterial P450)		>90% conversion was obtained at 70nm/min	[27]
Ni-NTA agarose bead immobilization	Benzaldehyde liase		>90% yields were obtained	[28]
His$_6$-tag affinity	Transketolase		Productivity was unchanged over 5 cycles of regeneration	[66]
Covalently immobilized on layer of γ-aluminum oxide	Thermostable β-glycosidase CelB		Similar conversion characteristics with batchwise stirred reactor	[67]

This reversible immobilization technique was also used for transketolase, which catalyses the synthesis of L-erythrulose [66]. Its productivity was unchanged over five regeneration cycles. Schwarz et al. reported transglycosilation using cellobiose and glycerol to produce β-glycosylglycerol [67]. The enzyme was immobilized onto the γ-alumina layer formed within a microchannel. However, the resulting microreactor showed almost similar conversion characteristics as the batchwise reaction.

Oxidation and Reduction

The application of enzyme-immobilized microreactors for oxidation and reduction was also reported (Table 6). Although the uses of oxidation reactions with the enzyme-immobilization techniques were mainly for analytical use, the oxidation of phenols with a horseradish peroxidaseimmobilized microreactor was recently reported [68].

Table 6: Oxidation, reduction and miscellaneous reactions in enzyme-immobilized microreactor

Immobilization technique	Enzyme	Reaction scheme	Results	Ref.
Covalently immobilized on gold patterned surface	Horseradish peroxidase	Oxidation reactions	Conversion with self-assembled monolayer approach was 1.5 time higher than physical adsorption	[68]
Surface modification by sol-gel technique/Ni-NTA immobilization	L-Lactic dehydrogenase		Crude enzyme can be used for immobilization Reversible immobilization was achieved by EDTA treatment Reaction was completed within 15 min	[47]
Entrapment of Novozym-435™ within microchannel	Lipase		Apparent rate of reaction is at least an order higher than that observed for batch reactors	[70]
CIM-disk epoxy monolith	Glycosyl-transferases		Immobilized enzyme is stable and exhibits good reproducibility	[37]
Entrapment of silica-immobilized enzymes within microchannel	Zinc Hydroxy-aminobenzene mutase Peroxidase		Used combinatorial synthesis of 2-aminophenoxyazin-3-one	[71]

We demonstrated the reduction of pyruvic acid to produce L-lactic acid by L-lactic dehydrogenase immobilized on microchannel surface through Ni-NTA group formed by sol-gel technique [47]. By this method, crude enzyme extract from bacterial lysed solution could be used for immobilization without further purification. Also, reversible immobilization was enabled to regenerate the microreactor upon enzyme denaturation. This reactor showed higher conversion rates than that of the batchwise reaction; however regeneration of co-enzyme still remains a major problem in this case. Yoon et al. reported an electrochemical microreactor for regeneration of coenzymes [69]. However, they used solution-phase reactions using enzyme solutions. Integration of enzyme immobilization techniques with this microreactor for co-enzyme regeneration might solve this problem

Miscellaneous Reactions

Enzymatic polymerization has been performed in microfluidic format. Entrapped Novozym-435TM was used for ring-opening polymerization of ε-caprolactone to produce polycaprolactone [70]. The microreactor showed improved reaction rates, higher than those observed in the batchwise reaction.

Glycosyltransferase-entrapped monolith was used for the preparation of oligosaccharides from monosaccharides [37]. The immobilized enzyme was stable and the resulting microreactor exhibited good reproducibility. The application of enzyme-immobilized microreactors for multistep synthesis was also demonstrated [71].

Three separate microfluidic devices, which possesed metallic zinc, silica-immobilized hydroxyaminobenzene mutase, and silica-immobilized peroxidase within a microchannel, were prepared and connected sequentially. These devices were used for combinatorial synthesis of 2-aminophenoxyazin-3-ones. These results open the door for the application of micro bioreactors for the enzymatic synthesis of bioactive natural products

CONCLUSIONS

Microchannel devices can be useful in imitating biological reaction apparatus, such as cellular surfaces and vascular systems, by providing the advantages of limited space and laminar flow compared with the conventional reaction apparatus. The quest for microreaction technologies will lead to better process intensification and efficient analytical methods. Increasingly, new findings are being achieved in microfluidics. Further investigation on microfluidics could provide novel mechanisms not observed in conventional systems, and better understanding of fluidics in microchannels might enable new reaction pathways not possible with conventional systems. The strong advantages offered by microreaction devices are useful, particularly in the development of microreaction systems for commercial purposes. Once a microreactor is optimized, it can be easily introduced into an industrial-scale plant. Parallel scale-out enables extension of reaction conditions optimized in a single reactor, and eliminates scale-up problems arising from conventional processes. Parallel operation of the same microreaction provides high throughput operation of different reagents at a single operation and serves as an excellent tool for combinatorial processing. Although several problems, such as connection, parallel control of fluid and reaction conditions, and monitoring, are common challenges, the benefits offered by microreaction technology accelerate the development of enzyme reaction devices.

As described here, few enzymes have been applied for microreaction process development, and not many patents describing the construction of micro enzyme reactors are published. These facts are an indication that the field is still in its initial stages. Efforts directed to the development, optimization and application of micro enzyme reactors will open a new era for biochemical processing in the synthetic organic chemistry field.

ACKNOWLEDGEMENTS

This work was supported by Grant-in-Aid for Basic Scientific Research (B: 23310092) and for Young Scientists (B: 23710153), from the Japan Society for the Promotion of Science (JSPS).

REFERENCES

1. Ehrfeld, W.; Hessel, V.; Lowe, H. Microreactors-New Technology for Modern Chemistry; Wiley-VCH: Weinheim, Germany, 2000.

2. Székely, L.; Guttman, A. New advances in microchip fabrication for electrochromatography. Electrophoresis 2005, 26, 4590-4604. 3. Ziaie, B.; Baldia, A.; Leia, M.; Guc, Y.; Siegelb, R.A. Hard and soft micromachining for BioMEMS: Review of techniques and examples of applications in microfluidics and drug delivery. Adv. Drug Delivery Rev. 2004, 56, 145-172.

3. Hessel, V.; Hardt, S.; Lowe, H. Chemical Micro Process Engineering; Wiley-VCH: Weinheim, Germany, 2004.

4. Watts, P.; Haswell, S.J. The application of micro reactors for organic synthesis. Chem. Soc. Rev. 2005, 34, 235-246.

5. Chován, T.; Guttman, A. Microfabricated devices in biotechnology and biochemical processing. Trends Biotechnol. 2002, 20, 116-122.

6. Andersson, H.; van den Berg, A. Microfluidic devices for cellomics: A review. Sens. Actuat. B 2003, 92, 315-325.

7. Cullen, C.J.; Wootton, R.C.; De Mello, A.J. Microfluidic systems for high-throughput and combinatorial chemistry. Curr. Opin. Drug Discov. Dev. 2004, 7, 798-806.

8. Wang, H.; Holladay, J.D. Microreactor Technology and Process Intensification, ACS Symposium Series, Vol. 914; American Chemical Society: Washington, DC, USA, 2005.

9. Miyazaki, M.; Maeda, H. Microchannel enzyme reactors and their applications for processing. Trends Biotechnol. 2006, 24, 463-470.

10. Schoemaker, H.E.; Mink, D.; Wubbolts, M.G. Dispelling the myths-biocatalysis in industrial synthesis. Science 2003, 299, 1694-1697.

11. García-Junceda, E.; García-García, J.F.; Bastida, A.; Fernández-Mayoralas, A. Enzymes in the synthesis of bioactive compounds: The prodigious decades. Bioorg. Med. Chem. 2004, 12, 1817-1834. 13. Schmid, A.; Dordick, J.S.; Hauer, B.; Kiener, A.; Wubbolts, M.; Witholt, B. Industrial biocatalysis today and tomorrow. Nature 2001, 409, 258-268.

12. Urban, P.L.; Goodall, D.M.; Bruce, N.C. Enzymatic microreactors in chemical analysis and kinetic studies. Biotechnol. Adv. 2006, 24, 42-57.

13. Tokeshi, M.; Minagawa, T.; Uchiyama, K.; Hibara, A.; Sato, K.; Hisamoto, H.; Kitamori, T. Continuous-flow chemical processing on a microchip by combining microunit operations and a multiphase flow network. Anal. Chem. 2002, 74, 1565-1571.

14. Hibara, A.; Nonaka, M.; Hisamoto, H.; Uchiyama, K.; Kikutani, Y.; Tokeshi, M.; Kitamori, T. Stabilization of liquid interface and control of two-phase confluence and separation in glass microchips by utilizing octadecylsilane modification of microchannels. Anal. Chem. 2002, 74, 1724-1728.

15. Zhao, B.; Moore, J.S.; Beebe, D.J. Principles of surface-directed liquid flow in microfluidic channels. Anal. Chem. 2002, 74, 4259-4268.

16. Yamaguchi, Y.; Ogino, K.; Takagi, F.; Honda, T.; Yamashita, K.; Miyazaki, M.; Nakamura, H.; Maeda, H. Partial Chemical Modification of a Microchannel and Stabilization of Water-Oil Phase Separation. In Proceedings of the 8th International Conference on Microreaction Technology (IMRET 8), Atlanta, GA, USA, 10–14 April 2005; American Institute of Chemical Engineers: New York, NY, USA, 2005.

17. Erickson, D.; Li, D. Influence of surface heterogeneity on electrokinetically driven microfluidic mixing. Langmuir 2002, 18, 1883-1892.

18. Seong, G.H.; Crooks, R.M. Efficient mixing and reactions within microfluidic channels using microbead-supported catalysts. J. Am. Chem. Soc. 2002, 124, 13360-13361.

19. Stroock, A.D.; Dertinger, S.K.W.; Ajdari, A.; Mezic, I.; Stone, H.A.; Whitesides, G.M. Chaotic mixer for microchannels. Science 2002, 295, 647-651.

20. Kim, J.-H.; Kin, B.-G.; La, M.; Yoon, J.-B.; Yoon, E. A Disposable Passive Microfluidic System Integrated with Micromixer and DNA Purification Chip for DNA Sample Preparation. In Micro Total Analysis System 2002; Baba, Y., Shoji, S., van den Berg, A., Eds.; Kluwer Academic Publishers: Dordrecht, The Netherlands, 2002; Volume 1, pp. 224-226.

21. Mengeaud, V.; Josserand, J.; Girault, H.H. Mixing processes in a zigzag microchannel: finite element simulations and optical study. Anal. Chem. 2002, 74, 4279-4286.

22. Yamaguchi, Y.; Ogino, K.; Yamashita, K.; Maeda, H. Rapid micromixing based on multilayer laminar flows. J. Chem. Eng. Jpn. 2004, 37, 1265-1270.

23. Chang, C.-C.; Yang, R.-J. Electrokinetic mixing in microfluidic systems. Microfluid. Nanofluid. 2007, 3, 501-525.

24. Richter, T.; Shultz-Lockyear, L.L.; Oleschuk, R.D.; Bilitewski, U.; Harrison, D.J. Bi-enzymatic and capillary electrophoretic analysis of non-fluorescent compounds in microfluidic devices determination of Xanthine. Sens. Actuat. B 2002, 81, 369-376.

25. Srinivasan, A.; Bach, H.; Sherman, D.H.; Dordick, J.S. Bacterial P450-catalyzed polyketide hydroxylation on a microfluidic platform. Biotechnol. Bioeng. 2004, 88, 528-535.

26. Dräger, G.; Kiss, C.; Kunz, U.; Kirschning, A. Enzyme-purification and catalytic transformations in a microstructured PASSflow reactor using a new tyrosine-based Ni-NTA linker system attached to a polyvinylpyrrolidinone-based matrix. Org. Biomol. Chem. 2007, 5, 3657-3664.

27. Nomura, A.; Shin, S.; Mehdi, O.O.; Kauffmann, J.-M. Preparation, characterization, and application of an enzyme-immobilized magnetic microreactor for flow injection analysis. Anal. Chem. 2004, 76, 5498-5502.

28. Li, Y.; Xu, X.; Yan, B.; Deng, C.; Yu, W.; Yang, P.; Zhang, X. Microchip reactor packed with metal-ion chelated magnetic silica microspheres for highly efficient proteolysis. J. Proteome Res. 2007, 6, 2367-2375.

29. Sakai-Kato, K.; Kato, M.; Ishihara, K.; Toyo'oka, T. An enzyme-immobilization method for integration of biofunctions on a microchip using a water-soluble amphiphilic phospholipid polymer having a reacting group. Lab Chip 2004, 4, 4-6.

30. Sakai-Kato, K.; Kato, M.; Toyo'oka, T. Creation of an on-chip enzyme reactor by encapsulating trypsin in sol-gel on a plastic microchip. Anal. Chem. 2003, 75, 388-393.

31. Kawakami, K.; Sera, Y.; Sakai, S.; Ono, T.; Ijima, H. Development and characterization of a silica monolith immobilized enzyme micro-bioreactor. Ind. Eng. Chem. Res. 2005, 44, 236-240.

32. Heule, M.; Rezwan, K.; Cavalli, L.; Gauckler, L.J. A miniaturized enzyme reactor based on hierarchically shaped porous ceramic microstruts. Adv. Mater. 2003, 15, 1191-1194.

33. Kawakami, K.; Abe, D.; Urakawa, T.; Kawashima, A.; Oda, Y.; Takahashi, R.; Sakai, S. Development of a silica monolith microbioreactor entrapping highly activated lipase and an experiment toward integration with chromatographic separation of chiral esters. J. Sep. Sci. 2007, 30, 3077-3084.

34. Krenkova, J.; Lacher, N.A.; Svec, F. Highly efficient enzyme reactors containing trypsin and endoproteinase LysC immobilized on porous polymer monolith coupled to MS suitable for analysis of antibodies. Anal. Chem. 2009, 81, 2004-2012.

35. Delattre, C.; Vijayalakshmi, M.A. Monolith enzymatic microreactor at the frontier of glycomic toward a new route for the production of bioactive oligosaccharides. J. Mol. Catal. B 2009, 60, 97-105.

36. Yang, C.; Zhang, Z.; Shi, Z.; Xue, P.; Chang, P.; Yan, R. Development of a novel enzyme reactor and application as a chemiluminescence flow-through biosensor. Anal. Bioanal. Chem. 2010, 397, 2997-3003.

37. Baeza, M.; López, C.; Alonso, J.; López-Santín, J.; Alvaro, G. Ceramic microsystem incorporating a microreactor with immobilized biocatalyst for enzymatic spectrophotometric assays. Anal. Chem. 2010, 82, 1006-1011.

38. Gleason, N.J.; Carbeck, J.D. Measurement of enzyme kinetics using microscale steady-state kinetic analysis. Langmuir 2004, 20, 6374-6381.

39. Mao, H.; Yang, T.; Cremer, P.S. Design and characterization of immobilized enzymes in microfluidic systems. Anal. Chem. 2002, 74, 379-385.

40. Holden, M.A.; Jung, S.-Y.; Cremer, P.S. Patterning enzymes inside microfluidic channels via photoattachment chemistry. Anal. Chem. 2004, 76, 1838-1843.

41. Ekström, S.; Onnerfjord, P.; Nilsson, J.; Bengtsson, M.; Laurell, T.; Marko-Varga, G. Integrated microanalytical technology enabling rapid and automated protein identification. Anal. Chem. 2000, 72, 286-293.

42. Miyazaki, M.; Kaneno, J.; Uehara, M.; Fujii, M.; Shimizu, H.; Maeda, H. Simple method for preparation of nanostructure on microchannel surface and its usage for enzyme-immobilization. Chem. Commun. 2003, 648-649.

43. Kaneno, J.; Kohama, R.; Miyazaki, M.; Uehara, M.; Kanno, K.; Fujii, M.; Shimizu, H.; Maeda, H. A simple method for surface modification of microchannels. New J. Chem. 2003, 27, 1765-1768.

44. Miyazaki, M.; Kaneno, J.; Kohama, R.; Uehara, M.; Kanno, K.; Fujii, M.; Shimizu, H.; Maeda, H. Preparation of functionalized nanostructures on microchannel surface and their use for enzyme microreactors. Chem. Eng. J. 2004, 101, 277-284.

45. Miyazaki, M.; Kaneno, J.; Yamaori, S.; Honda, T.; Briones, M.P.P.; Uehara, M.; Arima; K.; Kanno, K.; Yamashita, K.; Yamaguchi, Y.; et al.

Efficient immobilization of enzymes on microchannel surface through His-tag and application for microreactor. Protein Pept. Lett. 2005, 12, 207-210.

46. Qu, H.; Wang, H.; Huang, Y.; Zhong, W.; Lu, H.; Kong, J.; Yang, P.; Liu, B. Stable microstructured network for protein patterning on a plastic microfluidic channel: Strategy and characterization of on-chip enzyme microreactors. Anal. Chem. 2004, 76, 6426-6433.

47. Wu, H.; Tian, Y.; Liu, B.; Lu, H.; Wang, X.; Zhai, J.; Jin, H.; Yang, P.; Xu, Y.; Wang, H. Titania and alumina sol-gel-derived microfluidics enzymatic-reactors for peptide mapping: Design, characterization, and performance. J. Proteome Res. 2004, 3, 1201-1209.

48. Ji, J.; Zhang, Y.; Zhou, X.; Kong, J.; Tang, Y.; Liu, B. Enhanced protein digestion through the confinement of nanozeolite-assembled microchip reactors. Anal. Chem. 2008, 80, 2457-2463.

49. Thomsen, M.S.; Pölt, P.; Nidetzky, B. Development of a microfluidic immobilised enzyme reactor. Chem. Commun. 2007, 2527-2529.

50. Nakamura, H.; Li, X.; Wang, H.; Uehara, M.; Miyazaki, M.; Shimizu, H.; Maeda, H. A simple ethod of self-assembled nano-particles deposition on the micro-capillary inner walls and the reactor application for photo-catalytic and enzyme reactions. Chem. Eng. J. 2004, 101, 261-268.

51. Schilke, K.F.; Wilson, K.L.; Cantrell, T.; Corti, G.; McIlroy, D.N.; Kelly, C. A novel enzymatic microreactor with Aspergillus oryzae β-galactosidase immobilized on silicon dioxide nanosprings.Biotechnol. Prog. 2010, 26, 1597-1605.

52. Logan, T.C.; Clark, D.S.; Stachowiak, T.B.; Svec, F.; Fréchet, J.M.J. Photopatterning enzymes on polymer monoliths in microfluidic devices for steady-state kinetic analysis and spatially separated multi-enzyme reactions. Anal. Chem. 2007, 79, 6592-6598.

53. Koh, W.; Pishko, M. Immobilization of multi-enzyme microreactors inside microfluidic devices. Sens. Actuat. B 2005, 106, 335-342.

54. Gao, J.; Xu, J.; Locascio, L.E.; Lee, C.S. Integrated microfluidic system enabling protein digestion, peptide separation, and protein identification. Anal. Chem. 2001, 73, 2648-2655.

55. Hisamoto, H.; Shimizu, Y.; Uchiyama, K.; Tokeshi, M.; Kikutani, Y.; Hibara, A.; Kitamori, T. Chemicofunctional membrane for integrated chemical processes on a microchip. Anal. Chem. 2003, 75, 350-354.

56. Honda, T.; Miyazaki, M.; Nakamura, H.; Maeda, H. Immobilization of enzymes on a microchannel surface through cross-linking polymerization.

Chem. Commun. 2005, 5062-5064.

57. Honda, T.; Miyazaki, M.; Nakamura, H.; Maeda, H. Facile preparation of an enzyme-immobilized microreactor using a cross-linking enzyme membrane on a microchannel surface. Adv. Synth. Catal. 2006, 348, 2163-2171.

58. Honda, T.; Miyazaki, M.; Yamaguchi, Y.; Nakamura, H.; Maeda, H. Integrated microreaction system for optical resolution of racemic amino acids. Lab Chip 2007, 7, 366-372.

59. Wang, H.; Li, X.; Nakamura, H.; Miyazaki, M.; Maeda, H. Continuous particle self-arrangement in a long micro-capillary. Adv. Mater. 2002, 14, 1662-1666.

60. Cao, L.; Langen, L.V.; Sheldon, R.A. Immobilised enzymes: carrier-bound or carrier-free? Curr.Opin. Biotechnol. 2003, 14, 387-394.

61. Matsuura, S.; Ishii, R.; Itoh, T.; Hamakawa, S.; Tsunoda, T.; Hanaoka, T.; Mizukami, F. Immobilization of enzyme-encapsulated nanoporous material in a microreactor and reaction analysis. Chem. Eng. J. 2011, 167, 744-749.

62. Anuar, S.T.; Villegas, C.; Mugo, S.M.; Curtis, J.M. The development of flow-through bio-catalyst microreactors from silica micro structured fibers for lipid transformations. Lipids 2011, 46, 545-555.

63. Garcia, E.; Ferrari, F.; Garcia, T.; Martinez, M.; Aracil, J. Use of Microreactors in Biotransformation Processes: Study of the Synthesis of Diglycerol Monolaurate Ester. In Proceedings of the 4th International Conference on Microreaction Technology (IMRET 4), Atlanta, GA, USA, 5–9 March 2000; American Institute of Chemical Engineers: New York, NY, USA, 2000; pp. 5-9.

64. Matosevic, S.; Lye, G.J.; Baganz, F. Design and characterization of a prototype enzyme microreactor: Quantification of immobilized transketolase kinetics. Biotechnol. Prog. 2010, 26, 118-126.

65. Schwarz, A.; Thomsen, M.S.; Nidetzky, B. Enzymatic synthesis of β-glucosylglycerol using a continuous-flow microreactor containing thermostable β-glycoside hydrolase CelB immobilized on coated microchannel walls. Biotechnol. Bioeng. 2009, 103, 865-872.

66. Tudorache, M.; Mahalu, D.; Teodorescu, C.; Stan, R.; Bala, C.; Parvulescu, V.I. Biocatalytic microreactor incorporating HRP anchored on micro-/nano-lithographic patterns for flow oxidation of phenols. J. Mol. Catal. B 2011, 69, 133-139.

67. Yoon, S.K.; Choban, E.R.; Kane, C.; Tzedakis, T.; Kenis, P.J. a Laminar

flow-based electrochemical microreactor for efficient regeneration of nicotinamide cofactors for biocatalysis. J. Am. Chem. Soc. 2005, 127, 10466-10467.

68. Kundu, S.; Bhangale, A.S.; Wallace, W.E.; Flynn, K.M.; Guttman, C.M.; Gross, R.A.; Beers, K.L. Continuous Flow Enzyme-Catalyzed Polymerization in a Microreactor. J. Am. Chem. Soc. 2011, 6006-6011.

69. Luckarift, H.R.; Ku, B.S.; Dordick, J.S.; Spain, J.C. Silica-immobilized enzymes for multi-step synthesis in microfluidic devices. Biotechnol. Bioeng. 2007, 98, 701-705.

Chapter 12

IMMOBILIZATION OF PAPAIN ON CHITIN AND CHITOSAN AND RECYCLING OF SOLUBLE ENZYME FOR DEFLOCCULATION OFSACCHAROMYCES CEREVISIAE FROM BIOETHANOL DISTILLERIES

Douglas Fernandes Silva,[1] Henrique Rosa,[1] Ana Flavia Azevedo Carvalho,[2] andPedro Oliva-Neto[1]

[1]Department of Biological Science, University of State of São Paulo (UNESP), 19806-900 Assis, SP, Brazil

[2]Food Engineering Faculty, State University of Campinas (UNICAMP), 13083-970 Campinas, SP, Brazil

ABSTRACT

Yeast flocculation (Saccharomyces cerevisiae) is one of the most important problems in fuel ethanol production. Yeast flocculation causes operational difficulties and increase in the ethanol cost. Proteolytic enzymes can solve this problem since it does not depend on these changes. The recycling of soluble papain and the immobilization of this enzyme on chitin or chitosan were studied. Some cross-linking agents were evaluated in the action of proteolytic activity of papain. The glutaraldehyde ($0.1-10\% \, w \cdot v^{-1}$), polyethyleneimine ($0.5\% \, v \cdot v^{-1}$), and tripolyphosphate ($1-10\% \, w \cdot v^{-1}$) inactivated the enzyme in this range, respectively. Glutaraldehyde inhibited all treatments of papain immobilization. The chitosan cross-linked with TPP in 5 h of reaction showed the yield of active immobilized enzyme of 15.7% and 6.07% in chitosan treated with 0.1% PEI. Although these immobilizations have been possible, these levels have not been enough to cause deflocculation of yeast cells. Free enzyme was efficient for yeast deflocculation in dosages of 3 to $4 \, g \cdot L^{-1}$. Recycling of soluble papain by centrifugation was effective for 14 cycles with yeast suspension in time perfectly compatible to industrial conditions. The reuse of proteases applied after yeast suspension by additional yeast centrifugation could be an alternative to cost reduction of these enzymes.

INTRODUCTION

During the industrial process of fuel ethanol fermentation the contamination caused by bacteria and/or wild yeast is very common. The microbial contaminants cause cells flocculation or flakes of yeast and bacteria and this contamination causes settling of yeast cells at the bottom of the vats. The yeast flocculation is a serious current problem in fuel ethanol technology since this process uses cells recycle. The flocculation of yeast cells decreases the ethanol efficiency by some operational problems as the loss of yeast cells due to difficulties in yeast centrifugation and obstruction of bombs and pipes. Other important problem caused by yeast cells flocculation is the adhesion of bacterial cells on yeast cell surface in the yeast flake. This fact is responsible for the increase of lactic acid bacteria population. Consequently, organic acids are produced leading to yeast metabolism inhibition and ethanol production [1–3]. Furthermore, yeast flocculation increases the use of sulfuric acid and also increases the cost of fuel ethanol [1, 2]. Besides bacteria, wild yeasts and salts could be responsible for the phenomenon of yeast flocculation causing serious operational problems and economic losses in the processes [2–4]. Protein factors associated with minerals such as Ca^{+2} [3, 5], as well as mannans, have been proved to be involved in the process of flocculation.

The cell's flocculation also is responsible for the increase in the production of organic acids by bacteria causing inhibition of yeast metabolism and consumption of sugar by contaminants [6, 7]. All these problems result in a partial conversion of sugar into ethanol and CO_2 decreasing the ethanol yield and productivity and increasing the use of chemicals like sulfuric acid, antibiotics, and antifoam to control, respectively, yeast flocculation, microbial infection, and bubbles [8–10].

Conventional treatment of flocculated yeast cells using sulfuric acid leads to an immediate cell deflocculation, although this procedure allows the return of the flocculation due to pH change of the industrial process, when yeasts are returned to the fermenter (pH 4.0). The low pH used for yeast deflocculation can affect the yeast metabolism decreasing yeast cell viability, depending on the operational process. The residence times of the acid treatment are generally 0.5 to 2 hours, but an increase of these times and/or decrease of pH cause a detrimental impact on the yeast metabolism. Furthermore, younger and older yeast cells are less resistant to this treatment [11]. The action of proteases has been effective on yeast deflocculation [1]. A widely used proteolytic enzyme is papain, an alkaloid protein from the latex of papaya (Carica papaya), which is characterized as a cysteine endopeptidase that has strong proteolytic action [12] and it is relatively inexpensive.

The treatment with proteolytic enzymes for the control of flocculation could be an adequate alternative, since this method is less affected by the changes in the pH of the process [1] and it does not affect the yeast metabolism [13]. However, the application of these enzymes on industrial scale will only be economically viable if they are of low cost. For this reason, the reuse of proteases is required, which can be achieved with the recovery of free enzyme by centrifugation or ultrafiltration in the bioprocess [5, 14, 15] or, alternatively the immobilization of this enzyme on solid supports.

Some low cost products can be used in enzyme immobilization and reuse of papain. The chitin is a natural polysaccharide with acetamide group which has a positive charge [16], and it is possible to be obtained as residue from the food industry [17]. The chitosan is a natural polymer, of low cost, is renewable, and is obtained from chitin by deacetylation with alkali [16, 18]. The lower level of N-acetyl groups (<40%) provides greater solubility when chitosan solutions are in pH below 6.5; this polymer is nontoxic, available in different forms (powder, gel, fibers, and membranes), and easily derivatizable, demonstrating high protein affinity [19,20]. Therefore, a chemical treatment of chitosan is required in low pH conditions to maintain its insolubility, a key feature of the support for the success of the enzyme immobilization [21]. The treatment of chitosan with sodium tripolyphosphate ($Na_5P_3O_{10}$) is necessary to promote the formation of bonds between these molecules reducing its solubility in acid conditions by ionic cross-linking between them [22], preventing chitosan to be dissolved in this condition [23].

The cross-linking method by bifunctional agents (e.g., glutaraldehyde) [19] is one of the most common procedure for immobilization of enzymes since this method is of low cost, simple, and fast and is able to be widely applied [24]. The procedure involves a covalent bond by fixing the enzyme on the matrix or by cross-linking in the matrix, containing the enzyme and a bifunctional agent [4, 25, 26].

In this paper, the reuse of papain through the recovery of soluble enzyme by centrifugation or by its immobilization on polysaccharides was evaluated for S. cerevisiae cells deflocculation from fuel ethanol distilleries.

MATERIALS AND METHODS

Microorganisms and Reagents

Samples of flocculated Saccharomyces cerevisiae from fuel ethanol distillery (Raizen, Maracaí, SP, Brazil) were used in deflocculation test with commercial crude papain (Vetec Química Fina LTDA, EC 3.4.22.2) with $6000\,U \cdot mg^{-1}$ of proteolytic activity. This enzyme was dissolved in phosphate buffer $0.2\,mol \cdot L^{-1}$, pH 6.4. The PA (pure for analysis) reagents used for enzyme immobilization were 25% glutaraldehyde in water, polyethyleneimine (PEI) of high molecular weight (Sigma-Aldrich Co.), and sodium tripolyphosphate (TPP) ($Na_5P_3O_{10}$). Chitin was extracted from shrimp shells and high molecular weight chitosan was obtained from Aldrich (code 419419-50 G). The proteolytic activity of papain was determined by hydrolysis of sulfanilamide azocasein [27], and total protein was determined by Bradford [28].

Minimum Inhibitory Concentration of Papain for Yeast Deflocculation

The cells deflocculation of the industrial S. cerevisiae was proceeded in a range of $0–4\,g \cdot L^{-1}$ of soluble papain in a 30% $(w \cdot v^{-1})$ flocculated yeast suspension for 15 minutes of reaction, at 27°C, according to a method developed by Ludwig et al. [1] relating absorbance of yeast suspension and percentage of yeast deflocculation (Table 3).

Immobilization of Papain on Chitin or Chitosan

Immobilization of papain on chitin was conducted in two ways [20]: without any pretreatment of the support or pretreatment with $10\,mL$ of 2% $(v \cdot v^{-1})$ glutaraldehyde per gram of chitin at 27°C for 5 h. The enzyme immobilization was conducted by the addition of $1.0\,g$ of treated or untreated chitin into $10\,mL$ of 1% $(w \cdot v^{-1})$ papain solution in phosphate buffer ($0.2\,mol \cdot L^{-1}$) and/or by the addition of 0.1% or 0.5% $(v \cdot v^{-1})$ of PEI in papain solution. These protocols were performed in pH 7.0 at 27°C for 12 h in 125 mL Erlenmeyer flasks at shaker 80 rpm, totaling six different immobilization protocols (Table 1).

Table 1: Protocols used for papain immobilized on chitin

Treat-ment	Immobilization conditions	Pretreatment of support
A	1 g chitin + 1% papain[a]	—
B	1 g chitin + 1% papain + 0.1% PEI	—
C	1 g chitin + 1% papain + 0.5% PEI	—
D	1 g chitin + 1% papain	2% glutaraldehyde[b]
E	1 g chitin + 1% papain + 0.1% PEI[c]	2% glutaraldehyde
F	1 g chitin + 1% papain + 0.5% PEI	2% glutaraldehyde

The cross-linking process of chitosan by sodium tripolyphosphate (TPP) was processed according to Laus et al. [23], in which 1.0 g of chitosan was dissolved into 100 mL of 1% (w·v⁻¹) acetic acid solution. This viscous solution was dripped into 1% papain (w·v⁻¹) in a ratio of 1 : 2 (v·v⁻¹) in three different protocols for 5 or 12 h of reaction (Table 2) in order to evaluate the effect of two different times on the result of immobilization. Subsequently, the immobilized chitin and chitosan were filtered and washed two times with distilled water (100 mL and 1000 mL) and packed in 5°C in phosphate buffer (pH 6.4) for use in testing of the proteolytic activity.

Table 2: Protocols used for papain immobilized on chitosan

Treatment	Preparation of chitosan with TPP	Adsorption of papain on chitosan with TPP	Reaction time (h)
A′	1 g chitosan in 100 mL 1% acetic acid solution dripped into 200 mL 1% TPP	Washing and addition of the microspheres in 200 mL 1% papain solution (phosphate buffer pH 6.4)	5 and 12
B′	1 g chitosan in 100 mL 1% acetic acid solution dripped into a 200 mL 1% TPP and 1% papain to produce microspheres	—	5 and 12
C′	1 g chitosan in 100 mL 1% acetic acid solution dripped into a solution of 200 mL 1% TPP	Addition of the 1% papain solution instantly after the formation of microspheres	5 and 12

Table 3: Percentage of S. cerevisiae cells deflocculation, yeast precipitation, and turbidity measured in 600 nm by spectrophotometry

Total turbidity (600 nm)	Yeast deflocculation (%)	Yeast precipitation
≤12.0	0	p^2
27.0	25.0	np^3
40.5	37.5	np
54.0	50.0	np
67.5	62.5	np
81.0	75.0	np
94.5	87.5	np
≥120.0[1]	100	np

Yeast Deflocculation Using Immobilized Papain

The deflocculation of yeast cells through immobilized papain on chitin or chitosan was conducted in conical glasses with 50 mL of 30% $(w \cdot v^{-1})$ flocculated S. cerevisiae from ethanol distillery and 2 g of papain immobilized on chitin or chitosan. The reaction was processed at 25°C and pH 4.5 for 120 min.

Yield Calculations of the Immobilization Process

The yields of enzyme activity and immobilized protein were calculated according to Varavinit et al. [29] by equations

$$\text{Yield of active immobilized enzyme (\%)} = \frac{\text{Total activity of immobilized enzyme}}{(A - B)} \times 100,$$

$$\text{Yield of active immobilized enzyme (\%)} = \frac{\text{Total activity of immobilized enzyme}}{(A - B)} \times 100, \tag{1}$$

where A is total activity of the soluble papain added on the support; B is total activity of the enzyme remained in solution after immobilization process. Thus,

$$\text{Yield of immobilized protein (\%)} = \frac{C}{D} \times 100, \tag{2}$$

where C is total immobilized protein (g); D is total protein added on support (g), total protein remaining in solution after the immobilization process (g).

Yeast Viability

The viability of yeast cells was performed by counting living cells by light microscopy in a Neubauer chamber [30], comparing yeast deflocculation by papain ($4\,g\cdot L^{-1}$) against H_2SO_4 at minimum concentration for complete deflocculation. The mixture was resting for 2 hours. Yeast cells of both test solutions were stained with erythrosine [6]. The yeast viability was expressed by the percentage of live cells of the total number of cells.

Proteolytic Activity and Protein Determination

The proteolytic activity of the enzyme was determined by hydrolysis of azocasein sulfanilamide, according to the method of Leighton et al. [27], at 60°C for 30 minutes. The reaction was stopped with 10% ($w\cdot v^{-1}$) trichloroacetic acid (TCA), and the proteolytic activity was analyzed by spectrophotometer at 440 nm. One unit was defined as the absorbance change in 30 minutes of reaction per mL of solution or g support. The protein was determined according to Bradford [28], and the measurements were performed before and after immobilization procedures.

Recycled Soluble Enzyme in Yeast Deflocculation

The recycle of soluble enzyme was performed according to the best concentration of yeast deflocculation test using soluble papain with or without $0.1\,g\cdot L^{-1}$ sodium dodecyl sulfate (SDS) in 50 mL flocculated yeast cell solution (30% $w\cdot v^{-1}$) in 250 mL Erlenmeyer flask. They were incubated in shaker for 6–120 min at 25°C. After each cycle, 5% of enzyme was replaced by new enzyme in order to maintain the enzyme activity. The quantification of yeast deflocculation and phase separation were evaluated in conical glasses with flocculated S. cerevisiae suspension according to Ludwig et al. [1] (Table 3).

Statistic Treatment

The statistical analyses were processed in triplicate of results submitted to analysis of variance (ANOVA), while the means were compared using Student's t-test or Tukey test by the program GraphPad InStat version 3.05 (Rutgers University). The treatments were considered statistically significant at $P < 0.005$.

RESULTS AND DISCUSSION

Cell Deflocculation of S. cerevisiae by Soluble Papain and Yeast Viability

The minimum concentration of papain necessary to induce yeast cell deflocculation was $3\,g\cdot L^{-1}$ of papain inS. cerevisiae suspension after 15 min of reaction (Figure 1), but $4\,g\cdot L^{-1}$ of papain produced better results (ANOVA and Tukey $F = 162.32, P < 0.05$) reaching almost 100% of cells deflocculation.

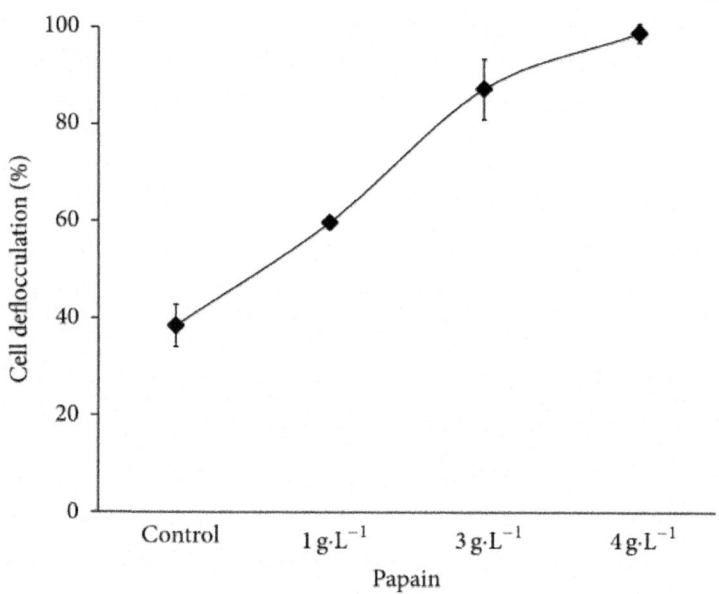

Figure 1: Yeast cell deflocculation on the S. cerevisiae suspension from fuel ethanol distillery treated with soluble papain in 15 minutes of reaction.

The current method of cell deflocculation of S. cerevisiae in industrial fuel ethanol production in Brazil is the use of H_2SO_4 in pH 2.5, and this method can decrease yeast cells viability. Table 4 shows a comparative effect of sulfuric acid and papain $(4\,g\cdot L^{-1})$ for 2 h of reaction to induce yeast cell deflocculation. The yeast cells treated with H_2SO_4 showed a lesser (P<0.005) yeast viability (67.2%) than papain (72.4%) in just one treatment of 2 h. However, the yeast cells are treated with H_2SO_4 about 2-3 times per day, and sometimes pH<2.5. According to these results papain does not affect cell viability and for this reason it is more convenient than the sulfuric acid method for the health of yeast.

Table 4: Viability of yeast cell treated with H_2SO_4 and papain in 2 h of reaction

Treatments	Cell viability (%)
Control	72.0 ± 1.0^a
Sulfuric acid[1]	67.19 ± 0.8^b
Papain[2]	72.4 ± 1.4^a

Inhibition of Soluble Papain by Glutaraldehyde, Polyethyleneimine, and Tripolyphosphate

Glutaraldehyde (0.1–10%) strongly inhibited (P<0.005) the papain activity compared to the untreated soluble enzyme. 1–10% sodium tripolyphosphate (TPP) reduced (P<0.005) the proteolytic activity of papain; however, TPP was less aggressive than glutaraldehyde, since 1% TPP reduced only 45% of the enzyme activity, while glutaraldehyde at this same concentration reduced this activity by 94.1%. Despite their inhibition, 1% TPP and 2% glutaraldehyde were evaluated as cross-linking agents in immobilization of papain on chitin (Table 5).

Table 5: Proteolytic activity of the papain in the presence of glutaraldehyde, polyethyleneimine, and tripolyphosphate, for 120 minutes at 27°C

Concentration (%)	Glutaraldehyde (U/mL)	Polyethyleneimine (PEI) (U/mL)	Sodium tripolyphosphate (U/mL)
Control	2.04 ± 0.04^a	2.68 ± 0.09^a	2.61 ± 0.07^a
0.1	0.18 ± 0.06^b	3.01 ± 0.04^b	1.68 ± 0.09^b
0.5	0.13 ± 0.01^b	2.25 ± 0.24^c	1.47 ± 0.10^c
1.0	0.12 ± 0.00^b	0.96 ± 0.07^d	1.43 ± 0.09^c
1.5	0.12 ± 0.02^b	0.81 ± 0.08^d	0.88 ± 0.10^d
2.0	0.11 ± 0.02^b	0.58 ± 0.06^d	0.55 ± 0.04^e
5.0	0.01 ± 0.01^c	0.74 ± 0.14^d	0.48 ± 0.09^e
10.0	0.00 ± 0.00^c	0.59 ± 0.27^d	0.34 ± 0.00^e

Polyethyleneimine (PEI) also was less inhibitor of papain activity than glutaraldehyde. Up to 84% of proteolytic activity of papain was maintained in concentrations less than 0.5% PEI. However, more than 70% reduction in the papain activity with 1% PEI was verified. Therefore, 0.5% PEI was considered the limit of concentration to use in the tests of immobilization process of papain. TPP and glutaraldehyde are considered excellent cross-linking agents for chitosan microspheres [31]; however, the inhibitory effect of this agent was proved inducing a total or partial loss of enzyme activity depending on each enzyme and chemical or physical treatments [20]. The action of glutaraldehyde in denaturation of aldehyde dehydrogenase has been showed [32]. The nature of the enzyme is primarily responsible for the denaturing action by glutaraldehyde. Enzymes rich in lysine are more resistant to degrading action of this compound [25]. Therefore, the concentrations of the enzyme and glutaraldehyde need to be carefully considered in order to obtain active derivatives via cross-linking. Low concentrations of enzyme and the bifunctional agent tend to induce intramolecular cross-linking [33]; however, the enzyme activity is inversely proportional to the concentration of glutaraldehyde [34]. Excess of cross-linking can result in a distortion in the enzyme structure [34], and this conformational change may induce loss of the catalytic site, thereby reducing enzyme activity.

Immobilization of Papain on Chitin and Chitosan

Immobilization on Chitin

The process of immobilization of papain on chitin with 0.1% PEI (protocol B) showed better yield of active immobilized enzyme (6%) when compared with protocol A (2.6 times lower), proving the efficiency of PEI in the process of immobilization of papain on chitin. However, although protocol B has shown better performance in yield of active immobilized enzyme, only less than 0.72 U was immobilized on chitin, or 0.11 U/g of support, and only 0.3% of the immobilized protein yield was obtained. Protocol A showed double (0.23 U/g) of support of protein immobilization (Tables 6 and 7).

Table 6: Total activity of papain in several steps, according to each protocol of immobilization

Immobilization protocol[a]	Total enzyme in solution	Enzyme solution after the immobilization process	1st pellets washing-water (100 mL)	2nd pellets washing-water (1000 mL)	Active immob. enzyme (U)[c]
A	355.14	215.88 (14)[b]	74.0	0.0	1.53 (6.5)[d]
B	481.34	333.48 (14)	136.0	0.0	0.72 (6.5)

C	416.86	122.46 (13)	148.0	66.7	0.43 (6.5)
D	316.00	132.66 (18)	48.0	0.0	0.00 (7.5)
E	360.94	240.79 (18)	70.3	31.0	0.00 (7.5)
F	424.46	225.54 (18)	87.0	0.0	0.10 (7.5)
A'-5 h	206.50	55.97 (39)	0.0	0.0	5.76 (2.78)
B'-5 h	180.65	155.61 (26)	0.0	0.0	3.94 (3.17)
C'-5 h	201.00	138.13 (33)	0.0	0.0	2.24 (2.70)
A'-12 h	184.75	34.59 (30)	0.0	0.0	2.21 (2.66)
B'-12 h	197.50	6.65 (25)	0.0	0.0	3.72 (2.86)
C'-12 h	197.75	29.92 (34)	0.0	0.0	2.34 (2.84)

Table 7: Determination of yield of active immobilized papain and other parameters

Immobilization protocols[a]	Enzyme activity after immobilization (in solution) (%)	Enzyme activity involved in immobilization (%)	Yield[c] of active immobilized enzyme (%)	Yield[d] of immobilized protein (%) and immob. protein on support (mg·g^{-1})	Enzymatic activity in support (U/g)
A	81.62	18.37	2.34	0.24 (0.12)	0.23
B	97.53	2.46	6.07	0.3 (0.148)	0.11
C	64.88	35.11	0.29	0.3 (0.15)	0.06
D	57.17	42.82	0.00	13.56 (6.73)	0.00
E	86.18	13.81	0.00	43.58 (21.63)	0.00
F	73.63	26.36	0.08	67.41 (33.46)	0.013
A'-5 h	27.10	72.89	3.82	35.00 (24.60)	2.07
B'-5 h	86.13	13.86	15.73	9.06 (4.50)	1.24
C'-5 h	68.72	31.27	3.56	45.9 (23.90)	0.82
A'-12 h	18.72	81.27	1.47	44.06 (30.80)	0.83

| B'-12 h | 3.36 | 96.63 | 1.94 | 45.9 (23.90) | 1.30 |
| C'-12 h | 15.13 | 84.86 | 1.39 | 46.64 (29.40) | 0.82 |

Depending on the enzyme, the method of chitin treated with glutaraldehyde had a higher immobilization of proteins, probably due to a better cross-link between enzyme-enzyme and enzyme-support [4]. The immobilization of papain on chitin using glutaraldehyde (Tables 6 and 7) has confirmed the inhibition of papain activity in tests with soluble enzyme (Table 5), even if a higher linkage of protein (67.41% or 33.46 mg·g−1 support) (Table 7) was obtained. However, only 0.012 U/g of support of the active enzyme was determined. For the protocols with glutaraldehyde (D, E, and F), although they have shown higher yield of immobilized protein than protocols untreated (A, B, and C), all of them did not show yield of active immobilized enzyme (Tables 6 and 7).

This fact could be explained due to the reactivity of glutaraldehyde with some groups of catalytic site of the enzyme, leading to a loss of proteolytic activity [35]. The enzyme was connected to support in various ways and its activity may be affected due to this link involving the catalytic site or preventing its availability to the substrate, leading to a loss of proteolytic activity.

On the other hand, the importance of glutaraldehyde for immobilization process of several enzymes is unquestionable. This agent promotes the formation of Schiff bases between aldehyde and amine groups of support and enzyme, resulting in a better adsorption of the enzyme on the support [4, 20, 36]. Gaspari et al. [36] working with inulinase of Kluyveromyces marxianus showed a low yield of immobilization on chitin without any cross-linking agent, and best results occurred when it was previously treated with glutaraldehyde. On the other hand, concentrations up to 0.4% of glutaraldehyde for immobilization procedures cause a significant decrease of papain activity [35]. Martins et al. [17] working with trehalase of Escherichia colireached a maximum value of 0.026% of enzyme even with the treatment of glutaraldehyde. Therefore, each enzyme may have different sensitivity depending on the support immobilization and concentration of glutaraldehyde. If lesser concentrations of glutaraldehyde were evaluated (e.g., less than 0.1%), maybe the concentration of active papain linkage on chitin would improve, due to less inhibition of the catalytic site of this enzyme and better protein immobilization, probably improving the yield of active enzyme immobilization.

Immobilization on Chitosan

The highest activities of papain immobilized on chitosan were in protocols A' and B' both in 5 h, respectively, 2.07 U/g and 1.24 U/g support (Tables 6 and 7).

The yield of active papain immobilized on chitosan (Table 7) cross-linked with 1% TPP (protocol B') was higher (15.7%) than in protocol B on chitin (6.07%) with 0.1% PEI, both at the same time (5 h). In protocol B' the high concentration of active enzymes in the enzyme solution after immobilization process (86.13% of the total activity) was evident, which could be used for additional immobilization.

Despite the best result of protocol B' (15.7%), Qiuhua et al. [37] obtained 50% of recovered activity of papain immobilized on microcrystalline chitosan when treated with 3% glutaraldehyde and Hong et al. [38] obtained 66.6%. These values were higher than those found in the present study probably due to the quality of papain and its resistance to glutaraldehyde. The purity of papain could influence the reactions during the immobilization process.

The yields of immobilized protein in protocols A' and B' were, respectively, 35% and 9.06% or 24.6 mg and 4.5 mg of protein·g−1 of support (Table 7). The time of immobilization process is also important. The yield of immobilized protein in protocol B' in 5 h and 12 h was, respectively, 9.06% and 45.9%. However if the time of immobilization is important to increase the yield of immobilized protein, the yield of active immobilized enzyme decreased with the time, from 15.73% in 5 h to only 1.94% in 12 h (Table 7). Therefore, the time of enzyme immobilization is critical and more studies are justified in order to improve this process.

In relation to the yield of active immobilized papain, Lorenzoni et al. [18] worked with fructooligosaccharides and inverted sugar by β-fructofuranosidase and β-fructosyltransferase in a covalent immobilization on chitosan spheres, using glutaraldehyde as a coupling agent. 42% of immobilization yield and 12% of immobilization efficiency (yield of active immobilized enzyme) were obtained in these conditions. This yield was lower than that obtained with the immobilization of papain on chitosan and TPP in the present work (15.7%). The best yields of immobilized protein (Table 7) were, respectively, for protocols A', B', and C' (44–46.6%) in 12 h of reaction. These are high yields if they are compared with the literature. Spagna et al. [20] working with α-L-arabinofuranosidase immobilized on chitosan obtained a lesser yield from 25% to 30%. Zappino et al. [39] working with immobilization of bromelain on microbial and animal chitosan treated with 25% of glycerol achieved the best result of 41% of protein immobilization yield in films of chitosan. In another study with β-galactosidase immobilized on chitosan, 60% of protein immobilization yield on support of alginate-chitosan was performed [40].

However, the yield of immobilized protein is not an efficient parameter to evaluate the immobilization of enzyme. The best protocols for protein immobilization (A', B', and C') in 12 h reaction showed lower yield of immobilized active papain when compared with the same protocols in 5 h of reaction (Table 7). Probably the catalytic site of this enzyme could be inactivated by enzyme denaturation due to higher time of contact between papain and 1% TPP, since in the test of evaluation of TPP concentration on the activity of soluble papain there was a decrease in the enzyme activity from 2.61 to 1.43 U/mL with the increase of TPP from 0 to 1.0% (Table 5).

Protocol B' showed the best yield of active immobilized enzyme, but the yield of active immobilized protein was only 9.06% (Table 7). However, this result was higher than that found by Hong et al. [38] also working with papain immobilization on chitosan microspheres, cross-linked with 0.5% glutaraldehyde (0.19%).

Yeast Deflocculation Using Immobilized Papain on Chitin and Chitosan

The immobilized papain was tested in flocculated S. cerevisiae suspension from ethanol distillery in 120 min of reaction using 2 g of support (chitin or chitosan) in 50 mL yeast suspension. In the case of chitin protocol B was selected (Tables 1 and 7) and for chitosan protocol B' (Tables 2 and 7).

Yeast suspension treated with the immobilized enzymes showed no significant differences () when compared to the control (flocculated). After 120 minutes of reaction, both treatments with the immobilized enzyme showed only 15% of deflocculation, while the yeast treated with 4 g·L−1 soluble papain showed more than 70% of yeast cells deflocculation (Figure 2). These results were expected since there was a relatively low concentration of active enzyme in both supports, since for chitosan (protocol B') there was a yield of active immobilized enzyme of 15.7% and for chitin (protocol B) only 6.07% (Table 7).

Figure 2: Effect of the soluble and immobilized papain in the suspension of flocculated yeast from fuel ethanol distillery.

The ideal concentration of active papain in immobilized support for yeast deflocculation can be estimated. If 10 g·L−1 papain solution showed 2.04–2.68 U/mL (Table 5), there is 0.200–0.268 U/mg papain. If 4 g·L−1 of papain in 50 mL of yeast suspension was necessary to obtain 70% deflocculation of yeast cells, it means 40 U to 53.6 U or 46.8 U in average of total proteolytic activity was needed to defloculate 50 mL of yeast suspension. 2 g of papain immobilized on chitin or chitosan was used in yeast deflocculation tests, with, respectively, 0.11 U/g (protocol B) and 1.24 U/g (protocol B') (Table 7). Therefore, 0.22 U for chitin or 2.48 U for chitosan of total papain activity was used in 50 mL of 30% yeast suspension. These values were not enough for yeast deflocculation, since they represent only 0.47% (for chitin) and 5.3% (for chitosan) of the total protease activity spent in the deflocculation tests with soluble enzymes.

The yeast deflocculation may be possible if papain immobilized in these supports was packaged in columns of a plug flow reactor and the flocculated yeast suspension was carried through the reactor. Using this process there will be an increase in the percentage of active immobilized papain to yeast cells deflocculation, but this condition needs to be tested.

The purity of enzyme is another important point to improve the efficiency in yield of immobilization. The higher the purity of the enzyme obtained the better the efficiency in the enzyme immobilization. In this work only 3.6% of protein was determined in the papain used for immobilization tests. In addition, the improvement in the concentration of the active enzyme in derivatives for yeast cell deflocculation is also possible by other pretreatments of supports, the use of more resistant protease for cross-linking agent, and other techniques as multipoint covalent immobilization of enzymes [41].

Reuse of the Soluble Papain by Centrifugation

The flocculated S. cerevisiae suspension was reacted with a papain solution and SDS in 15 to 120 min (Figure3) of reaction time. The results were very interesting since yeast deflocculation was possible in a short time and at the temperature of the industrial process (25–30°C). 14 recycles of papain on yeast cells deflocculation was possible with this proposed method. In addition, just only 5% of new enzyme solution after each cycle of deflocculation was needed to add on the yeast cell suspension. Figure 3: Yeast cell deflocculation with the recycle of soluble papain by centrifugation of yeast suspension and enzyme recovery.

In the treatment using soluble papain (4 g·L^{-1}) and 0.10 g·L^{-1} of SDS for yeast deflocculation a slightly better performance of SDS was observed when compared to the treatment with soluble papain (4 g·L^{-1}). However, there was no significant difference () in these treatments, therefore not justifying the use of SDS in cell deflocculation by soluble papain (Figure 3). The use of low concentrations of SDS in this work is justified by the toxicity of this product to yeast metabolism in concentrations higher than 0.30 g·L^{-1}, as observed by de Oliva-Neto and Yokoya [10].

The acid treatment could be excessively abrasive on the yeast wall, which is essential for cell viability and production of ethanol, and according to Paterson et al. [42] the intensity of the acid treatment changes with the level of contamination and flocculation. On the other hand, the use of sulfuric acid as the agent for disinfecting the yeast cream and yeast deflocculation constitutes one of the most efficient and economic practices to the production of fuel ethanol. However, it should not be indiscriminately used as it currently is but carefully monitored during fermentation process [43].

Although some researchers advocate the use of sulfuric acid to decontaminate the alcoholic fermentation process, for others this practice is not efficient, raising the cost of ethanol due to the increase in the use of this product and damaging the environment since the acid vinasse is placed in the soil [44].

An alternative treatment with the use of enzymes has been proposed by Ludwig [13] with two enzymes, protease (Novozyme 642) and carbohydrase (Novozyme SP 299), which have been proved to be effective in yeast deflocculation and further they did not affect the yeast viability. However, these enzymes are not economically viable for industrial application due to the need for high dosages. Therefore, the development of technology responsible for cost reduction of enzymes in specific applications could contribute to become economically viable the use of these biocatalysts in the industry. Currently, the yeast cells are centrifuged one time in fed-batch or continuous process with cell recycles in Brazilian distilleries [2]. The concept of this purposed new method is to improve the industrial process of fuel ethanol technology with the removal of acid treatment of yeast cells or its reduction replacing the acid by protease treatment in a new double yeast centrifugation. Firstly, yeast suspension (after the first centrifugation) will be treated with soluble proteases for a certain time (15–120 min). After yeast deflocculation, the yeast cells will be separated by the second centrifugation and enzyme solution will be recovered to be used in another treatment with another yeast suspension. The proposed method was demonstrated to be effective at yeast deflocculation in 14 recycles of papain. However, a test on an industrial scale is necessary to prove the efficiency of this method. After an economic analysis, maybe a new alternative to the acid treatment of yeast will be possible, especially when the yeast viability is low and the acid treatment is not recommended.

CONCLUSION

The inhibitory effect of some cross-linking agents to papain activity was evaluated. The glutaraldehyde and polyethyleneimine in concentrations, respectively, 0.10% and 1%, strongly inhibited the activity of soluble papain. However, this enzyme was partly inhibited by tripolyphosphate (TPP) at the minimum concentration required for cross-linking with chitosan. The treatment of soluble papain in S. cerevisiae cells was efficient in yeast cell deflocculation in a short time, maintaining yeast cell viability, but only in a high dosage. This method could replace the current method used in industry with sulfuric acid, but the reuse of the enzyme is needed for an economical process on an industrial scale.

Several protocols of papain immobilization were performed. The protocols with papain on chitin and glutaraldehyde had a higher protein immobilization, but the great inhibition of papain by this agent was confirmed. Chitosan cross-linked with tripolyphosphate showed higher yield of active immobilized enzyme than chitin treated with polyethyleneimine. Although these immobilizations have been possible, these levels have not been enough to cause yeast cells deflocculation in a reaction by batch process; therefore higher active papain pellets are required. Recycling of free enzyme by centrifugation was effective for 14 cycles with yeast suspension in a time perfectly compatible to industrial conditions. The reuse of proteases applied in yeast suspension by additional yeast centrifugation could be a new alternative to sulfuric acid, especially when the yeast viability is low and the treatment with sulfuric acid is not recommended.

ACKNOWLEDGMENTS

The authors thank Fundação de Amparo à Pesquisa do Estado de São Paulo (FAPESP) and Conselho Nacional de Desenvolvimento Científico e Tecnológico (CNPq) for financial support.

REFERENCES

1. K. M. Ludwig, P. Oliva-Neto, and D. F. Angelis, "Quantification of Saccharomyces cerevisiae flocculation by bacterial contaminants of alcoholic fermentation," Ciência e Tecnologia Alimentos, vol. 21, no. 1, pp. 63–68, 2001.

2. P. de Oliva-Neto and F. Yokoya, "Evaluation of bacterial contamination in a fed-batch alcoholic fermentation process," World Journal of Microbiology and Biotechnology, vol. 10, no. 6, pp. 697–699, 1994.

3. P. Oliva-Neto, C. Dorta, A. F. A. Carvalho, V. M. G. Lima, and D. F. Silva, "The Brazilian technology of fuel ethanol fermentation—yeast inhibition factors and new perspectives to improve the technology," in Materials and Processes for Energy: Communicating Current Research and Technological Developments, A. Méndez-Vilas, Ed., vol. 1, pp. 371–379, Formatex, Badajoz, Spain, 1st edition, 2013.

4. I. Migneault, C. Dartiguenave, M. J. Bertrand, and K. C. Waldron, "Glutaraldehyde: behavior in aqueous solution, reaction with proteins,

and application to enzyme crosslinking," BioTechniques, vol. 37, no. 5, pp. 790–802, 2004.

5. P. Oliva Neto, K. M. Ludwig, C. Dorta, A. F. A. Carvalho, D. F. Silva, and V. M. G. Lima, "Microbial contamination of the alcholic fermentation for fuel ethanol production," in Bioenergy: Developing, Research and Inovation, N. Stradioto and E. Lemos, Eds., vol. 1, pp. 407–488, Cultura Acadêmica, São Paulo, Brazil, 2012.

6. M. T. Santos and F. Yokoya, "Characteristics of yeast cell flocculation by Lactobacillus fermentum,"Journal of Fermentation and Bioengineering, vol. 75, no. 2, pp. 151–154, 1993.

7. F. Yokoya and P. Oliva-Neto, "Characterization of yeast flocculation by Lactobacillus fermentum,"Revista de Microbiologia, vol. 22, pp. 12–16, 1991.

8. S. H. Hynes, D. M. Kjarsgaard, K. C. Thomas, and W. M. Ingledew, "Use of virginiamycin to control the growth of lactic acid bacteria during alcohol fermentation," Journal of Industrial Microbiology and Biotechnology, vol. 18, no. 4, pp. 284–291, 1997.

9. S. P. Meneghin, F. C. Reis, P. G. De Almeida, and S. R. Ceccato-Antonini, "Chlorine dioxide against bacteria and yeasts from the alcoholic fermentation," Brazilian Journal of Microbiology, vol. 39, no. 2, pp. 337–343, 2008.

10. P. de Oliva-Neto and F. Yokoya, "Effect of 3,4,4′-trichlorocarbanilide on growth of lactic acid bacteria contaminants in alcoholic fermentation," Bioresource Technology, vol. 63, no. 1, pp. 17–21, 1998.

11. C. Dorta, P. Oliva-Neto, M. S. de-Abreu-Neto, N. Nicolau-Junior, and A. I. Nagashima, "Synergism among lactic acid, sulfite, pH and ethanol in alcoholic fermentation of Saccharomyces cerevisiae (PE-2 and M-26)," World Journal of Microbiology and Biotechnology, vol. 22, no. 2, pp. 177–182, 2006.

12. M. Azarkan, A. El Moussaoui, D. van Wuytswinkel, G. Dehon, and Y. Looze, "Fractionation and purification of the enzymes stored in the latex of Carica papaya," Journal of Chromatography B: Analytical Technologies in the Biomedical and Life Sciences, vol. 790, no. 1-2, pp. 229–238, 2003.

13. K. M. Ludwig, Flocculation of Saccharomyces cerevisiae—characterization and the action of deflocculation enzymes [Master in Microbiology], Instituto de Biociências, Universidade Estadual Paulista, Rio Claro, Brazil, 1998.

14. S. Luque, J. M. Benito, and J. Coca, "The importance of specification

sheets for pressure-driven membrane processes," Filtration and Separation, vol. 41, no. 1, pp. 24–28, 2004. ·

15. P. de Oliva-Neto and P. T. P. Menão, "Isomaltulose production from sucrose by Protaminobacter rubrum immobilized in calcium alginate," Bioresource Technology, vol. 100, no. 18, pp. 4252–4256, 2009.

16. N. Kubota, N. Tatsumoto, T. Sano, and K. Toya, "A simple preparation of half N-acetylated chitosan highly soluble in water and aqueous organic solvents," Carbohydrate Research, vol. 324, no. 4, pp. 268–274, 2000.

17. A. S. Martins, D. N. Peixoto, L. M. C. Paiva, A. D. Panek, and C. L. A. Paiva, "A simple method for obtaining reusable reactors containing immobilized trehalase: characterization of a crude trehalase preparation immobilized on chitin particles," Enzyme and Microbial Technology, vol. 38, no. 3-4, pp. 486–492, 2006.

18. A. S. G. Lorenzoni, L. F. Aydos, M. P. Klein, M. A. Z. Ayub, R. C. Rodrigues, and P. F. Hertz, "Continuous production of fructooligosaccharides and invert sugar by chitosan immobilized enzymes: comparison between in fluidized and packed bed reactors," Journal of Molecular Catalysis B: Enzymatic, 2014.

19. H. Chen, Q. Zhang, Y. Dang, and G. Shu, "The effect of glutaraldehyde cross-linking on the enzyme activity of immobilized β-galactosidase on chitosan bead," Advance Journal of Food Science and Technology, vol. 5, no. 7, pp. 932–935, 2013.

20. G. Spagna, F. Andreani, E. Salatelli, D. Romagnoli, and P. G. Pifferi, "Immobilization of α-L-arabinofuranosidase on chitin and chitosan," Process Biochemistry, vol. 33, no. 1, pp. 57–62, 1998.

21. M. W. Anthonsen, K. M. Vårum, and O. Smidsrød, "Solution properties of chitosans: conformation and chain stiffness of chitosans with different degrees of N-acetylation," Carbohydrate Polymers, vol. 22, no. 3, pp. 193–201, 1993.

22. B. Krajewska, "Chitin and its derivatives as supports for immobilization of enzymes," Acta Biotechnologica, vol. 11, no. 3, pp. 269–277, 1991.

23. R. Laus, M. C. M. Laranjeira, A. O. Martins, et al., "Chitosan microspheres crosslinked with tripolyphosphate used for the removal of the acidity, iron (III) and manganese (II) in water contaminated in coal mining," Química Nova, vol. 29, no. 1, pp. 34–39, 2006.

24. M. Koudelka-Hep, N. F. Rooij, and D. J. Strike, "Immobilization of enzymes on microelectrodes using chemical crosslinking," in Immobilization of Enzimes and Cells, F. G. Bickerstaff, Ed., vol. 10,

pp. 83–85, Humana Press, Totowa, NJ, USA, 1997.

25. G. B. Broun, "Chemically aggregated enzymes," Methods in Enzymology, vol. 44, pp. 263–280, 1976. ·

26. R. Dalla-Vecchia, M. D. G. Nascimento, and V. Soldi, "Synthetic applications of immobilized lipases in polymers," Química Nova, vol. 27, no. 4, pp. 623–630, 2004. ·

27. T. J. Leighton, R. H. Doi, R. A. J. Warren, and R. A. Kelln, "The relationship of serine protease activity to RNA polymerase modification and sporulation in Bacillus subtilis," Journal of Molecular Biology, vol. 76, no. 1, pp. 103–122, 1973.

28. M. M. Bradford, "A rapid and sensitive method for the quantitation of microgram quantities of protein utilizing the principle of protein-dye binding," Analytical Biochemistry, vol. 72, no. 1-2, pp. 248–254, 1976.

29. S. Varavinit, N. Chaokasem, and S. Shobsngob, "Immobilization of a thermostable alpha-amylase,"Science Asia, vol. 28, pp. 247–251, 2002.

30. A. Johnstone and R. Thorpe, Immunochemistry in Practice, Blackwell Scientific, Oxford, UK, 1987.

31. F. C. Vasconcellos, G. A. S. Goulart, and M. M. Beppu, "Production and characterization of chitosan microparticles containing papain for controlled release applications," Powder Technology, vol. 205, no. 1–3, pp. 65–70, 2011.

32. R. S. Lima, G. S. Nunes, T. Noguer, and J.-L. Marty, "Enzymatic biosensor for the detection of dithiocarbamate fungicides. Kinetic study of aldehyde dehydrogenase enzyme and biosensor optimization," Química Nova, vol. 30, no. 1, pp. 9–17, 2007. ·

33. O. Zaborsky, Immobilized Enzymes, CRC Press (Chemical Rubber Co.), Cleveland, Ohio, USA, 1973.

34. W. K. Chui and L. S. C. Wan, "Prolonged retention of cross-linked trypsin in calcium alginate microspheres," Journal of Microencapsulation, vol. 14, no. 1, pp. 51–61, 1997.

35. Y. F. Li, F. Y. Jia, J. R. Li, G. Liu, and Y. Z. Li, "Papain immobilization on a nitrilon fibre carrier containing primary amine groups," Biotechnology and Applied Biochemistry, vol. 33, no. 1, pp. 29–34, 2001.

36. J. W. Gaspari, L. H. Gomes, and F. C. A. Tavares, "Immobilization of inulinase from Kluyveromyces marxianus for the hydrolysis of extracts of Helianthus tuberosus L," Scientia Agricola, vol. 56, pp. 1135–1140, 1999.

37. L. Qiuhua, M. Jingesu, G. Gangjun, C. Xu, and X. Hong, "Immobilization

of Papain on microcrystalline chitosan," Journal of Branch Campus of the First Military Medical, vol. 1, 1998.

38. L. Hong, W. W. Jun, and X. F. Cai, "Studies on preparation of chitosan microspheres and immobilization of papain," Journal of South China Agricultural University, vol. 2, 2000.

39. M. Zappino, I. Cacciotti, I. Benucci et al., "Bromelain immobilization on microbial and animal source chitosan films, plasticized with glycerol, for application in wine-like medium: microstructural, mechanical and catalytic characterisations," Food Hydrocolloids, vol. 45, pp. 41–47, 2015.

40. E. Taqieddin and M. Amiji, "Enzyme immobilization in novel alginate-chitosan core-shell microcapsules," Biomaterials, vol. 25, no. 10, pp. 1937–1945, 2004.

41. J. M. Guisán, A. Bastida, R. M. Blanco, R. Fernández-Lafuente, and E. García-Junceda, "Immobilization of enzymes on glyoxyl agarose," in Immobilization of Enzymes and Cells, G. F. Bickerstaff, Ed., pp. 277–287, Humana Press, Totowa, NJ, USA, 1997.

42. M. Paterson, J. M. M. Borba, F. A. D. Melo, and J. I. Moraes, "Evaluating the performance of ethanol fermentation in different situations of industrial process," Brasil Açucareiro, vol. 106, no. 516, pp. 27–32, 1988.

43. F. E. Alves da Silva, Ethanolic fermentation: influence of sulfuric acid on yeast viability and bacterial and yeast contaminants [M.S. thesis], Instituto de Biociências, Unesp, Rio Claro, Brazil, 1993.

44. M. H. Otênio, Comparative evaluation of the effect of removal of acid treatment with sulfuric acid in yeast during the recycles in the Bandeirantes Distillery (Brazil) [Master degree], Instituto de Biociências de Rio Claro, Universidade Estadual Paulista, Rio Claro, Brazil, 1998.

Chapter 13

ENZYME INHIBITION: MECHANISMS AND SCOPE

Rakesh Sharma[1,2,3]

[1]Center of Nanomagnetics Biotechnology, Florida State University, Tallahassee, FL

[2]Innovations and Solutions Inc. USA, Tallahassee, FL

[3]Amity University, NOIDA, UP [1,2]USA [3]India

INTRODUCTION

Enzyme is a protein molecule acting as catalyst in enzyme reaction. Enzyme inhibition is a science of enzyme-substrate reaction influenced by the presence of any organic chemical or inorganic metal or biosynthetic compound due to their covalent or non-covalent interactions with enzyme active site. It is well known that all these inhibitors follow same rule to interplay in enzyme reaction. Present chapter introduces beginners with basic tenets of classic presumptions of enzyme inhibition, types of enzyme inhibitors, different models of enzyme inhibition with established examples cited in literature, and scientific basis of emerging immobilized enzyme technology in different applications. In the end, limitations of using classic presumptions and variants of enzyme inhibition are highlighted with new challenges to achieve best results. Present time, best approach is 'customize new technology with detailed analysis to make it highly efficient' in both drug discovery and enzyme biosensor industry. However, other applications are described in following chapters on pesticides, herbicides.

WHAT ARE ENZYME INHIBITORS?

The enzyme inhibitors are low molecular weight chemical compounds. They can reduce or completely inhibit the enzyme catalytic activity either reversibly or permanently (irreversibly). Inhibitor can modify one amino acid, or several side chain(s) required in enzyme catalytic activity. To protect enzyme catalytic site from any change, ligand binds with critical side chain in enzyme. Safely, chemical modification can be done to test inhibitor for any drug value.

In drug discovery, several drug analogues are chosen and/or designed to inhibit specific enzymes. However, detoxification or reduced toxic effect of many antitoxins is also accomplished mainly due to their enzyme inhibitory action. Therefore, studying the aforementioned enzyme kinetics and structure-function relationship is vital to understand the kinetics of enzyme inhibition that in turn is fundamental to the modern design of pharmaceuticals in industries [Sami et al. 2011]. Enzyme inhibition kinetics behavior and inhibitor structure-function relationship with enzyme active site clarify the mechanisms of enzyme inhibition action and physiological regulation of metabolic enzymes as evidenced in following chapters in this book. Some notable classic examples are: drug and toxin action and/or drug design for therapeutic uses e.g., iodoacetamide deactivates cys amino acid in enzyme side chain; methotrexate in cancer chemotherapy through semi-selectively inhibit DNA synthesis of malignant cells; aspirin inhibits the synthesis of the proinflammatory prostaglandins; sulfa drugs inhibit the folic acid synthesis essential for growth of pathogenic bacteria and so many other drugs. Many life-threatening poisons, e.g., cyanide, carbon monoxide and polychlorinated biphenols are all enzyme inhibitors. Conceptually, enzyme inhibitors are classified into two types: non-specific inhibitors and specific inhibitors.

The enzyme inhibition reactions follow a set of rules as mentioned in following rules. Presently, computer based enzyme kinetics data analysis softwares are developed using following basic presumptions.

- Enzyme interacts with substrate in 1:1 ratio at active site to catalyze the reaction.
- Enzyme binds with substrate at active site in the form of a lock-key 3D arrangement for induced fit.
- Inhibitor active groups compete with substrate active groups and/or active groups at enzyme allosteric catalytic site in a synergistic manner or first cum first preference (competition) to make enzyme-inhibitor-substrate/enzyme-substrate/enzyme-inhibitor complexes.
- Enzyme-inhibitor-substrate complex formation depends on active free energy loss and thermodynamic principles.
- Enzyme and substrate or inhibitors react with each other as active masses and reaction progresses in kinetic manner of forward or backward reaction.
- Kinetic nature of inhibitor or substrate binding with enzyme is expressed as kinetic constants of a catalytic reaction.
- Enzyme reaction(s) are highly depend on physiological conditions

such as pH, temperature, concentration of reactants, reaction period to determine the rate of reaction.

• Substrate and inhibitor molecules arrange over enzyme active site on specific sub unit(s) in 3D manner. As a result enzyme-substrate-inhibitor exhibit binding rates depend on allosteric sites or subunit-subunit homotropic or heterotropic interactions.

• Intermolecular forces between enzyme subunits, substrate or inhibitor active group interactions, physical properties of binding nature: electrophilic, hydrophilic, nucleophilic and metalloprotein nature; hydrogen bonding affect the overall enzyme reaction rates and mode of inhibition (3D orientation of inhibitor molecule on enzyme active site)

Other factors are also significant in determining enzyme inhibition reaction as described in each individual inhibitor in following sections. For basic principles of enzyme units (apoenzyme, holoenzyme, co-factor, co-enzyme) in enzyme catalysis, active energy loss, Michaelis-Menton Equations, LeChatelier's principle, Lineweaber-Burk and semi-log plots, apparent and actual plots, readers are requested to read text books [Schnell et al. 2003, Nelson, et al. 2008, Jakobowski 2010a, Strayer et al. 2011]. Our focus is enzyme inhibition mechanisms with examples in following description. For multisubstrate enzymes, pingpong mechanism, allosteric mechanisms, and diffusion kinetics, readers are requested to read original papers [Pryciak 2008, Bashor 2008, Jakobowski 2010b]

These inhibitors may act in reversible or irreversible manner. Non-specific irreversible noncompetitive inhibitors include all protein denaturing factors (physical and chemical denaturation factors). The specific inhibitors attack a specific component of the holoenzyme system. The action depends on increased amount of substrate or by other means of physiological conditions, toxins. Specific inhibitors can be described in several forms including; 1) coenzyme inhibitors: e.g., cyanide, hydrazine and hydroxylamine that inhibit pyridoxal phosphate, and, dicumarol that is a competitive antagonist for vitamin K; 2) inhibitors of specific ion cofactor: e.g., fluoride that chelates Mg^{2+} of enolase enzyme; 3) prosthetic group inhibitors: e.g., cyanide that inhibits the heme prosthetic group of cytochrome oxidase; and, 4) apoenzyme inhibitors that attack the apoenzyme component of the holoenzyme; 5) physiological modulators of reaction pH and temperature that denature the enzyme catalytic site.

The apoenzyme inhibitors are of two types; i) Reversible inhibitors; their inhibitory action is reversible because they make reversible association with the enzyme, and, ii) Irreversible inhibitors; because they make inactivating irreversible covalent modification of an essential residue of the enzyme.

Apoenzyme inhibitors show effect on Km and Vmax. The reversible apoenzyme inhibitors are also called metabolic antagonists. They are of three subtypes; a) competitive, b) uncompetitive and c) non-competitive or mixed type. For example: enzyme inhibitors are used in drug design.

Discovery of useful new enzyme inhibitors used to be done by trial and error through screening a huge library of compounds against a target enzyme at allosteric catalytic site. This approach is still in use for compounds with combinatorial chemistry and highthroughput screening technology as described in following description based on recent concepts [El-Metwally et al. 2010]. However, rational drug design as an alternative approach uses the three-dimensional structure of an enzyme's active site or transition-state conformation to predict which molecules might be ideal inhibitors as given an example of urease in chapter 11 in this book. 3D-structure shortens the long screening list towards a right set of novel inhibitor which kinetically characterizes and allows specific structural changes in amino acids of catalytic site chain to optimize inhibitor-enzyme binding.

Alternatively, molecular docking and molecular mechanics are computer-based methods that predict the affinity of an inhibitor for an enzyme. In following description, a glimpse of these mechanisms is given on different types of inhibitors based on recent classic book [ElMetwally et al. 2010]. Readers are requested to read other classic details from advanced text books [Dixon and Webb, 1979].

IRREVERSIBLE INHIBITION

The irreversible apoenzyme inhibitors have no structural relationship to the substrate and bind covalently. They also bind stable non-covalently with the active site of the enzyme or destroy an essential functional group of active site. So, irreversible inhibitors are used to identify functional groups of the enzyme active sites at which location they bind. Although inhibitors have limited therapeutic applications because they are usually act as poisons. A subset of irreversible inhibitors called suicide irreversible inhibitors, are relatively inactive compounds. They get activated upon binding with the active site of a specific enzyme. After such binding, the suicide irreversible inhibitor is activated by the first few intermediary steps of the biochemical reaction - like the normal substrate. However, it does not release any product because of its irreversible binding at the enzyme active site. Inhibitors make use of the normal enzyme reaction mechanism to get activated and subsequently inactivate the enzyme. Due to this very nature, suicide irreversible inhibitors are also called mechanismbased inactivators or transition state analog inhibitors.

Thus, inhibitor exploits the transition state stabilizing effect of the enzyme, resulting in a better binding affinity (lower Ki) than substrate-based designs. An example of such a transition state inhibitor is active form of the antiviral drug oseltamivir (Tamiflu; see Figure 1); this drug mimics the planar nature of the ring oxonium ion in the reaction of the viral enzyme neuraminidase [El-Metwally et al. 2010]. After drug activation in the liver, the drug replaces sialic acid as the normal substrate found on the surface proteins of normal host cells. It prevents the release of new viral particles from infected cells. It has been used to treat and prevent Influenza virus A and Influenza virus B infections. Most of such inhibitors are classified as tight-binding competitive inhibitors in other references of enzymes. However, their reaction kinetics is essentially irreversible.

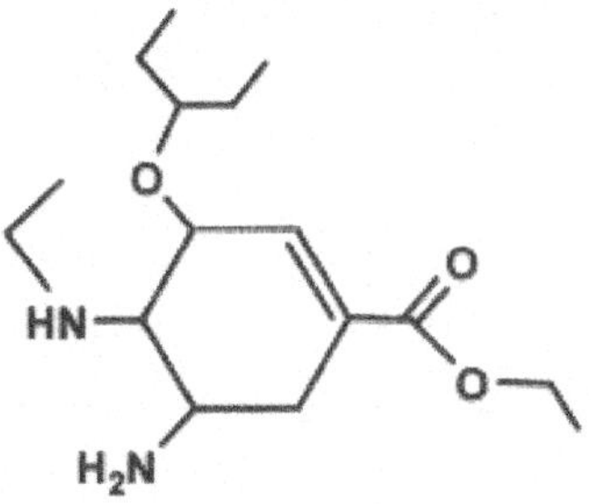

Figure 1: The transition state analog oseltamivir - the viral neuraminidase inhibitor.

The present art of drug discovery and design of new drugs is based on suicidal irreversible inhibitors. Chemicals are synthesized based on knowledge of 3D conformation of substrateactive site binding at specific binding rates in presence of co-factors, co-enzyme (enzyme reaction mechanisms) to inhibit at specific enzyme active site with minimal side-effects due to its non-specific binding nature. Transition state analogs are extremely potent and specific inhibitors of enzymes because they have higher affinity and stronger binding to the active site of the target enzyme than the natural substrates or products. However, exact design of drugs that precisely mimic the transition state is a challenge because of unstable structure of transition state in the free-state. Prodrugs undergo initial reaction(s) to form an overall electrostatic and three-dimensional intermediate transition state complex form with close similarity to that of the substrate. These prodrugs serve as guideline for drug development to form transition state suitable for stable modification; or, using the transition state analog to design a complementary catalytic antibody; called Abzyme. Example: Abzymes are used in catalytic antibodies and ribozymes in catalytic ribosomes [El-Metwally et al. 2010].

- Abzymes are antibodies generated against analogs of the transition state complex of a specific chemical. The arrangement of amino acid side chains at the abzyme variable regions is similar to the active site of the enzyme in the transition state and work as artificial enzymes. For example, an abzyme was developed against analogs of the transition state complex of cocaine esterase, the enzyme that degrades cocaine in the body [El-Metwally et al. 2010]. Thus, this abzyme has similar esterase activity that is

- used as injection drug to rapidly destroy cocaine in the blood of addicted individuals to decreasing their dependence on it.

- Thrombin inhibition is common in saliva of leeches and other blood-sucking organisms. They contain the anticoagulant hirudin that irreversibly inhibits thrombin, and, to regain thrombin action synthesis of new thrombin molecules is required. This made it unsafe as an anticoagulation drug. However, based on hirudin structure, rational drug design synthesized 20-amino acids peptide known as bivalirudin that is safe for longterm use because of its reversible effects on thrombin; despite its high binding affinity and specificity for thrombin.

- Ornithine decarboxylase by difluoromethylornithine is used to treat African trypanosomiasis (sleeping sickness). The enzyme initially decarboxylates difluoromethylornithine instead of ornithine and releases a fluorine atom, leaving the rest of the molecule as a highly electrophilic conjugated imine. The later reacts with either a cysteine or lysine residue in the active site to irreversibly inactivate the enzyme.

- Inhibition of thymidylate synthase by fluoro-dUMP. Imidazole antimycotic drugs are examples of such group that inhibit several subtypes of cytochrome P450 [Sharma, 1990]. The mechanisms of toxicities and antidotes of irreversible inhibitors are of medical pathological importance. Because of the irreversible inactivation of the enzyme, irreversible inhibition is of long duration in the biological system because reversal of their action requires synthesis of new enzyme molecules at the enzyme genetranscription-translation level.

- Inhibition of acetylcholine esterase (ACE) by diisopropylfluorophosphate (DPFP), the ancestor of current organophosphorus nerve gases (e.g., Sarin and Tabun) and other organophosphorus toxins (e.g., the insecticides Malathion and Parathion and chlorpyrifos). ACE hydrolyzes the acetylcholine into acetate and choline to terminate the transmission of the neural signal form the neuromuscular excitatory acetylcholine presynaptic cell to somatic neuromuscular junction (see Figure 2). DPFP as a potent neurotoxin inhibits ACE and acetylcholine hydrolysis.

Failure of hydrolysis leads to persistent acetylcholine excitatory state and improper vital function particularly respiratory muscles that may lead to suffocation; with a lethal dose of less than 100 mg. DPFP inhibits other enzymes with the reactive serine residue at the active site, e.g., serine proteases such as trypsin and chymotrypsin, but the inhibition is not as lethal as that of acetylcholine esterase. Similar to DPFP, malaoxon the toxic reactive derivative from Malathion (after its metabolism by the liver) binds initially reversibly and then irreversibly (after dealkylation of the inhibitor) to the active site serine and inactivates ACE and other enzymes. Lethal doses of oral Malathion are estimated at 1 g/kg of body weight for humans.

- Inhibition of ACE by these poisons leads to accumulation of acetylcholine that overstimulates the autonomic nervous system (including heart, blood vessels, and glands), thereby accounting for the poisoning symptoms of vomiting, abdominal cramps, nausea, salivation, and sweating. Acetylcholine is also a neurotransmitter for the somatic motor nervous system, where its accumulation resulted in poisoning symptom of involuntary muscle twitching (muscle fasciculation), convulsions, respiratory failure and coma. Intoxication of Malathion is treated by the antidote drug Oxime that reactivates the acetylcholine esterase and by intravenous injection of the anticholinergic (antimuscarinic) drug atropine to antagonize the action of the excessive amounts of acetylcholine [El-Metwally et al. 2010].

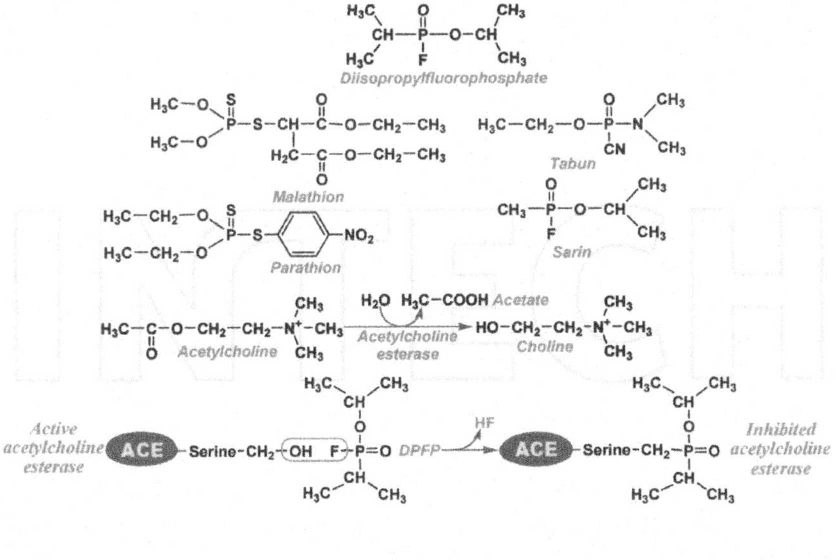

Figure 2:. Organophosphorus compounds and the suicidal irreversible mechanism-based inhibition of the enzyme acetylcholine esterase by diisopropylfluorophosphate. Malathion and parathion are organophosphorus insecticides. The nerve gases Tabun and Sarin are other organophosphorus compounds.

Another example of irreversible inhibition is iodoacetate inhibition of the glycolytic glyceraldehyde-3-phosphate dehydrogenase (GPD). Iodoacetate is a sulfhydryl compound that covalently alkylates and blocks the sulfhydryl group at the active site of the enzyme. Iodoacetate also inhibits other enzymes with -SH at the active site (Figure 3).

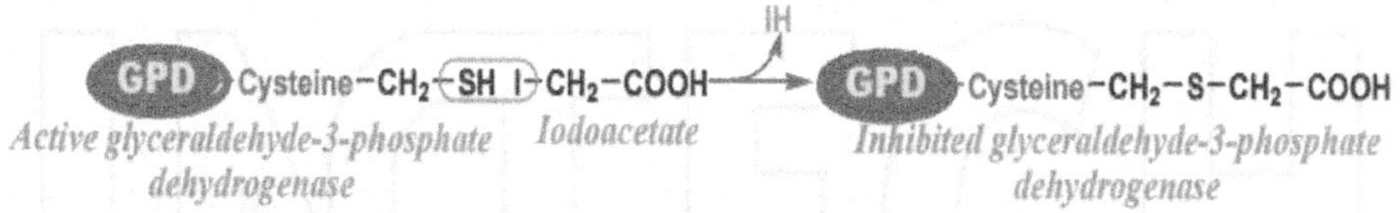

Figure 3: The suicidal irreversible mechanism-based inhibition of the enzyme glyceraldehyde- 3-phosphate dehydrogenase by iodoacetate.

- Allopurinol - the anti-gout drug - is a suicidal irreversible mechanism-based inhibitor of the enzyme xanthine oxidase that works as oxidase or dehydrogenase. The enzyme commits suicide by initial activating allopurinol into a transition state analog - oxypurinol - that bind very tightly to molybdenum-sulfide (Mo-S) complex at the active site (Figure 4). This enzyme accounts for the human dietary requirement for the trace mineral molybdenum. The molybdenum-sulfide (Mo-S) complex binds the substrates and transfers the electrons required for the oxidation reactions.

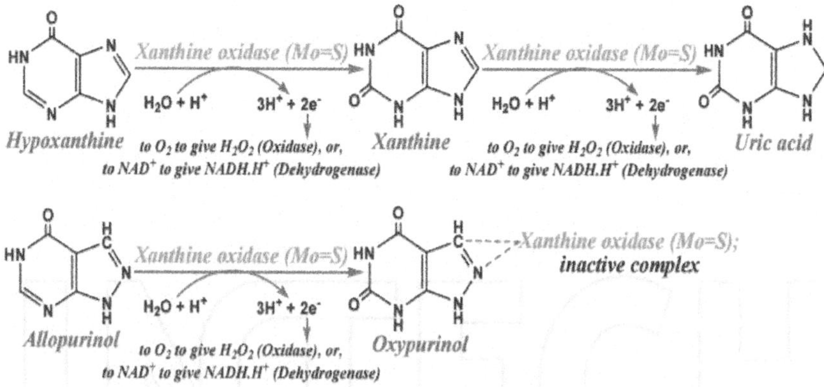

Figure 4: The suicidal irreversible mechanism-based inhibition of the enzyme xanthine oxidase by allopurinol.

- Guanosine analogue antiviral drug aciclovir - acycloguanosine (2-amino-9-((2- hydroxyethoxy)methyl)-1H-purin-6(9H)-one), as one of the most commonly-used antiviral drugs, it is primarily used for the treatment of herpes simplex and herpes zoster (shingles) viral infections. Aciclovir (see Figure 5) started a new era in antiviral therapy, as it is extremely selective and low in cytotoxicity. Aciclovir as a prodrug differs from previous nucleoside analogues in that it contains only a partial nucleoside structure: the sugar ring is replaced by an open-chain structure. It is selectively converted into acyclo-guanosine monophosphate (acyclo-GMP) by viral thymidine kinase, which is far more effective (3000 times) in phosphorylation than cellular thymidine kinase. Subsequently, the monophosphate form is further phosphorylated into the active triphosphate form, acyclo-guanosine triphosphate (acyclo-GTP), by cellular kinases. Acyclo-GTP is a very potent inhibitor of viral DNA polymerase; it has approximately 100 times greater affinity for viral than cellular polymerase. As a substrate, acyclo-GTP is incorporated into viral DNA, resulting in chain termination. Acyclo-GTP is fairly rapidly metabolized within the cell, possibly by cellular phosphatases.

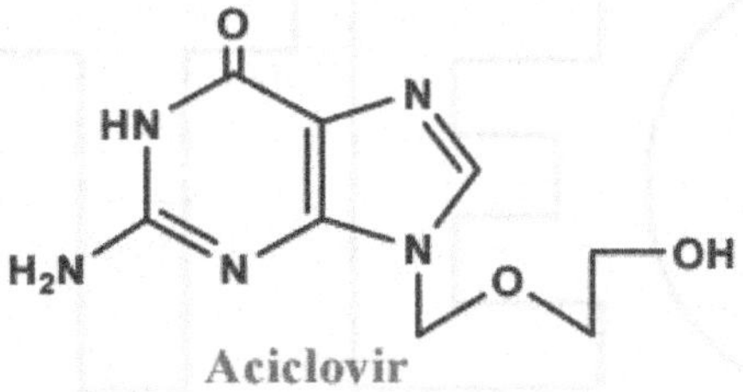

Aciclovir

Figure 5: Aciclovir; the prodrug for the suicidal irreversible inhibition of the viral DNA polymerase.

- The antibiotic penicillin is another transition state analog suicidal inhibitor that binds irreversibly covalently to serine at the active site of the bacterial enzyme glycopeptide transpeptidase. The enzyme is a serine protease required for synthesis of the bacterial cell wall and is essential for bacterial growth and survival. It normally cleaves the peptide bond between two D-alanine residues in a polypeptide. Penicillin structure contains a strained peptide bond within the β-lactam ring that resembles the transition state of the normal cleavage reaction, and thus penicillin binds very readily to the enzyme active site. The partial reaction to cleave the imitating penicillin peptide bond activates penicillin to bind irreversibly covalently to the active site serine (Figure 6).

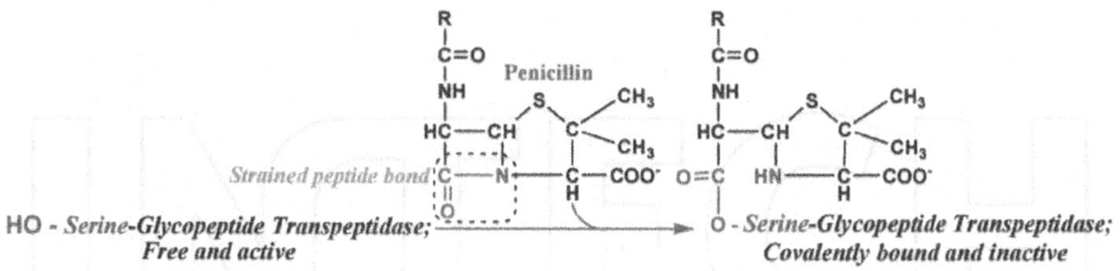

Figure 6: The suicidal irreversible mechanism-based inhibition of the bacterial enzyme glycopeptide transpeptidase by the antibiotic penicillin.

- Aspirin (acetylsalicylic acid) provides an example of a pharmacologic drug that exerts its effect through the covalent acetylation of an active site serine in the enzyme cyclooxygenase (prostaglandin endoperoxide synthase). Aspirin resembles a portion of the prostaglandin precursor that is a physiologic substrate for the enzyme.

- Heavy metal toxicity is caused by tight binding of a metal such as mercury, lead, aluminum, or iron, to a functional group at the active site of an enzyme. At high concentration of the toxin, heavy metals are relatively nonspecific for the enzymes they inhibit and inhibit a large number of enzymes. For example, it is impossible to specify which particular enzyme is implicated in mercury toxicity that binds reactive -SH groups at the active sites. Lead developmental and neurologic toxicity is caused by its ability to replace the normal functional metal in target enzymes; particularly Ca^{2+} in important enzymes, e.g., Ca^{2+}-calmodulin and protein kinase C. Because of their irreversible effect, heavy metals are routinely use as fixatives in histological preparations.

Kinetically, the irreversible inhibitors decrease the concentration of active enzyme and in turn decrease the maximum possible concentration of ES complex with ultimate reduction in the reaction rate of the inactivated individual enzyme molecules. The remaining unmodified enzyme molecules are normally functional considering their turnover number and K_m. For example: Natural poisons act as Enzyme inhibitors and Inhibitory enzymes In nature, animals and plants are rich in poisons as secondary metabolites, peptides and proteins that can act as enzyme inhibitors. Natural toxins are small organic molecules and act as natural inhibitors for enzymes in metabolic pathways and non-catalytic proteins.

- Neurotoxins are natural inhibitors, toxic but valuable for therapeutic uses at lower doses. For example, glycoalkaloids from Solanaceae family plants (potato, tomato and eggplant) act as acetylcholinesterase inhibitors to increase the acetylcholine neurotransmitter, muscular paralysis and then death. Many natural toxins are secondary metabolites. These neurotoxins also include peptides and proteins. An example of a toxic peptide is alpha-amanitin, found in death cap mushroom and acts potent enzyme inhibitor, in this case preventing the RNA polymerase II enzyme from transcribing DNA. The algal toxin microcystin is also a peptide and is an inhibitor of protein phosphatases. This toxin can contaminate water supplies after algal blooms and is a known carcinogen that can also cause acute liver hemorrhage and death at higher doses. Proteins can also be natural poisons or antinutrients, such as the trypsin inhibitors that are found in some legumes, potato, and tomato. Several invertebrate and vertebrate venoms contain protein and peptide enzyme inhibitors for, e.g., plasmin, renin and angiotensin converting enzymes. Inhibitory enzymes are enzymes that irreversibly inhibit other enzymes by chemically modifying them. In the broad sense, they include all proteases and lysosomal enzymes. Some of them are toxic plant products, e.g., ricin, a glycosidase that is an extremely potent protein toxin found in castor oil beans. It inactivates ribosomes by cleavage the eukaryotic 28S rRNA and reduces protein synthesis and a single molecule of ricin is enough to kill a cell.

REVERSIBLE INHIBITION

Reversible inhibitors may be competitive, noncompetitive, or uncompetitive inhibitors relative to a particular substrate. Products of enzymatic reactions are reversible inhibitors of the enzymes. A decrease in the rate of an enzyme caused by the accumulation of its own product plays an important role in the balance and most economic usage of metabolic pathways. It prevents one enzyme in a sequence of reactions from generating a new product more than the capacity of the next enzyme in that sequence, e.g., inhibition of hexokinase by accumulating glucose 6-phosphate. With the reduction in the inhibitor concentration, the enzyme activity is regenerated due to the non-covalent association and the reversible equilibrium with the enzyme. The equilibrium constant for the dissociation of enzyme inhibitor complexes is known as K_i that equals $[E][I]/[EI]$ [Cheng et al. 1973]. The inhibition effect of K_i on the reaction kinetics is reflected on the normal K_m and or Vmax observed in Lineweaver-Burk plots; in a pattern dependent on the type of the inhibitor [Nelson et al. 2008]. The inhibitor is removable by several ways. The three common types of reversible inhibitions are:

- Competitive reversible inhibition.
- Uncompetitive reversible inhibition.
- Mixed reversible inhibition (or non-competitive inhibition).

Competitive reversible inhibition

The competitive inhibitor is structurally related to the substrate and binds reversibly at the active site of enzyme and occupies it in a mutually exclusive manner with the substrate. Therefore, the competitive inhibitor competes with the substrate for the active site. The binding is mutually exclusive because of their free competition. According to the law of mass action, relatively higher inhibitor concentration prevents the substrate binding. Since the reaction rate is directly proportional to [ES], reduction in ES formation for EI formation lowers the rate. Increasing substrate towards a saturating concentration alleviates competitive inhibition. In the time enzyme-substrate complex releases the free enzyme and a product, the enzyme-inhibitor complex does release neither free enzyme nor a product.

Reversible inhibition is of short duration in the biological system because it depends on substrate availability and/or rate of the catabolic clearance of the inhibitor (Figure 7).

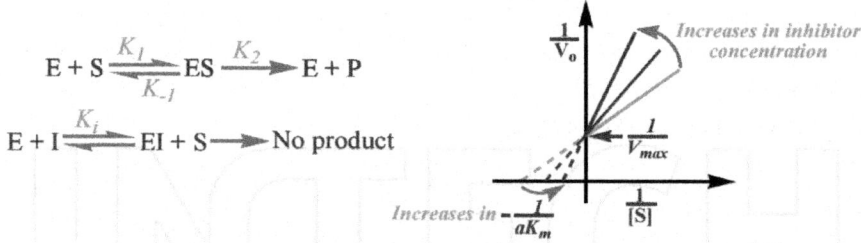

Figure 7: The equation and the effect of the competitive inhibitor on the double reciprocal plot of the substrate-reaction rate relationship.

Kinetically, the inhibitor (I) binds the free enzyme reversibly to form enzyme inhibitor complex (EI) that is catalytically inactive and cannot bind the substrate. The competitive inhibitor reduces the availability of free enzyme for the substrate binding. Thus, the K_m of the normal reaction is increased to a new K_m (aK_m) as a function of the inhibitor concentration (expressed in the "a" factor - apparent K_m in presence of the inhibitors), where the substrate concentration at $Vo = \frac{1}{2} V_{max}$ is equal to aKm. The "a" can be calculated from the change in the slope of the line at a given inhibitor concentration;

$$a = 1 + \frac{[I]}{K_I}, \text{where}, K_I = \frac{[E][I]}{[EI]}$$

(1)

Therefore, competitive inhibitors do not affect the turnover number (active site catalysis per unit time) or the efficiency of the enzyme because once enzyme is free, enzyme behaves normally. The Michaelis-Menten equation for competitive inhibitors becomes

$$V_o = \frac{V_{max}[S]}{aK_m + [S]}$$

(2)

Consequently, the double reciprocal form of the equation is also modified so as the line slope becomes $\dfrac{aK_m}{V_{max}}$ and the intercept with y-Axis stays at $\dfrac{1}{Vmax}$ but the intercept with the x-axis at $\dfrac{-1}{aK_m}$ will differ according to the concentration of the competitive inhibitor. The later property is characteristic for competitive inhibitors.

Examples include the classical competitive inhibitory effect of malonic acid on succinate dehydrogenase (SD) of the Krebs' cycle that reversibly dehydrogenates succinate into fumarate. Other less potent competitive inhibitors of succinate dehydrogenase include; oxalate, glutamate and oxaloacetate. The common molecular geometric feature of these compounds is the presence of two negatively charged -COOH groups suggesting that the active site of the flavoprotein SD has specifically positioned two positively charged binding groups (Figure 8).

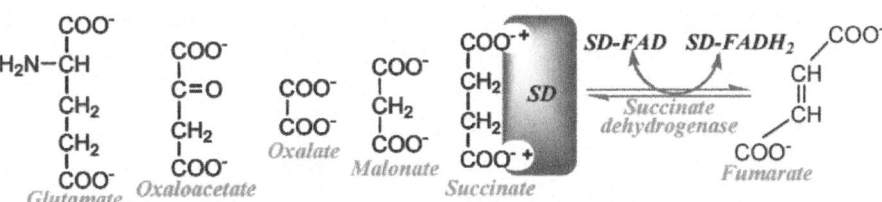

Figure 8: The substrate and different competitive inhibitors of succinate dehydroge-nase (SD).

Methotrexate - competitive inhibitor of dihydrofolate reductase (DHFR) is another example. The drug is used as anticancer antimetabolite chemotherapy particularly for pediatric leukemia. It hinders the availability of tetrahydrofolate as a carrier for one-carbon moieties important for anabolic pathways -particularly synthesis of purine nucleotides for DNA replication (Figure 9).

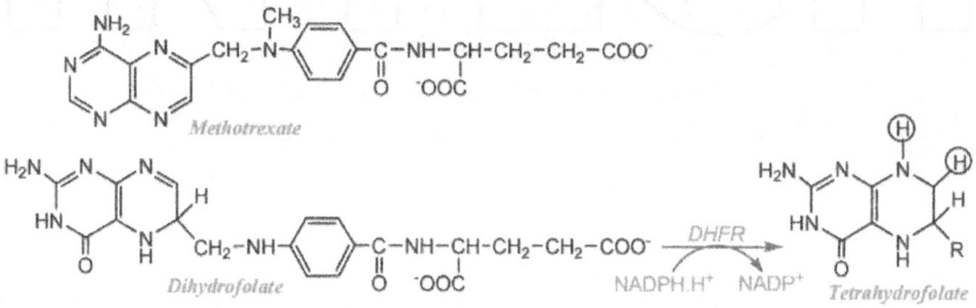

Figure 9: The substrate and methotrexate as a competitive inhibitor for dihydrofolate reductase.

Sulfanilamides - the simplest form of Sulfa drugs - were among earliest antibacterial chemotherapeutic drugs classified as enzyme inhibitors. They are competitive inhibitors of the bacterial folic acid synthesizing enzyme system from p-aminobenzoic acid. Bacterial cannot absorb pre-made folate that is necessary to be synthesized de novo. Structural similarity of sulfanilamide (and other sulfas derived from it) to p-aminobenzoic acid made them competitive inhibitors to the enzyme (Figure 10).

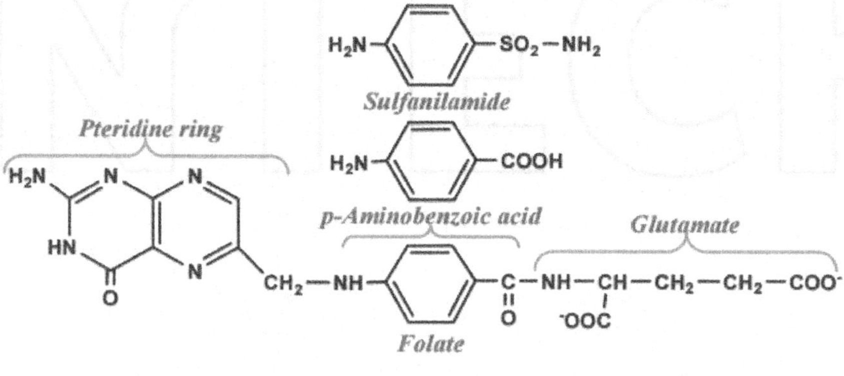

Figure 10: The p-aminobenzoic acid substrate and sulfanilamide as a competitive inhibitor during the bacterial folate synthesis.

Male erectile impotence was a major medical problem. Now a group of chemicals with molecular structural similarity to cGMP is promising that competitively inhibit the cGMPphosphodiesterase-5. They include sildenafil citrate (Viagra; Figure 11), vardenafil (Levitra) and tadalafil (Cialis). The inhibition of this enzyme that has a limited tissue distribution including the penile cavernous tissue spares cGMP. Accumulation of cGMP leads to smooth muscle relaxation (vasodilation) of the intimal cushions of the helicine arteries, resulting in increased inflow of blood and an erection.

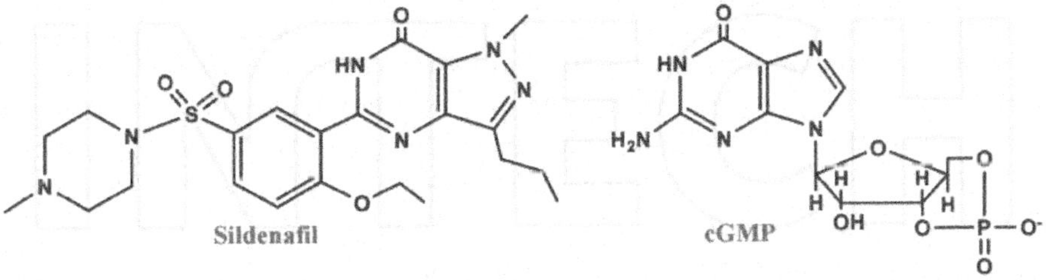

Figure 11: The cGMP substrate and sildenafil a competitive inhibitor of the cGMP-phosphodiesterase-5.

Another example of these substrate mimics competitive inhibitors are the peptide-based protease inhibitors, a very successful class of antiretroviral drugs used to treat HIV, e.g., ritonavir that contains three peptide bonds (see Figure 12).

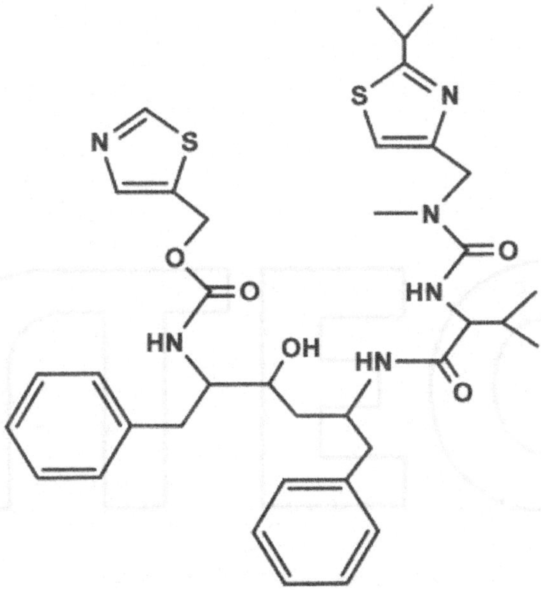

Figure 12: The peptide-based competitive protease inhibitor ritonavir.

Reversible competitive inhibitors of acetylcholinesterase, such as edrophonium, physostigmine, and neostigmine, are used in the treatment of myasthenia gravis and in anesthesia. The carbamate pesticides are also examples of reversible acetylcholinesterase inhibitors.

Uncompetitive Reversible Inhibition

Uncompetitive inhibitor has no structural similarity to the substrate. It may bind the free enzyme or enzyme substrate complex that exposes the inhibitor binding site (ESI). Its binding, although away from the active site, causes structural distortion of the active and allosteric sites of the complexed enzyme that inactivates the catalysis. This leads to a decrease in both K_m and V_{max}. Increasing substrate towards a saturating concentration does not reverse this type of inhibition and reversal requires special treatment, e.g., dialysis. This type of inhibition is also encountered in multi-substrate enzymes, where the inhibitor competes with one substrate (S2) to which it has some structural similarity and is uncompetitive for the other (S1). The reaction without the inhibitor would be; $E + S_1 Û ES_1 + S2 Û ES_1S_2 Þ E + Ps$ and with uncompetitive inhibitor becomes; $E + S_1 Û ES_1 + I Þ ES_1I$ (prevents S2 binding) Þ no product. It is a rare type and the inhibitor may be the reaction product or a product analog.

Kinetically, uncompetitive inhibition modifies the Michaelis-Menten equation by (a') factor that proportionates with the inhibitor concentration to be:

$$V_o = \frac{V_{max}[S]}{K_m + a'[S]}$$

(3)

and in the double-reciprocal equation to be:

$$\frac{1}{V_o} = \frac{a'}{V_{max}} + \frac{K_m}{V_{max}} X \frac{1}{[S]}$$

(4)

while y-intercept is at $\frac{a'}{V_{max}}$ and x-intercept is at $-\frac{a'}{K_m}$, whereas, the line slope stays $\frac{K_m}{V_{max}}$. This gives a number of lines in the Lineweaver-Burk plot that are parallel to the normal line with decreased $1/V_{max}$ and $-a'/K_m$ proportional to concentrations of the uncompetitive inhibitor. The later is characteristic to uncompetitive inhibition (Figure 13).

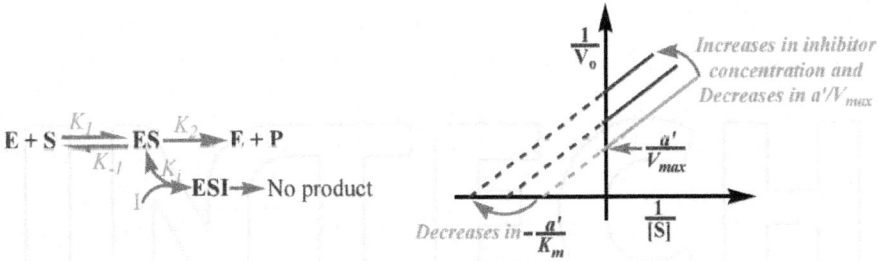

Figure 13: The equation and the effect of the uncompetitive inhibitor on the double reciprocal plot of the substrate-reaction rate relationship

Uncompetitive reversible inhibition is rare, but may occur in multimeric enzymes. Examples of uncompetitive reversible inhibitors include; inhibition of lactate dehydrogenase by oxalate; inhibition of alkaline phosphatase (EC 3.1.3.1) by L-phenylalanine, and, inhibition of the key regulatory heme synthetic enzyme; δ-aminolevulinate synthase and dehydratase and heme synthetase by heavy metal ion, e.g., lead. Heavy metals, e.g., lead, form mercaptides with -SH at the active site of the enzyme (2 R-SH + Pb Þ R-S-Pb-S-R + 2H).

Oxidizing agents, e.g., ferricyanide also oxidizes -SH into a disulfide linkage (2 R-SH Þ R-SS-R). Reversion here requires treatment with reducing agents and/or dialysis.

Mixed (noncompetitive) inhibition

The mixed type inhibitor does not have structural similarity to the substrate but it binds both of the free enzyme and the enzyme-substrate complex. Thus, its binding manner is not mutually exclusive with the substrate and the presence of a substrate has no influence on the ability of a non-competitive inhibitor to bind an enzyme and vice versa. However, its binding - although away from the active site - alters the conformation of the enzyme and reduces its catalytic activity due to changes in the nature of the catalytic groups at the active site. EI and ESI complexes are nonproductive and increasing substrate to a saturating concentration does not reverse the inhibition leading to unaltered K_m but reduced V_{max}. Reversal of the inhibition requires a special treatment, e.g., dialysis or pH adjustment. Some classifications differentiate between non-competitive inhibition as defined above and mixed inhibition in that the EIS-complex has residual enzymatic activity in the mixed inhibition. Kinetically, mixed type inhibition causes changes in the Michaelis-Menten equation so as

$$V_o = \frac{V_{max}[S]}{aK_m + a'[S]}$$

(5)

Mixed type inhibition - as the name imply - has a change in the denominator with Km modified by factor (a) as in competitive inhibition, and [S] modified by factor (a') as in uncompetitive inhibition. In the double reciprocal equation 6,

$$\frac{1}{V_o} = \frac{a'}{V_{max}} + \frac{aK_m}{V_{max}} X \frac{1}{[S]}$$

(6)

A line slope is $\frac{aK_m}{V_{max}}$, and the intercept with y-axis is at $\frac{a'}{V_{max}}$ and with x-axis is at $\frac{a'}{aK_m}$. This results in progressive decreases in V_{max} and progressive increases in K_m proportional to the increase in the mixed inhibitor concentration. The double reciprocal plot shows a number of lines reflecting decreases in V_{max}/increases in Km but their intercept is to the left of the yaxis. Mixed type inhibitor would be called non-competitive only if [a = a'], where, it will only lower V_{max} without affecting the K_m (Figure 14).

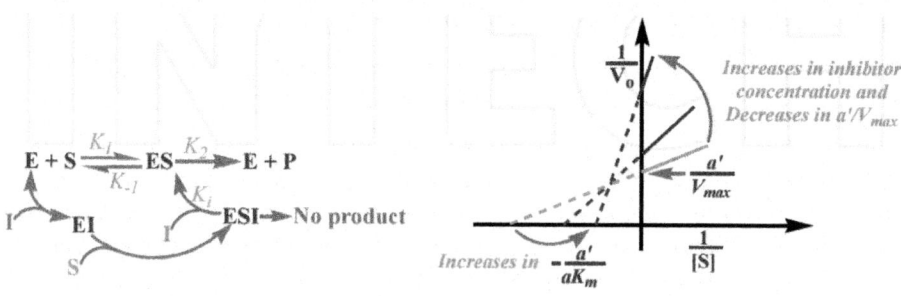

Figure 14: The equation and the effect of the mixed type (noncompetitive) inhibitor on the double reciprocal plot of substrate-reaction rate relationship.

Examples of noncompetitive inhibitors are mostly poisons because of the crucial role of the targeted enzymes. Cyanide and azide inhibits enzymes with iron or copper as a component of the active site or the prosthetic group, e.g., cytochrome c oxidase (EC 1.9.3.1). They include the inhibition of an enzyme by hydrogen ion at the acidic side and by the hydroxyl ion at the alkaline side of its optimum pH. They also include inhibition of; carbonic anhydrase by acetazolamide; cyclooxygenase by aspirin; and, fructose-1,6- diphosphatase by AMP. Cyanide binds to the $Fe3+$ in the heme of the cytochrome aa3 component of cytochrome c oxidase and prevents electron transport to $O2$. Mitochondrial respiration and energy production cease, and cell death rapidly occurs. The central nervous system is the primary target for cyanide toxicity. Acute inhalation of high concentrations of cyanide (e.g., smoke inhalation during a fire and automobile exhaust) provokes a brief central nervous system stimulation rapidly followed by convulsion, coma, and death. Acute exposure to lower amounts can cause lightheadedness, breathlessness, dizziness, numbness, and headaches. Cyanide is present in the air as hydrogen cyanide (HCN), in soil and water as cyanide salts (e.g., NaCN), and in foods as cyanoglycosides. Comparison of the three types of the reversible enzyme inhibitors is presented in Table 1.

In a special case, the mechanism of partially competitive inhibition is similar to that of noncompetitive, except that the EIS complex has catalytic activity, which may be lower or even higher (partially competitive activation) than that of the enzyme-substrate (ES) complex. This inhibition typically displays a lower Vmax, but an unaffected Km value. We compare three main types of inhibitors in terms of reaction properties as shown in Table 1 and Figure 15.

Table 1: Comparison of the different types of reversible inhibition is shown in Table with a quick view of mechanism in sketches as below.

Competitive inhibitor	Uncompetitive inhibitor	Mixed (noncompetitive inhibitor)
• The inhibitor binds the catalytic/substrate binding site. • It competes with substrate for binding. • Inhibition is reversible by increasing substrate concentration. • V_{max} constant, the substrate concentration has to be increased as reflected on increased K_m.	• Substrate binding exposes the inhibitor binding site away from the catalytic/substrate binding site. • Increasing substrate concentration does not reverse the inhibition. • The inhibited reaction rate parallel the normal one as reflected on decreased both V_{max} and K_m.	• The inhibitor binds each of the free enzyme and the substrate-enzyme complex away from the catalytic/substrate binding site. • Increasing substrate concentration does not reverse the inhibition. • Only V_{max} is decreased proportionately to inhibitor concentration, • K_m is unchanged since increasing substrate concentration is ineffective.

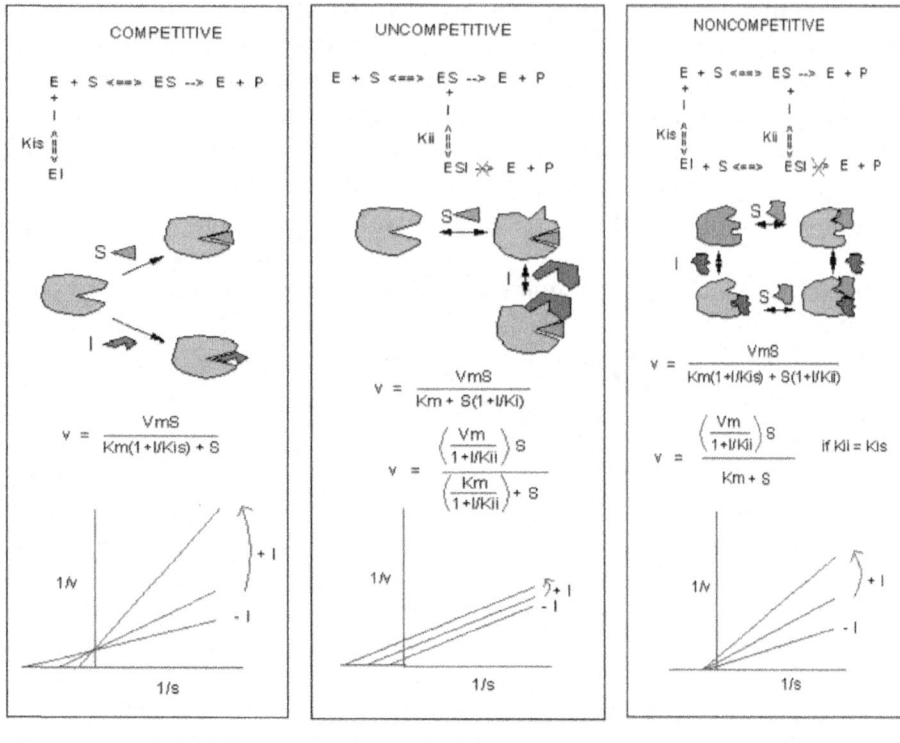

Figure 15: Sketch of three different enzyme inhibition by competitive, uncompetitive and noncompetitive types are shown with illustration of enzyme-substrate or inhibitor binding, kinetics and graphs.

In last decade, role of membrane receptors was explored in relation with enzyme inhibition. Membrane receptors or transmembrane proteins bind with natural ligands such as hormones, neurotransmitters in tissue membranes. Receptor-ligand binding modulates the binding of drugs with enzyme. Such ligand binding behavior also influences the analysis of competitive, uncompetitive and noncompetitive inhibition by biological effect of prodrugs on enzymes. It usually involves a shape change in the receptor, a transmembrane protein, which activates intracellular activities. The bound receptor usually does not directly express biological activity, but initiates a cascade of events which leads to expression of intracellular activity. However, occupied receptor actually expresses biological activity itself. For example, the bound receptor can acquire enzymatic activity, or become an active ion channel with similar competitive, noncompetitive behavior. Drugs targeted to membrane receptors can have biological effects similar to the natural ligands, they are called agonists, or conversely they may inhibit the biological activity of the receptor, they are called antagonists [Jakobowski 2010a].

Agonist

An agonist or test drug or substrate is similar to natural ligand and binds with receptor to produce a similar biological effect as the natural ligand. Agonist binds at the same binding site in competition with natural ligand to show full or partial response. So, it is called partial agonist. If receptor has a basal (or constitutive) activity in the absence of a bound ligand, it is called inverse agonist. If either the natural ligand or an agonist binds to the receptor site, the basal activity is increased. If an inverse agonist binds, the activity is decreased. Ro15- 4513 and benzodiazepines (Valium) bind with the GABA receptor. As a result, GABA receptor is "activated" to become a ion channel allowing the inward flow of Cl- into a neural cell, inhibiting neuron activation. Ro15-4513 binds to the benzodiazepine site, which leads to the opposite effect of valium, the inhibition of the receptor bound activity - a chloride channel as shown in Figure 16

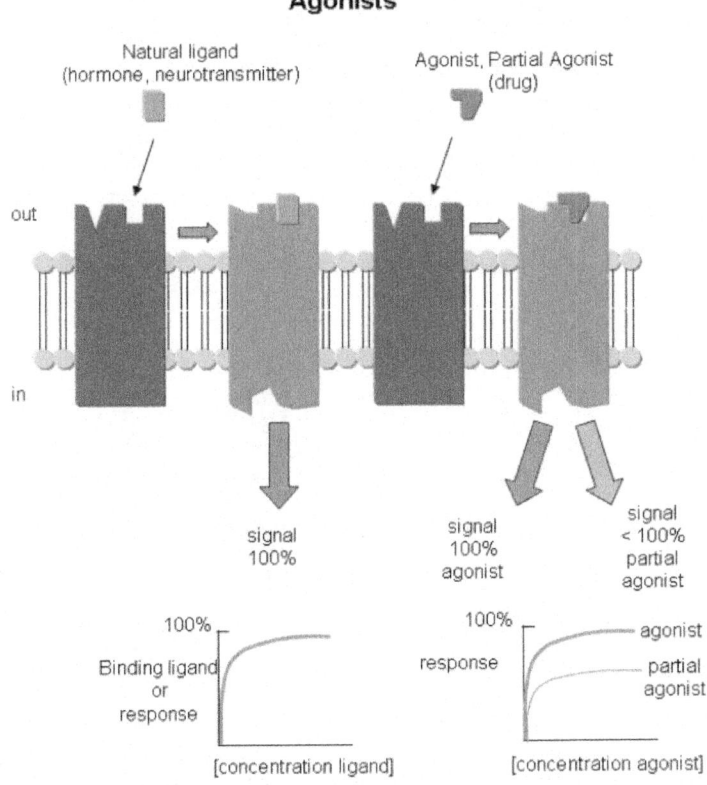

Figure 16: A sketch is shown for membrane receptor binding with ligand (agonist) acting like as enzyme. Reproduced with permission [Jakobowski 2010a].

Antagonist

Antagonist or test inhibitor can inhibit the effects of the natural ligand (hormone, neurotransmitter), agonist, partial agonist, and inverse agonists. We can think of them as inhibitors of receptor activity behaving as competitive, noncompetitive and irreversible antagonists as shown in Figure 17. For further details, readers are requested to read advanced text book [Nelson et al. 2008, Dixon and Webb 1979]

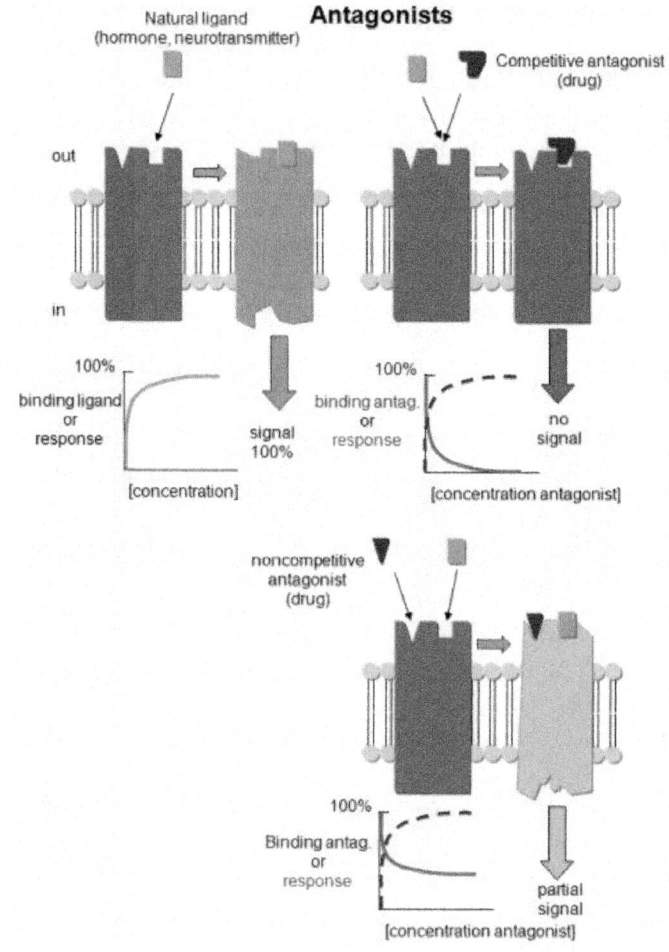

Figure 17: Sketch is shown for membrane receptor binding with ligand (acting as agonist) and antagonist (acting as inhibitor) in competition with agonist to bind with enzyme. Reproduced with permission [Jakobowski 2010a]

INHIBITION BY PHYSIOLOGICAL MODULATORS

Temperature of reaction

Some endothermic or exothermic chemical compounds change the temperature of reaction. Enzyme reaction experiences inhibition at higher or lower than optimal physiological temperature. For example, human body optimal temperature of human body is 37 oC. For most of the enzyme reactions, enzyme activity usually increases at 0 to about 40-50 oC in the absence of catalysts. As a general rule of thumb, reaction velocities double for each increment of 10oC rise. At higher temperatures, the activity decreases dramatically as the enzyme denatures as shown in Figure 18.

Temperature and enzyme activity

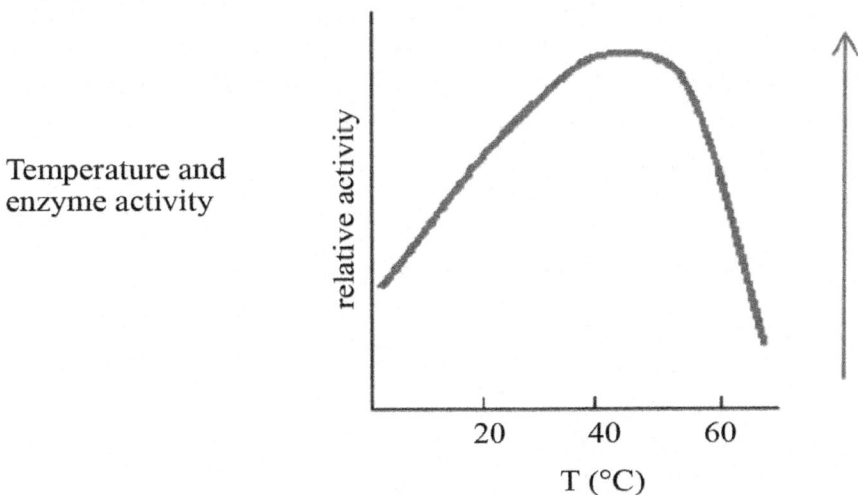

Figure 18: Figure shows the effect of temperature change on the rate of enzyme reaction. Notice the initial rise of rate of reaction and sudden fall near to optimal temperature 37-42 °C.

Hydrogen ion concentration or pH of reaction

Think of all the things that pH changes might affect. Many chemicals such as acids or alkaline chemical compounds if mixed in enzyme reaction medium can change the pH. As a result, reaction rate changes. It might

- affect E in ways to alter the binding of S to E, which would affect K_m
- affect E in ways to alter the actual catalysis of bound S, which would affect k_{cat}

- affect E by globally changing the conformation of the protein
- affect S by altering the protonation state of the substrate

The easiest assumption is that certain side chains necessary for catalysis must be in the correct protonation state. Thus, some side chain, with an apparent pK_a of around 6, must be deprotonated for optimal activity of trypsin which shows an increase in enzyme activity with the increase in range centered at pH 6. Which amino acid side chain would be a likely candidate to participate in enzyme inhibition? It all depends on net charge on active group of each amino acid in the active site chain. The pH of reaction thus depends on net pK_a value of amino acids and presence of acid or alkaline nature of substrate effects on enzyme kinetics by formation of EH, ESH as shown in Figure 19. It can be modeled at the chemical and mathematical level to calculate velocity(v), Vm(apparent) and K_m(apparent) as shown in Equations 7-9. Different enzymes show different behavior of enzyme catalyzed reactions such as chymotrypsin, cholinesterase, papain, and papsin show distinct graphs (see Figure 20). For further details, readers are requested to read text books [Nelson et al. 2008, Berg et al. 2011]

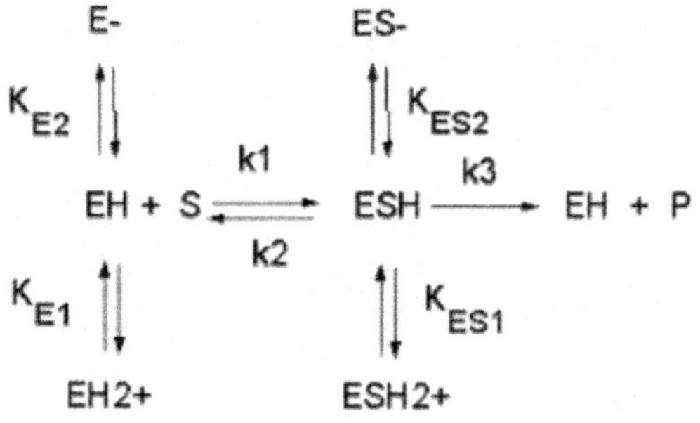

$$V = \frac{Vm\,app\,S}{Km\,app + S} \tag{7}$$

$$Vm\,app = \frac{Vm}{1 + H+ / Kes1 + Kes2 / H+} \tag{8}$$

$$Km\,app = \frac{Km(1 + H+ / Ke1 + Ke2 / H+)}{1 + H+ / Kes1 + Kes2 / H+} \tag{9}$$

Figure 19: Chemical equations showing the mechanism of pH effects on enzyme catalyzed reactions. Different mathematical equations 7-9 illustrate the modeling pH effects on enzyme catalyzed reactions.

Three dimensional nature of enzyme-inhibitor complex at enzyme active site

The role of non-covalent interactions such as hydrogen bonding, hydrophobic interaction and orientation of inhibitor and enzyme in an organized fashion was well described in classic paper [Amtul et al., 2002]. 3D nature of enzyme reaction can be understood as following. There are two sites on enzyme molecule: 1. at allosteric site, inhibitor binds with enzyme, and 2. at active site, substrate binds with enzyme. However, substrate and inhibitor interact with each other by non-covalent interactions of their chemical groups. Inhibitors interact at allosteric site and known as 'pharmacohores'. Presently, structure-based design and testing, mechanistic biological approach is a state-of-art to develop new pharmacohores. The non-covalent interactions determine the chemoselectivity of the substrate and enzymes during formation of the ESI complex. In other words, ESI complex provides enzyme as a platform to perform catalysis. 3D geometrical shape and topology of active site match with orientation of chemical groups in substrate molecule that fit together in a 'lock and key' arrangement. Several possibilities happen to make enzyme-inhibitor complexes such as bidentate, tri-, tetra- and polydentate, trigonal, pyramidal, tetrahedral, polyhedral charge transfer complexes due to co-ordinate interactions between metallic co-factor with hydrophilic groups on inhibitor(s). In this process, geometry of amino acid side chains at allosteric site changes due to hydrogen bonding between amino acid residues.

Suboptimal interactions of metal-solvent, oxygen-water molecular bridge, free energy content loss, subunit-subunit biophysical interactions as a result play a significant role in inhibitorenzyme complex formation and completion of enzyme catalysis.

For more details, readers are requested to read recent reference papers on 3D mechanistic studies on enzymes. Specific example on urease is cited in chapter 11 in this book. Now science is shifting to develop crystallized enzyme molecules, better structural-functional relationship in enzyme catalysis and immobilized enzyme chips. In following description, factors are discussed on different practical considerations that influence the enzyme reaction rates, enzyme inhibition kinetics, % binding efficiency on enzyme solid support with a glimpse of known theories and concepts on real-time, cheaper, economic, user-friendly immobilized enzyme technology. When actual and practical considerations are analyzed to work in enzyme reactor, the scenario becomes complicated. Several factors such as inhibitor chemical state, substrate structure, enzyme 3D conformation or peptide subunit interactions, physiological reaction conditions in reactor and enzyme carrier supports also contribute in inhibition kinetics and rates of reaction to form ES,ESI and P.

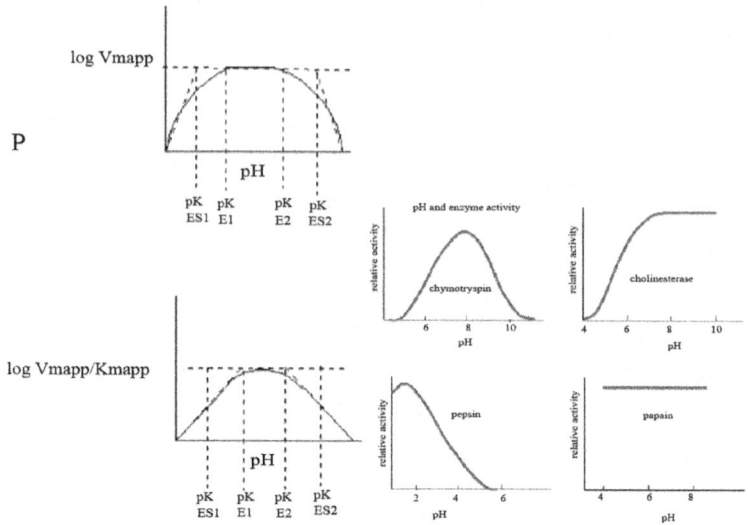

Figure 20: Graphs of different pH effects on enzyme catalyzed reactions as log $V_m(app)$ and $V_m/K_m(app)$ are shown on left. Different enzymes such as chymotrypsin, cholinesterase, pepsin and papain are illustrated with different rates of enzyme reaction. Reproduced with permission [Jakobowski 2010a]

Every year list of new factors grows in new enzyme systems. Author believes that more and more contributory factors introduced, will influence enzyme reaction rate kinetics and more and more additive kinetic constants are introduced with new variants to define the action of inhibitors on enzyme catalysis.

Other factors to keep in mind for new possibilities are:

- enzyme autoinhibition and enzyme molecular structural-functional factors affecting 3D conformation of active site compatible with active groups of substrate or inhibitor
- porosity and diffusion across the enzyme support material and availability of exposed active sites to react
- real-time recording the instant formation of ESI or ES or EP or EI on solid phase enzyme support organic chip
- sustrate-inhibitor interactions, % binding of active site with each additive

- computer based semi-corrected or averaged calculations of kinetic constants of inhibition kinetics

- thermodynamic states of the enzyme reaction in reactor and fluctuating physiological and physical states of substrate, inhibitor, enzyme complexes in reactor.

- synergy of inhibitors, substrate, subunits in enzyme on active site

For all these factors and details, readers are expected to read advanced text books on enzyme inhibition and enzyme engineering. Readers will experience a wide variation in the scientific analysis of enzyme inhibition data in different enzyme reactors used in different studies. High efficiency with desired results of enzyme inhibitors is the new challenges to optimize reaction, scale-up, and phase out unwanted physiological factors from reaction. In following section, these issues are addressed. Author believes that above mentioned description is just iceberg from a large hidden treasure or unknown factors contributing enzyme inhibition to give desired outcome.

IMMOBILIZED ENZYME SYSTEMS

In search of economic, efficient and practical enzyme platforms to test enzyme inhibitors, new user-friendly immobilized enzyme technology is available now. It is based on principle that an enzyme molecule is contained within confined space for the purpose of retaining and re-using enzyme on solid medium in processing system or equipment. There are many advantages of immobilized enzymes and methods of immobilization such as low cost, suitability of reusable model system in membrane-bound enzymes in cell. However, some disadvantages are expansive methods of adsorption or covalent bound or matrix trapping or membrane trapping immobilization methods, low measurement of enzyme activity with mass transfer limitations. For knowledge sake, the entrapment of enzyme molecules on matrix, diffusion phenomenon and kinetics are important to understand. A brief description is given for interested readers on classic concepts and scientific basis of porous or nonporous enzyme supports, theory of enzyme immobilization and efficiency of reaction outcome. For more details of each aspect, readers are requested to read individual research papers.

Matrix entrapment is done by mixing enzyme solution with polymer fluid in matrices such as Ca-alginate, agar, polyacrylamide, collagen. Membrane entrapment is done by confining enzyme solutions between semi-permeable membrane hollow fibers made of nylon, cellulose, polysulfone, polyacrylate etc. Surface immobilization by adsorption is done by attaching enzymes on stationary solids such as alumina, porous glass, cellulose, ionexchange resin, silica, ceramic, clay, starch etc. by physical forces keeping active sites intact. Covalent bonding is done by enzyme retention on support surfaces by covalent binding between functional groups such as amino, carboxylic, sulfhydryl, hydroxyl groups on the enzyme and those on the support surface keeping enzyme active site(s) free (see Figure 21) [Laider et al. 1980].

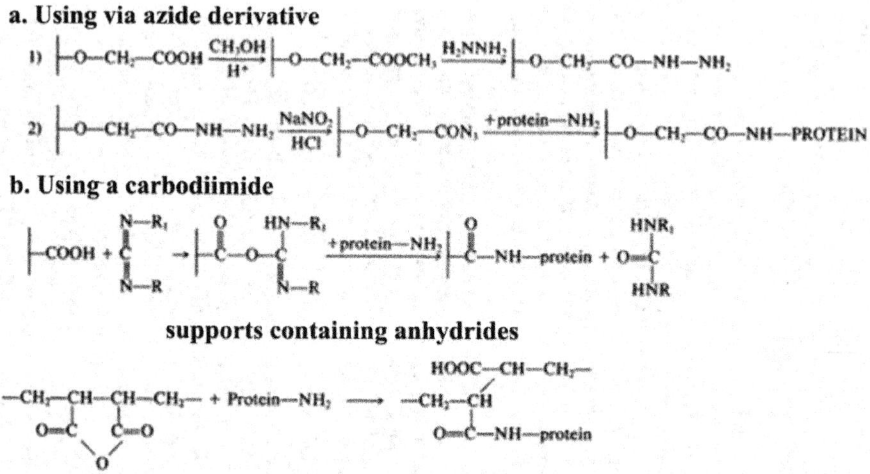

a. Using via azide derivative

1) $\left|-O-CH_2-COOH \xrightarrow[H^+]{CH_3OH} \right|-O-CH_2-COOCH_3 \xrightarrow{H_2NNH_2} \left|-O-CH_2-CO-NH-NH_2\right.$

2) $\left|-O-CH_2-CO-NH-NH_2 \xrightarrow[HCl]{NaNO_2} \right|-O-CH_2-CON_3 \xrightarrow{+protein-NH_2} \left|-O-CH_2-CO-NH-PROTEIN\right.$

b. Using a carbodiimide

$$\left|-COOH + \underset{N-R}{\overset{N-R_1}{\underset{\|}{C}}} \rightarrow \underset{N-R}{\overset{O\quad HN-R_1}{C-O-C}} \xrightarrow{+protein-NH_2} \overset{O}{C}-NH-protein + O=\underset{HNR}{\overset{HNR_1}{C}}$$

supports containing anhydrides

$$-CH_2-CH-CH-CH_2- + \text{Protein}-NH_2 \longrightarrow -CH_2-CH \overset{HOOC-CH-CH_2-}{\underset{O=C-NH-protein}{}}$$

$$\underset{O}{\overset{O=C\quad C=O}{\diagdown\diagup}}$$

Figure 21: Scheme of immobilization of enzyme is shown with chemical groups involved in binding of enzyme on solid surface. Reproduced with permission from reference Lieder et al.1980.

Diffusional limitations are observed to various degrees in all immobilized enzyme systems. This occurs because substrate must diffuse from the bulk solution up to the surface of the immobilized enzyme prior to reaction. The rate of diffusion relative to enzyme reaction rate determines whether limitations on intrinsic enzyme kinetics is observed or not as shown in Figures 22 [Laider et al.1980]. However, rate of diffusion across and within matrix is determinant of immobilized enzyme reaction as shown in Figure 22 and 23.

In immobilized enzyme reaction, two major effects due to diffusion and product inhibition are first observed by Lineweaber-Burk plots in classic study [Rees, 1984]. The diffusional effects and product inhibition both influenced the shape of Lineweaver-Burk plot (see Figure 22). In case of substrate inhibition effects binding of more than one substrate molecule(s) lead to inhibition showing same type of curved Lineweaver-Burk plot as those observed for diffusional limitation and product inhibition in immobilized enzymes. Combination of these two effects lead to intermediate behavior, such as normal Michaelis-Menten kinetics as shown in Figure 24, 25 by curves [Rees, 1984]. However, immobilized enzyme system also suffers from both diffusion and product inhibition effects. As a consequence, it is important to consider diffusion effects and product inhibition effects while extracting catalytic parameters from kinetic data for immobilized enzyme systems. Use of non-porous support in enzyme immobilization minimizes the diffusion effects to some extent.

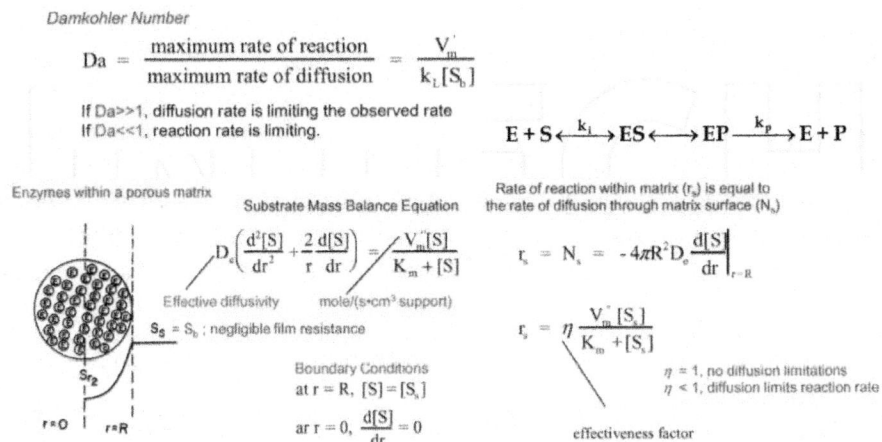

Figure 22: A sketch of porous matrix is shown (on left) and a scheme of substrate mass balance Equation to calculate rate of immobilized enzyme reaction rs is shown (on right)

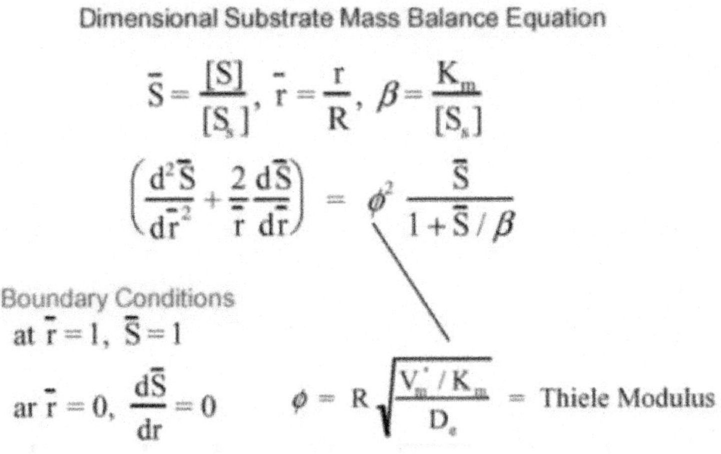

Dimensional Substrate Mass Balance Equation

$$\bar{S} = \frac{[S]}{[S_s]}, \quad \bar{r} = \frac{r}{R}, \quad \beta = \frac{K_m}{[S_s]}$$

$$\left(\frac{d^2 \bar{S}}{d \bar{r}^2} + \frac{2}{\bar{r}} \frac{d \bar{S}}{d \bar{r}} \right) = \phi^2 \frac{\bar{S}}{1 + \bar{S} / \beta}$$

Boundary Conditions
at $\bar{r} = 1$, $\bar{S} = 1$

ar $\bar{r} = 0$, $\frac{d\bar{S}}{dr} = 0$

$\phi = R \sqrt{\frac{V_m^{'} / K_m}{D_e}}$ = Thiele Modulus

Figure 23: A scheme of substrate mass balance is shown to calculate S with boundary conditions.

Enzyme kinetics predicts the efficiency of reaction. Kinetics of immobilized enzymes depends on conformational alterations within the enzyme due to the immobilization procedure, or the presence and nature of the immobilization support. Immobilization can greatly affect the stability of an enzyme such as any strain into the enzyme will inactivate the enzymes under denaturing conditions (e.g. higher temperatures or extremes of pH). An example of unstrained multipoint binding between the enzyme and the support to cause substantial stabilization is illustrated in Figure 20. From mechanistic standpoint, a lesser conformational change within the protein structure will initiate enzyme inactivation. As a result, covalent immobilization processes involve an initial freely-reversible stage. Covalent links may form, break and re-form till an unstrained covalently-linked structure is created. However, additional stabilization is derived from maximum enzyme-support compatibility, least enzyme molecule interactions, least proteolytic and microbiological attacks.

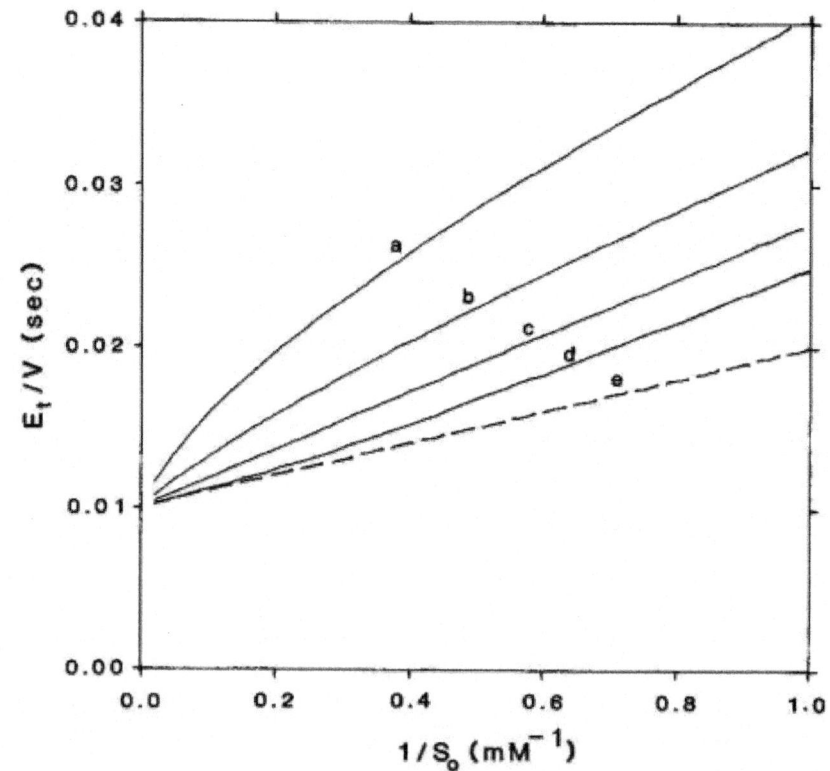

Figure 24: Effect of one or more inhibitor molecules on enzyme kinetics and their inhibition effect dependent on 1/So. Reproduced with permission from Rees et al. 1984.

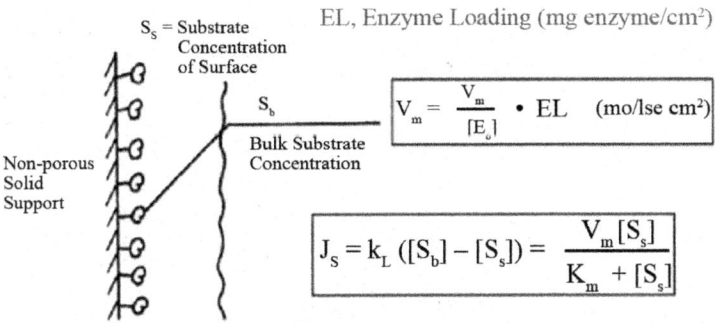

Figure 25: A scheme of immobilized enzyme action is shown on non-porous solid support. Notice the dependence of V_m on available immobilized enzyme active sites (E_L).

The kinetic constants (e.g. K_m, V_{max}) of immobilized enzymes may be altered by the process of immobilization due to internal structural changes and restricted access to the active site. Thus, the intrinsic specificity ($k./K_m$) of such enzymes may well be changed relative to the soluble enzyme. An example of trypsin is illustrated in Figure 21, where the freely soluble enzyme hydrolyses fifteen peptide bonds in the protein pepsinogen but the immobilized enzyme hydrolyses only ten. The apparent value of these kinetic parameters, when determined experimentally, may differ from the intrinsic values. This fact may be due to changes in the properties of the solution in the immediate vicinity of the immobilized enzyme, or the effects of molecular diffusion within the local environment. The relationship between these intrinsic and apparent parameters is shown below in Figure 26. Typically, nonporous microenvironment consists of the internal solution plus part of the surrounding solution which is influenced by the surface characteristics of the immobilized enzyme. Partitioning of substances occurs between these two environments.

Substrate molecule (S) diffuses through the surrounding layer (external transport) in order to reach the catalytic surface and gets converted to product (P). In order for all immobilized enzyme to be utilized, substrate must diffuse within the pores in the surface of the immobilized enzyme particle (internal transport) [Pryciak 2008]. The degree of stabilization is determined by strength of the gel, and hence the number of non-covalent interactions. As a result, intrinsic parameters of enzyme result with specific apparent parameters dependent on partition and diffusion as shown in Figure 27.

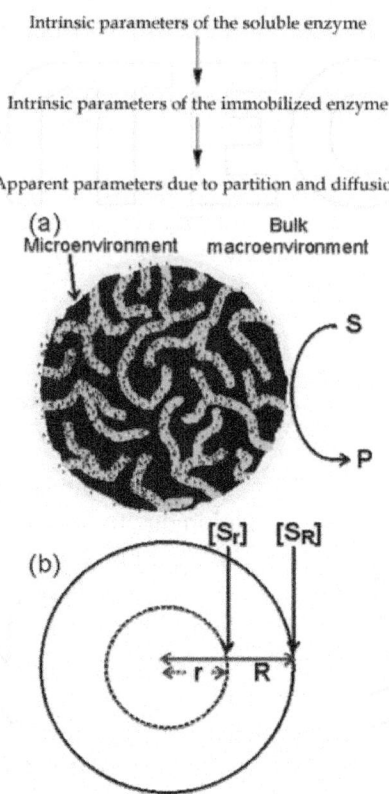

Intrinsic parameters of the soluble enzyme

Intrinsic parameters of the immobilized enzyme

Apparent parameters due to partition and diffusion

(a) Microenvironment   Bulk macroenvironment

(b)   [S$_r$]   [S$_R$]

r   R

Figure 26: A schematic cross-section of an immobilized enzyme particle (a) shows the macroenvironment and microenvironment. Triangular dots represent the enzyme molecules. Courtesy: Pangandai V. Pennirselvam, Ph.D UFRN, Lagoa Nova–Natal/ RN Campus Universitário. North East, Brazil.

- The porosity (e) of the particle can be expressed as ratio of the volume of solution contained within the particle to the total volume of the particle. The tortuosity (t) is the average ratio of the path length, via the pores, between any points within the particle to their absolute distance apart.

- The tortuosity, which is always greater than or equal to unity, depends on the pore geometry. The diagram exaggerates dimensions for the purpose of clarity.

- The concentration of the substrate at the surface of the particles [Sr] depends on radius R or internal concentration [Si] at any smaller radius (r) is the lower value.

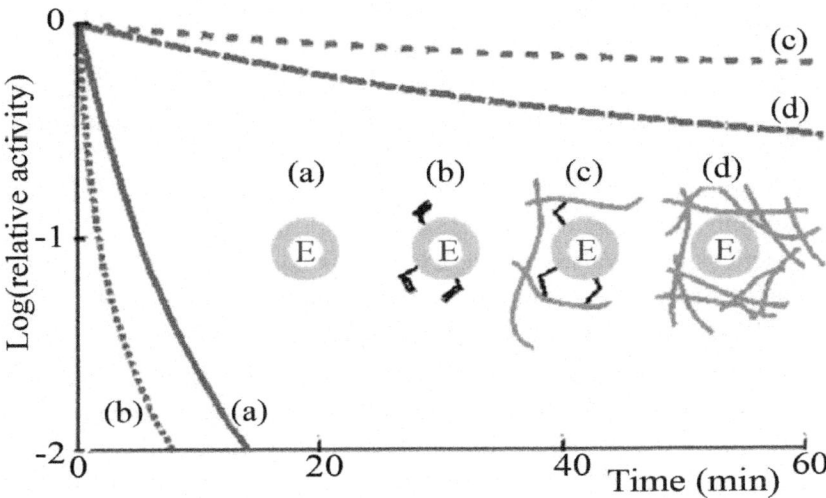

Figure 27: Illustration of the use of multipoint interactions for the stabilization of enzymes. (a) -------- activity of free un-derivatized chymotrypsin. (b) ..... activity of chymotrypsin derivatized with acryloyl chloride. (c) -- -- -- activity of acryloyl chymo-trypsin copolymerized within a polymethacrylate gel. Up to 12 residues are covalently bound per enzyme molecule. Lower derivatization leads to lower stabilization. (d) --- -- activity of chymotrypsin noncovalently entrapped within a polymethacrylate gel. All reactions were performed at 60°C using low molecular weight artificial substrates. The immobilized chymotrypsin preparations showed stabilization of up to 100,000 fold, most of which is due to their multipoint nature although the consequent prevention of autolytic loss of enzyme activity must be a significant contributory factor. Reproduced with permission from Martinek et al, 1977a,b.

In general, the use of immobilized enzyme can be divided into two major categories of applications: in biosensors and bioreactors. However, list is growing in the other fields of ecological, environmental, agriculture, health, oceanic, space and earth sciences.

NEW DEVELOPMENTS IN ART OF ENZYME INHIBITION

Now a day, immobilized enzymes are used in industries and have value as medicinal and industrial enzyme products. Good examples of industrial enzymes are amylase, glucoamylase, trypsin, pepsin, rennet, glucose isomerase, penicillinase, glucose oxidase, lipase, invertase, pectinase, cellulase in medicinal use. With emergence of new inhibitors in the quest of drug discovery, several new inhibition mechanisms are expected in case of new substrate analogues. New substrate–enzyme active site interactions are envisaged due to different binding intricacies. Some examples of emerging concepts are outlined in following description and readers are expected to read advanced literature on these applications.

- *Slow-tight inhibition*: Slow-tight inhibition occurs when the initial enzyme-inhibitor complex EI undergoes isomerizing conformational change to a more tightly binding complex. However, the overall inhibition process is reversible. This manifests itself as slowly increasing enzyme inhibition. Under these conditions, traditional MichaelisMenten kinetics gives a false value of a time-dependent Ki. The true value of Ki can be obtained through more complex analysis of the on (kon) and off (koff) rate constants for inhibitor association.

- *Substrate and product inhibition*: Substrate and product inhibition is where either the substrate or product of an enzyme reaction inhibits the enzyme's activity. This inhibition may follow the competitive, uncompetitive or mixed patterns. In substrate inhibition there is a progressive decrease in activity at high substrate concentrations. This may indicate the existence of two substrate-binding sites in the enzyme. At low substrate, the high-affinity site is occupied and normal kinetics is followed. However, at higher concentrations, the second inhibitory site becomes occupied, inhibiting the enzyme. Product inhibition is often a regulatory feature in metabolism and can also be a form of negative feedback; see allosteric regulation [Pryciak 2008, Bashor 2008].

- *Antimetabolites*: They are chemicals that interfere with the normal metabolism of normal biochemical metabolite(s). This in most of case is due to their structural similarity to such physiological substrates and therefore works as competitive enzyme inhibitors. They include antifolates such as methotrexate, hydroxyurea and purine and pyrimidine analogues. They are mainly used as cytotoxic anticancer drugs through inhibiting DNA and RNA synthesis and cell division. An example of nitroimidazole is described in detail on its metabolic effects at cellular level in this book [Sharma 2012a].

- *Antienzyme:* Intestinal parasites, e.g., Ascaris, protect themselves from digestion by expressing on their surface substances that are protein in nature which inhibit the action of digestive enzymes, e.g., pepsin and trypsin. The blood plasma and extracellular fluids are containing several types of protease inhibitors particularly important in controlling the blood clot formation and dissolution and matrix and cytokine homeostasis. Most of these inhibitors are peptides and several of them are also isolated from raw egg white, potatoes, tomatoes and Soya bean and other plant sources. Most of the natural peptide protease inhibitors are similar in structure to the amino acid sequence of the peptide substrates of the enzyme. Designed peptide protease inhibitors are important drugs, e.g., captopril that is a metalloprotease angiotensin-converting enzyme peptide inhibitor. Inhibiting this enzyme prevent activation of angiotensin and therefore prevent vasoconstriction to lower blood pressure. Crixivan is an antiretroviral aspartyl protease peptide inhibitor used in the treatment of Human Immunodeficiency Virus (HIV)-induce acquired immunodeficiency syndrome (AIDS). It inhibits the HIV protease that cleaves the large multidomain viral protein into active enzyme subunits. Because these peptide inhibitors may not be specific, they have several side-effects as drugs.

- Antibodies against several nonfunctional plasma enzymes have clinical diagnostic importance since they are longer living than the enzyme itself and hence reflect the disease history better. In this respect, autoimmune antibodies are clinically important in diagnosis of autoimmune diseases, e.g., anti-glutamic acid decarboxylase antibodies in type 1 diabetes mellitus.

- Biosensors: Light inhibits most enzyme activity although some enzymes, e.g., amylase are activated by red or green light and also specific DNA repairing enzymes (e.g., UVspecific endonuclease) are activated by the blue and UV light. Ultraviolet rays and ionizing radiations cause denaturation of most enzymes. Most enzymes contain sulfhydryl (-SH) groups at their active sites which upon oxidation by oxidants and free radicals by oxidants and free radicals inactivate the enzyme. Examples: Effect of radiations, light and oxidants on the rate of the enzyme catalyzed reaction.

- Other application of membrane bound redox enzymes constitutes them as a scaffolding enzyme arrangement into systems for multi-step catalytic processes. The reconstruction of portions of this redox catalytic machinery, interfaced to an electrical circuit leads to novel biosensing devices or biosensors. An example of nitric oxide synthase enzyme is cited in this book [Sharma, 2012b].

- In EzNET® water purifying system, nitrate pollution is eliminated. Enzyme is immobilized on "beads" with an electron-carrying dye as shown in Figure 28. Reduction of nitrate to environmentally safe nitrogen gas is driven by a low voltage direct current.

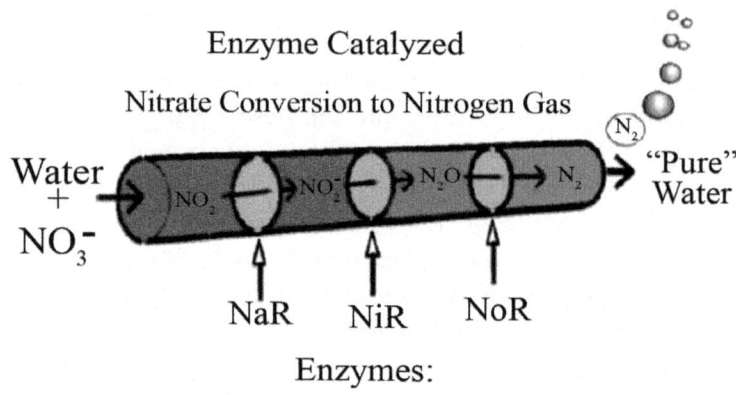

Figure 28: EzNET® system shows immobilized enzyme on "beads" with an electron-carrying dye. In this system, reduction of nitrate generates environmentally safe nitrogen gas driven by a low voltage direct current. Source: The Nitrate Elimination Co., Inc. 2000.

In biolumescence detection for toxicity of HPV chemicals or drug development, 62 kDa MW oxygenase (yellow green light emitted at 560 nm) enzyme gives 88 photon/cycle light output proportional to [ATP] according to:

Luciferin + luciferase + ATP → luciferyl adenylate-luciferase + pyrophosphate

Luciferyl adenylate-luciferase + O2 → Oxyluciferin + luciferase + AMP + light

Strong inhibition of luciferase by chloroform or HPV chemicals indicates the efficiency of immobilized recombinant luciferase enzyme system as shown in Figure 20. Inhibition by chloroform is much reduced in the mutant Luciferase compared to the wild type Luciferase as shown in Figures 29, 30.

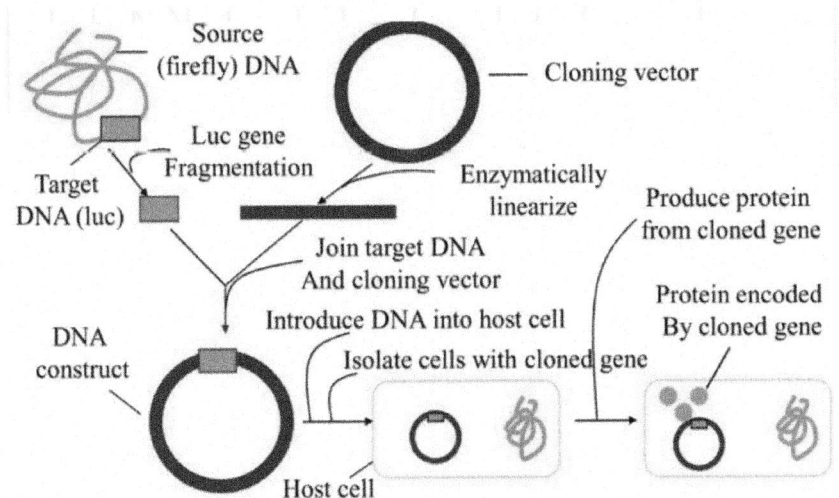

Source: Kim et al. AIChEngg Annual Meeting 2003, San Francisco, CA.

Figure 29: A sketch of recombinant luciferase is shown illustrating the gene clone.

- In the search for new therapeutics, the high throughput screening (HTS) of ligands for key target proteins, enzymes represent the principal hit identification tool for early drug discovery [Bartolini et al. 2009]. However, output depends on cost-based or amount-based limitation of target availability, need of speed, automation and easy coupling of the enzyme assay with separation systems (affinity chromatography of immobilized proteins) and appropriate detectors. Good example is targeting in drug discovery represented by enzyme inhibition mechanism in monolithic immobilized enzyme reactors (IMERs) to represent different phases of the drug discovery pathway-starting with active compounds (hit) identification, through drug development and lead optimization, early ADMET (absorption, distribution, metabolism, excretion, toxicity) studies and quality control of protein drugs. Some details are described in chapters in this book [Bartolini et al. 2005, 2007]. Interested readers are requested to read advanced text books on these aspects. Different IMER have own requirements for optimal performances to show an increased data output, reliability and stability to translate into cost reduction for potential applications in pharmacy industry [Bartolini et al. 2005, 2007].

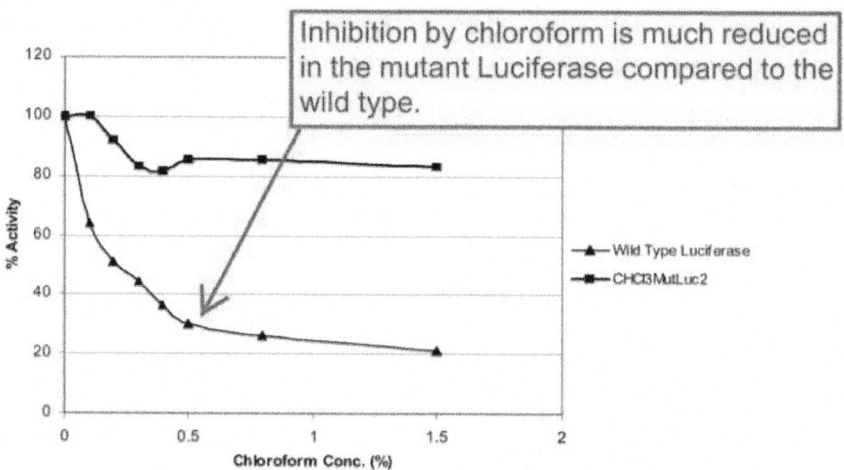

Source: Kim et al. AIChEngg Annual Meeting 2003, San Francisco, CA.

Figure 30: Inhibition of luciferase activity by increasing the concentration of chloroform.

SOFTWARES AND COMPUTERIZATION IN ENZYME INHIBITION KINETICS

Recently softwares have popped up to visualize custom visual interface to see curve fits in real-time, graph transforms, equations using kinetic data entry in terms of substrate, inhibitor, activator, velocity, and standard deviation of the velocity. Data tables are directly generated linked to the Fitting Panel of software. The data and results analysis is transferred in userfriendly lay-out, ANOVA window, % inhibition using Monte-Carlos fits, and receptor or ligand binding calculator. For interested readers, VISUALENZYMICS 2010® is available for statistical analysis for enzyme kinetics.[http://www.softzymics. com/visualenzymics.htm].

LIMITATIONS AND CHALLENGES

Above mentioned description on mechanism and applications shows a clear issue on need of careful analysis for enzyme inhibition factors, presumptions of enzyme reaction, use of new immobilized enzyme support and enzyme recording/monitoring methods. Challenge is that most of times, basic presumptions do not hold true in enzyme reactors and addition of new factors further complicate the calculation of reactor outcome. Most of the times, computer based kinetic calculations average out outcome as less realistic with more chances of variants. Other major challenge is that each time enzyme reactor outcome depends on individual inhibitor and individual enzyme reactor at different times. It is less reproducible and unpredictable because of synergy, interplay of known and unknown physical, physiological, biological, molecular factors affecting reaction kinetics.

IMPACT OF ENZYME INHIBITION SCIENCE IN BUSINESS

The major current and emerging therapeutic markets for enzyme inhibitors used in human therapeutics are very high. New information is available on biochemistry for enzyme inhibitors and classes of enzyme inhibiting products with broad current or potential therapeutic applications in large markets. However, more than 100 enzyme inhibitors are currently marketed and double than those are under development. A better understanding of the emerging enzyme inhibitors on enzyme mechanism is main key. These include selected indications for asthma and chronic obstructive pulmonary disease (COPD), cardiovascular diseases, erectile dysfunction, gastrointestinal disorders, hepatitis B virus infection, hepatitis C virus infection, herpesvirus infections, human immunodeficiency virus (HIV)/acquired immune deficiency syndrome (AIDS) and rheumatoid arthritis and related inflammatory diseases. Key information from the business literature and thorough enzyme inhibition research is the basis of expert opinion on commercial potential and market sizes from enzyme industry professionals. Since initial reports on chemical immobilization of proteins and enzymes first appeared ~30 years ago, immobilized proteins are now widely used for the processing of products in industries from food business to environmental control. In recent years, use of chemical immobilization was extended to immobilized antibodies or antigens in bioaffinity chromatography. In coming years, it is speculated that immobilization techniques of proteins and enzymes will have greater impact on point-ofcare medical and health business.

CONCLUSION

Enzyme inhibition is significant biological process to characterize the enzyme reaction, extraction of catalysis parameters in bio-industry and bioengineering. Conceptual models of inhibition define the interactions of substrate-enzyme or inhibitor-enzyme or both substrateenzyme-inhibitor in the moiety of active site. In recent years, application of enzymes and enzyme inhibition science have gone in healthcare, pharmaceutical, bio-industries, environment, and biochemical enzyme chip industries with great impact on healthcare and medical business. Last decade has shown the measurement and accuracy of enzyme detection up to the scale of picometer and enzyme industry is entering in the area of picotechnology. Immobilized enzyme technology has given a new way of economic tools in drug discovery and biosensor industry. Every year new enzyme inhibitors are discovered useful as drugs but success still needs to minimize challenges.

ACKNOWLEDGEMENTS

Author acknowledges the suggestions of Dr Pagandai V. Pannirselvam, MTech, Ph.D at Centro de Technologia, UFRN, Lagoa Nova–Natal/RN Campus Universitário. North East, Brazil. Author contributed to explain intriguing issues on enzyme inhibition and highlighted the need of better understanding on mechanism of inhibitors before applying them in industries.

REFERENCES

1. Amtul, Z. Atta, Ur. R., Siddiqui, R.A., Choudhary, M. I. (2002). Chemistry and mechanism of urease inhibition. Current Medicinal Chemistry, Vol 9, pp 1323-1348.

2. Bartolini M., Cavrini V., Andrisano V. (2005) J. Chromatogr A, Choosing the right chromatographic support in making a new acetylcholinesterase microimmobilized enzyme reactor for drug discovery. Vol 1065, pp 135-144.

3. Bartolini M, Greig NH, Yu QS, Andrisano V. (2009) Immobilized butyrylcholinesterase in the characterization of new inhibitors that could ease Alzheimer's disease. J Chromatogr A. Vol 1216(13), pp 2730-38.

4. Bartolini M., Cavrini V., Andrisano V. Characterization of reversible and irreversible acetylcholinesterase inhibitors by means of an immobilized enzyme reactor. J. Chromatogr. A (2007) Vol 1144, pp 102 –10.

5. Bartolini M, Andrisano V. (2009) Immobilized enzyme reactors into the drug discovery process: The Alzheimer's Disease case. Web Source: http://www.farm.unipi.it/npcf3/pdf/BartoliniManuela.pdf

6. Bashor, C.J., Helman,N.C., Yan, S., Lim, W.A. Using Engineered Scaffold Interactions to Reshape MAP Kinase Pathway Signaling Dynamics. Science.Vol 319 (5869), pp1539- 1543

7. Berg, J.M., Tymoczko, J.L., Stryer, L. (2011) Biochemistry ISBN-13: 978-1429231152, Freeman WH and Company. Cleland, W.W.(1979) Substrate inhibition, Methods Enzymol. Vol 63, pp 500-513.

8. Dixon,M., Webb,E.C. (1979) Enzymes, 3rd ed., Academic Press, New York.

9. El-Metwally, T.H., El-Senosi, Y. (2010) Enzyme Inhibition. Medical Enzymology: Simplified Approach.Chapter 6, Nova Publishers, NY. pp 57-77.

10. Jakbowski H. (2010a) Personal communication. Online study. Chapter 6- Transport and Kinetics. C. Models of Enzyme Inhibition and D. More complicated Enzymes. Internet source. http://employees.csbsju.edu/ hjakubowski/classes/ch331/transkinetics/olcompli catedenzyme.html

11. --ibid- (2010b) http://employees.csbsju.edu/hjakubowski/classes/ch331/ transkinetics/olinhibiti on.html

12. Laider, K., Bunting, P. (1980) The kinetics of immonbilized enzyme systems. Methods Enzymol. Vol 64, pp 227-248.

13. Martinek, K., Klibanov, A.M., Goldmacher, V.S. & Berezin, I.V. (1977a) The principles of enzyme stabilization 1. Increase in thermostability of enzymes covalently bound to a complementary surface of a polymer support in a multipoint fashion. Biochimica et Biophysica Acta, Vol 485, pp 1-12.

14. Martinek, K., Klibanov, A.M., Goldmacher, V.S., Tchernysheva, A.V., Mozhaev, V.V., Berezin, I.V. & Glotov, B.O. (1977b) The principles of enzyme stabilization 2. Increase in the thermostability of enzymes as a result of multipoint noncovalent interaction with a polymeric support. Biochimica et Biophysica Acta Vol 485, pp 13-28.

15. Nelson, D.L., Cox, M.M. (2008) Lehninger Principles of Biochemistry. 5th Edition ISBN-13: 978071677108, Freeman W.H. and Company.

16. Pryciak, P. (2008) Customized Signaling Circuits. Science 319, pg 1489.

17. Rees, D.C. (1984) A general solution for the steady state kinetics of immobilized enzyme systems. Bulletin of Mathematical Biology, Vol 46, 2,pp 229-234.

18. Sami, A.J., Shakoor, A.R.. (2011) Cellulase activity inhibition and growth retardation of associated bacterial strains of Aulacophora foviecollis by two glycosylated flavonoids isolated from Mangifera indica leaves. Journal of Medicinal Plants Research (2011) Vol. 5(2), pp. 184-190.

19. Sharma,R. (1990) The effect of nitroimidazoles on isolated liver cell metabolism during development of amoebic liver abscess. Dissertation submitted to Indian institute of Technology, Delhi and CCS University.

20. Sharma, R. (2012a) Mechanisms of Hepatocellular Dysfunction and Regeneration: Enzyme Inhibition by Nitroimidazole and Human Liver Regeneration. In: Enzyme Inhibition: Concepts and Bioapplications. Chapter 7, InTech Web Publishers, Croatia. ISBN 979- 953-307-301-8.

21. Sharma, R. (2012b) Inhibition of Nitric Oxide Synthase Gene Expression: In Vivo Imaging Approaches of Nitric Oxide with Multimodal Imaging. In: Enzyme Inhibition: Concepts and Bioapplications. Chapter 8, InTech Web Publishers, Croatia. ISBN 979- 953-307-301-8.

CITATION

CHAPTER 1

Fu J, Reinhold J, Woodbury NW (2011) Peptide-Modified Surfaces for Enzyme Immobilization. PLoS ONE 6(4): e18692. doi:10.1371/journal.pone.0018

\CHAPTER 2

Dong-Hao Zhang, Li-Xia Yuwen, and Li-Juan Peng, "Parameters Affecting the Performance of Immobilized Enzyme," Journal of Chemistry, vol. 2013, Article ID 946248, 7 pages, 2013. doi:10.1155/2013/946248

CHAPTER 3

Devi, R. , Kirthiga, O. and Rajendran, L. (2015) Analytical Expression for the Concentration of Substrate and Product in Immobilized Enzyme System in Biofuel/Biosensor. *Applied Mathematics*, **6**, 1148-1160. doi: 10.4236/am.2015.67105.

CHAPTER 4

Gan Z, Zhang T, Liu Y, Wu D (2012) Temperature-Triggered Enzyme Immobilization and Release Based on Cross-Linked Gelatin Nanoparticles. PLoS ONE 7(10): e47154. doi:10.1371/journal.pone.0047154

CHAPTER 5

Ang LF, Por LY, Yam MF (2013) Study on Different Molecular Weights of Chitosan as an Immobilization Matrix for a Glucose Biosensor. PLoS ONE 8(8): e70597. doi:10.1371/journal.pone.0070597

CHAPTER 6

Bau-Yen Hung, Yaswanth Kuthati, Ranjith Kumar Kankala, Shravankumar Kankala, Jin-Pei Deng, Chen-Lun Liu and Chia-Hung Lee, Utilization of Enzyme-Immobilized Mesoporous Silica Nanocontainers (IBN-4) in Prodrug-Activated Cancer Theranostics, doi:10.3390/nano5042169.

CHAPTER 7

Jian Li, Jun Ma, Tao Jiang Yanhuan Wang 1, Xuemei Wen and Guozhu Li, Constructing Biopolymer-Inorganic Nanocomposite through a Biomimetic Mineralization Process for Enzyme Immobilization, doi: 10.3390/ma8095286

CHAPTER 8

Jakub Zdarta, Łukasz Klapiszewski, Marcin Wysokowski, Małgorzata Norman, Agnieszka Kołodziejczak-Radzimska, Dariusz Moszyński, Hermann Ehrlich, Hieronim Maciejewski, Allison L. Stelling and Teofil Jesionowski, Chitin-Lignin Material as a Novel Matrix for Enzyme Immobilization, doi: 10.3390/md13042424

CHAPTER 9

Raushan Kumar Singh, Manish Kumar Tiwari, Ranjitha Singh and Jung-Kul Lee, From Protein Engineering to Immobilization: Promising Strategies for the Upgrade of Industrial Enzymes, doi: 10.3390/ijms14011232

CHAPTER 10

Quan Feng, Bin Tang 2, Qufu Wei, Dayin Hou, Songmei Bi and Anfang Wei, Preparation of a Cu(II)-PVA/PA6 Composite Nanofibrous Membrane for Enzyme Immobilization, doi: 10.3390/ijms131012734.

CHAPTER 11

Yuya Asanomi ,Hiroshi Yamaguchi, Masaya Miyazaki, and Hideaki Maeda, Enzyme-Immobilized Microfluidic Process Reactors, doi: 10.3390/molecules16076041.

CHAPTER 12

Douglas Fernandes Silva, Henrique Rosa, Ana Flavia Azevedo Carvalho, and Pedro Oliva-Neto, "Immobilization of Papain on Chitin and Chitosan and Recycling of Soluble Enzyme for Deflocculation of Saccharomyces cerevisiae from Bioethanol Distilleries," Enzyme Research, vol. 2015, Article ID 573721, 10 pages, 2015. doi:10.1155/2015/573721

CHAPTER 13

Rakesh Sharma (2012). Enzyme Inhibition: Mechanisms and Scope, Enzyme Inhibition and Bioapplications, Prof. Rakesh Sharma (Ed.), ISBN: 978-953-51-0585-5, InTech, DOI: 10.5772/39273.

INDEX